HISTORY OF
GENERAL PHYSIOLOGY

600 B.C. to A.D. 1900

VOLUME ONE
FROM PRE-SOCRATIC TIMES
TO THE ENLIGHTENMENT

THOMAS S. HALL

The University of Chicago Press
Chicago and London

This book was published originally under the title
Ideas of Life and Matter: Studies in the History of
General Physiology, 600 B. C.–A. D. 1900

The University of Chicago Press, Chicago 60637
The University of Chicago Press, Ltd., London

ISBN: 0–226–31360–3 (clothbound, 2 vols.); 0–226–31353–0
(paperback, vol. 1); 0–226–31354–9 (paperback, vol. 2)
Library of Congress Catalog Card Number: 69–16999

HISTORY OF GENERAL PHYSIOLOGY

CONTENTS

VOLUME II

THE PHYSICAL BASIS OF LIFE
(1860–1900)

AUTHOR'S NOTE

When I was a graduate student of biology at Yale, my professor and mentor, the late Alexander Petrunkevitch, admonished me to temper my ambitions and not to hope to "discover the difference between living and nonliving things." While accepting this advice, I was tantalized by it and made up my mind to find out, when time should allow, what major thinkers of the past had thought about this most central of biological questions.

It was only decades later, with administrative duties behind me, that I finally set off to explore, and then only in part, the history of what I had come to call the life-matter problem. This book is in effect a preliminary account of that voyage of exploration.

No one could be more acutely aware than I am that having traveled so far in so short a time I have nowhere penetrated the "back country" as deeply as needs to be done. Since completing this preliminary account, I have begun to revisit and explore selected areas—and new areas not visited before—in considerably greater detail. The results have begun to appear in a series of shorter publications. In the meantime, if some readers are stimulated to investigate on their own, to fill in the gaps and to deepen—and sharpen—the perspective, then even these tentative impressions may perform some useful service.

PREFACE TO
THE PHOENIX EDITION

This book tells the story of the long effort in Western thought to define life—and to interpret it—scientifically. Underlying that effort, from its beginnings in Greek antiquity, has been the assumption that life in its familiar manifestations (nutrition, reproduction, and the like) is an activity of matter possessing a distinctive organization. What we here chiefly explore, therefore, is the attempt of physiologists to portray that organization correctly—to solve what we call the life-matter problem. In one sense that problem is central to all others of physiological interest, for whether a physiologist investigates respiration, assimilation, developmental differentiation, or some other activity of the organism, his approach and his conclusions will to some degree reflect, explicitly or implicitly, a point of view about the essential difference between living and nonliving things. Each separate inquiry is thus linked interpretively to that global exploration of life which it seems proper to call general physiology. Many of our chapters begin by detailing a solution to the life-matter problem and then go on to show the implications of that solution for the interpretation of familiar physiological functions.

The story told here commences in the early sixth century B.C., since the scientific tradition to which we today still belong was born at about that time in Greek Asia Minor. It ends with the final years of the nineteenth century A.D., partly because biology about then began a new phase of its development. Around 1900 biologists began to be less concerned with broad theoretical concepts and more concerned with incisive "slices" of general biological questions. After 1900, so many workers began to explore the separate facets of the life-matter problem that a volume the size of this one would be needed to detail their investigations and conclusions. In order not to leave out the present era entirely, a few notes—scarcely more— are included, in the final chapter on the state of the problem today and on efforts being made to solve it in the second half of the twentieth century.

Chapters 1 and 5 present, and other chapters illustrate, a general thesis about the nature of theory-building in physiology and about factors affecting its flow. This thesis is rather often restated in the text (at the risk of repetitiveness) to permit the use of the book as a reference without its having to be read in its entirety. Greater emphasis is placed on the internal structure of particular theories, several score in all, than on the detailed events that connected these theories in time—but not to the exclusion of the latter. The author is grateful that the reception of the original edition* justified the present paperback version, which reproduces the original except for typographical corrections and a few textual and bibliographic revisions and additions.

---

* *Ideas of Life and Matter: Studies in the History of General Physiology, 600 B.C.—1900 A.D.*, 2 vols. Chicago (University of Chicago Press, 1969).

Acknowledgments

Where interpretive problems were encountered in the research for this book, direct help was sought from specialists in the relevant personalities or periods. Chapters treating Greek science owe much, but not their residual shortcomings, to the suggestions of the late Ludwig Edelstein and to Mrs. Margaret T. May, Phillip De Lacy, Norman De Witt, and especially Charles A. Kahn and Richard Durling. Some but not all chapters dealing with later periods were read critically by Walter Pagel, Robert P. Multhauf, Alistair Crombie, Everett Mendelsohn, Charles Gillispie, Thomas Kuhn, Mrs. Audrey Davis, and Bentley Glass. Their advice was not always followed, however, and responsibility for defects is the author's.

Parts of the research were underwritten by Washington University, and parts by the National Science Foundation whose grants numbers GS-143 and GS-462 permitted the help of an undergraduate assistant, Michael Millard; three graduate assistants, Robert Patricelli, Linda Levy, and Jane Christie; and a postdoctoral associate, Dr. Heliane Kraus-Sutter. Their bibliographic research and literary "scouting" activities were indispensable at various stages. Particular thanks are due Mrs. James J. Hanks, National Library of Medicine, whose bibliographic contributions went far beyond ordinary library reader service. The author's wife, Mary T. Hall, read critically and helpfully certain chapters on seventeenth- and eighteenth-century French writers. The book in its present condition would have been impossible without the tireless attention of Shirley Funk who checked innumerable details and typed each chapter many times as it moved through successive revisions.

LIST OF ABBREVIATIONS
Used in the Footnotes, Chapters 1–10

Arist.   Aristotle. Followed by the name of the work abbreviated.

Abbreviations of Aristotle's works: *Cael., De Caelo; De An., De Anima; GA, De Generatione Animalium; GC, De Generatione et Corruptione; HA, Historia Animalium; Juv., De Juventute; MA, De Motu Animalium; Metaph., Metaphysica; Mete., Meterologia; PA, De Partibus Animalium; Ph., Physica; Resp., De Respiratione; Sens., De Sensu.*

Bailey   Cyril Bailey, *The Greek Atomists and Epicurus* (Oxford, 1964, and New York, 1964).

JB (EGP)   John Burnet, *Early Greek Philosophy* (Oxford, 1892; 4th ed., London, 1930).

D-K   Herman Diels, *Die Fragmente der Vorsokratiker,* 7th ed., ed. with additions by W. Kranz (a photographic reprint of the 5th ed.) (Berlin, 1954).

DL   Diogenes Laertius, *De vitis . . . philosophorum,* trans. R. D. Hicks, *Lives of Eminent Philosophers* (Cambridge and London, 1942).

DNR   Lucretius, *De Natura Rerum,* trans. W. H. D. Rouse (London and New York, 1924; Cambridge, 1947). See also the editions of C. Giussani (Turin, 1896–97), C. Bailey (Oxford, 1899, 1910, 1947); H. A. J. Munro (London, 1900, 1929); W. E. Leonard (Madison, 1942; New York, 1947); R. C. Trevelyan (Cambridge, 1937).

EZ (HGP)   Eduard Zeller, *A History of Greek Philosophy* (London, 1881) (trans. of a part of the following).

EZ (PG)   Eduard Zeller, *Die Philosophie der Griechen in ihrer geschichtlichen Aufgabe,* etc. (7th ed., Leipzig, 1923).

Abbreviations of Galen's works: *An in Art., An in Arteriis Sanguis Contineatur; De Plac., De Hippocratis et Platonis Placitis; De Sem., De Semine; El., De Elementis ex Hippocrate; Form. Fet., De Foetuum Formatione; N.F., De Naturalibus Facultatibus; U.P. De*

Usu Partium; U. Puls., De Usu Pulsuum; U.R., [De] Utilitate
Respirationis; Temp., De Temperamentis.

KF (PSP)   Kathleen Freeman, *The Pre-Socratic Philosophers: A*
            Companion to Diels (Cambridge, 1959).

WKCG   William K. C. Guthrie, *A History of Greek Philosophy,*
            vol. 1 (Cambridge, 1962), vol. 2 (Cambridge, 1965).

Kahn   Charles H. Kahn, *Anaximander and the Origins of*
            Greek Cosmology (New York, 1960).

K-R   Geoffrey S. Kirk and John E. Raven, *The Presocratic*
            Philosophers (Cambridge, 1957).

K   Karl Gottlob Kühn, *Medicorum Graecorum Opera Quae*
            Exstant, vols. 1–20 (Leipsig, 1821–33) [cited by vol-
            ume and page no., e.g., K3:265 for Kühn, vol. 3, p.
            265].

LH   "Epicurus to Herodotus, Greetings, etc." (the letter to
            Herodotus, in *DL*).

HISTORY OF GENERAL PHYSIOLOGY

· 1 ·

The Life-matter Problem and
the Strategy of Biology

There is a mood of optimism today in biology stemming largely from insights rapidly being gained into the physics and chemistry of life. Stubborn questions are beginning to yield to biochemical probing—questions about such subtle, and central, phenomena as the supply of energy for vital action, the conversion of food into tissue, and, most enigmatic of all, the transmission and use of hereditary "information." None of our current ideas on these subjects is fully final or complete; all are certain to be elaborated and revised. But recent accounts are more convincing—more precisely explanatory and predictive, than earlier accounts have been. The result is a cautious new confidence of biologists in the power of their discipline to deal scientifically with life in its most unique and hitherto elusive manifestations.

These hopeful developments hold a special interest for the historian because he sees them as the latest phase in an enterprise as old as science itself: namely, the effort to associate life with certain kinds or conditions of matter. We shall discover, as we proceed, that even where life has been viewed as an immaterial entity or agent, as something separate from and "superadded" to the body, it has generally been thought to require for its active expression a proper material vehicle or substrate.

The Life-matter Problem

Ideas of life and matter form, historically, two streams of related scientific concepts, and our primary interest will be in the interactions of these two conceptual currents. Such interaction may be said to have occurred whenever scientists have asked What is matter? What is life? and What condition of matter permits the effective appearance of life? Taken together, these questions constitute what we shall call the *life-matter problem*. Our story tells, in effect, of the long effort of science to solve this problem. Perhaps also by studying the influence which ideas of life and matter have exerted on each other we may be lucky enough to learn something about the role of conceptual interaction in scientific thinking more generally.

A Nineteenth-Century Example

In 1835, the French microanatomist Felix Dujardin made an easy experiment that was to alter the course of biological inquiry. By mechanically crushing certain microscopically small animals, Dujardin obtained from them something which he called a "living jelly" and which he described as a "glutinous, diaphanous substance, insoluble in water, contracting into spherical masses, sticking to the dissecting needles and letting itself be drawn out like slime, . . ."[1] Dujardin called this substance *sarcode*.

The rest of the story is familiar—how other observers found an apparently similar substance in other animals, and in plants; how the term *sarcode* gave way to the more generally accepted *protoplasm;* how the idea grew that living things owe their vitality to this substance; and how efforts were made from then on to find out its character and constitution. Protoplasm was not only necessary for life, in certain sectors of mid-nineteenth-century opinion; it was itself alive. It was the actually living part of the organism.

Protoplasm in Perspective

Before protoplasm. The name protoplasm was new in the nineteenth century, but the idea it entailed was an old one. Scientists of every era have understood that life implies special physical arrangements and in every era have tried to find out what physical arrangements make life possible, or permit its effective expression. Historically, each solution to this problem has had ultimately to yield to a partly new solution—one that squared better with new knowledge and new assumptions. In perspective, then, protoplasm appears as a nineteenth-century solution—but only one in a series of solutions—to an old and stubborn scientific problem.

After protoplasm. Subsequent events have subjected the concept of protoplasm to the fate of its predecessors. In fact, the view that

[1] Felix Dujardin, "Sur les prétendus estomacs des animalcules infusoires et sur une substance appelée Sarcode," *Recherches sur les organismes inférieurs,* part 3, in *Annales des Sciences Naturelles: Zoologie* (Paris, 1841), 2d ser., vol. 4, p. 343.

would limit life to a structureless semifluid substance was challenged from the beginning. Dujardin himself admitted that modified sarcode appeared *as structure* in the more highly organized animals. After that for a hundred years, as light-microscopes and the techniques of using them improved, an increasing wealth of organized structure was discovered. And when, in our century, the electron microscope was perfected, a new world of microstructural detail came into view. Where structurelessness was formerly described, there now appeared, extending in every direction through the cell, a wonderful architecture of granules, fibers, and membranes which are not only needed for cellular life, but take part actively in it. The old protoplasm has been succeeded, in effect, by a new "protoplasm," some of whose components possess the remarkable property of being at once architecturally stable yet chemically active.

To be sure, these stable structures are chemically dependent upon and in some instances even move about in the complex moist milieux that bathe them. But few modern biologists would limit life to the cell's fluid (or optically homogeneous semifluid) component. Rather, "living matter" currently appears as a highly heterogeneous organization of chemically active, and interactive, components ranging from fluid to visibly structured.

Necessary Precautions

To speak of an early solution to the life-matter problem is not to suggest that its author saw the problem as we do. The unwary historian may be tempted to ascribe to early theorists ideas that were actually only introduced at a later period. Life meant something different, though not completely different, to certain Pythagoreans (for whom the whole cosmos was an animate being) from what it meant to, say, Bichat (for whom it was "the sum of those processes that offer resistance to death"), or to certain twentieth-century theorists who view life as a kind of evasion of the second law of thermodynamics.

And as with life, so with matter. Before Leucippus introduced the atomic theory into Greek thought (ca. 440 B.C.), there probably was no concept of matter in the sense of a homogeneous, impenetrable substance carved up into tiny fragments. Many classical scholars have suggested that the focus of Greek physics in general was less on matter than on nature, that is, on the really existent. Indeed, it was not until nearly a hundred years after Leucippus that Aristotle first used a word for matter as such. Once proposed, matter was accepted by some thinkers, rejected by others, and by still others subjected to a whole series of transformations until modern physics changed it almost beyond recognition.

Here we shall use *matter* in a very comprehensive sense. The world studied by physics is a differentiated world, and alternate ideas were early advanced concerning the nature of its differentiations. According to one idea, the cosmos is essentially diphasic, one of its phases being empty or incorporeal and the other corporeal or full. In dealing with this view, we shall limit the term *matter* to the supposed corporeal phase. According to the other idea, the cosmos is a multiphase system all of whose phases are corporeal with no empty space between them. In dealing with this view, we shall refer to all phases as varieties of matter.

Proponents of the diphase conception, in successive eras, have raised a host of questions concerning the character of the non-corporeal or intercorporeal phase. Greek physicists inquired, for example, whether this phase was a complete nonentity (absolutely nothing) or whether it was not, rather, an undifferentiated entity of some sort (we shall hear different theorists express different views on this question). Again, in post-Renaissance physics, the matrix of bodies was viewed by many thinkers as, though not corporeal, nevertheless differentiated and even substantial, the substance in question being referred to as *ether*. Thus wave-theorists viewed the intercorporeal matrix as susceptible of differentiation into waves. And certain gravitation-theorists saw it as a force-field differing from place to place. From the mid-nineteenth century the non-corporeal or intercorporeal phase became a medium for energy-transmission. For this and other reasons, physical differentiations were regarded, at this period, as instances of matter on the one hand and energy on the other. And even these distinctions were to require reformulation when near the beginning of our century matter and energy proved to be interconvertible.

From today's perspective, to apply corporeality too boldly to the subvisible organization of matter appears an uncritical extrapolation from gross sensory experience with apparently concrete systems termed "bodies." When Aristotle first introduced a term for matter he used a word that had previously meant wood (though his own "matter" was highly dynamic and far from wooden). The point is that a vast difference exists between the internally homogeneous, impenetrable, and indestructible atoms of a Leucippus and the differentiations (particles) still collectively termed matter in modern physics.

We stand warned, therefore, that we must be sensitive to the changing content of terms like "life" and "matter," whose meaning is altered by each generation of users.

The Interpretive Strategy of Biology

This book is organized around a central idea—a general thesis about the way in which physiology has developed. Even though this thesis can only become fully meaningful as our study unfolds, it may be well to state it at the outset. It is: that *progress in the solution of the life-matter problem, and in physiology as a whole, has occurred through the application over the centuries of analytical procedures laid down in ancient Greek science.* Building on Babylonian antecedents, on myth and religion, and very largely on everyday observation, early Greek thinkers introduced a master strategy of interpretation that proved so fruitful and so flexible as to have been used in all later periods up to the present. We shall sketch the main features of this strategy here and add depth and substance to the sketch as our narrative evolves.

1. Classic questions. Physiology, commencing with the Greeks, has concerned itself with certain stubbornly recurrent questions— *classic questions,* we shall call them—to which the answers given in successive generations have been only partially new. The content of these questions will be apparent almost as soon as our story begins, and we shall find that one of them—the life-matter question —is in specific ways central to all the others. *Questions* as we use the term here may be distinguished from *puzzles* as that term is used by Thomas Kuhn in his important study, *The Structure of Scientific Revolutions. Puzzles* in Kuhn's sense are exercises entailed in squaring certain tentatively accepted broad conceptual schemes, or paradigms, with arrays of perceptual data. *Classic questions* in the present sense stem from everyday facts of life—motion, generation, nutrition—whose superficial phenomenal expressions demand, or seem to the scientist to demand, some further and deeper understanding (see below, section 2). The distinctive feature about classic questions is their generally long duration. Paradigms and the puzzles to which they give rise may come and go; classic questions remain, though the content of each question changes with time.

2. Percept and concept. In treating classic questions, physiologists from Greek times have studied the organism with a dually directed concern. They have examined both (*a*) the sensible aspects of living things and (*b*) their insensible, but inferable or conceivable, aspects. One kind of scientific "explanation" is, in effect, a reformulation of "sensibles," as Aristotle sometimes called them, in terms of underlying insensibles, a restatement of percept in terms of concept. De Santillana has said (1961) that science means a search for the impersonal invariants *behind* apparent events. A fundamental task for the historian is to discover why such restatements tend to be tentative and to need to be replaced. About this we shall have much to say.

3. Cosmology and biology. The two sciences of cosmology and biology have developed from the beginning not separately but hand in hand. The connection between them is reflected in the habit of Greek and later thinkers of referring to the organism as a "little universe" or *microcosm* comparable in important ways to the *macrocosm* or universe at large. When Greek science began, it inherited a prescientific tendency to think of the cosmos and its contents in biological terms; the whole universe was, even for certain sophisticated thinkers in those days, an actually living being. Almost at once, however, this situation reversed itself and cosmology acquired the leading edge. Increasingly, new cosmological—and physical and chemical—ideas began to alter biological thought, rather than vice versa. After that, a perennial feature of interpretive strategy in biology was the use of cosmological and physical and chemical innovations as a basis for innovation in the interpretation of living things. We might state this in Kuhnian terms by saying that periodic reconstruction is forced upon biology by paradigm shifts occurring in cosmology and in chemistry and physics.

4. Thematic character of physiology. From the beginning, physiology has developed under the influence of certain organizing ideas or themes. Life is a flame, or is like a flame. It is a ferment, or like a ferment. The organism resembles a crystal. Life is a double process of degeneration and regeneration. As strategic devices, these and many other orienting ideas have exerted at times a liberating (but also at times a limiting) effect on the physiologist's effort to answer classic questions. Note, however, that a theme (e.g., "life is like fire") or classic question ("What is the nature of assimilation?") may mean different things when reintroduced, even in the same words, in successive scientific eras.

The foregoing four observations are not offered as new or profound. Yet, however obvious, they have sometimes been neglected in the historian's portrayal of the biological endeavor. Used with caution, they can help us understand the *why* of the biologist's interpretive behavior. And they alert us to ask to what extent his concern with the details of life is informed by a broader concern, implicit or expressed, with the nature and physical conditions of life in general.

The Boundaries of our Problem

The relation of life to matter is not an exclusively scientific problem. It extends, in its implications, to philosophy and religion, to social thought and conduct, and to literature and the arts. Our present concern is with the problem in its scientific aspects insofar as these can be isolated from other aspects. Moreover, since even the scientific aspects will prove immensely broad in their ramifications, our treatment must be selective rather than fully connected. We must be content to find out what *can* happen when streams of thought about life and matter come together, and cannot hope to discover everything that *did* happen.

As the drama develops, more than eighty personae will cross our stage. This number places a boundary on the amount of detail with which each can be presented. Our perspective cannot include any thinker's whole biological or biomedical theory; it must focus on the way in which his ideas of life and matter affected his interpretation of physiological function. This point merits special emphasis, since it distinguishes the purposes of this study from the legitimate aims of much historical scholarship. An analogy may perhaps be drawn between our present intent and that of the biologist who studies a single but not simple phenomenon (say, energetics or nervous integration) as represented in an array of animals without trying to consider the evolutionary relations among the systems he studies and without describing any one animal in its entirety. Such a study could perhaps be termed comparative—and in a partial sense causal-analytical—rather than primarily phylogenetic.

Finally, for reasons suggested in the preface, our study will be bounded by two dates, ca. 600 B.C. to ca. 1900 A.D., though concluding notes will be added on the life-matter problem as it exists in the third quarter of the twentieth century.

CONCLUSION

One must be impressed in tracing the life-matter problem by the wealth and ingenuity of solutions offered by successive generations of thinkers. Each new solution reflects the new knowledge and assumptions prevalent in its particular period. Each solution stands in a kind of metaphorical relation to those which preceded it in the series. Each bears also a metaphorical relation to the basic reality which all would discover. How close or what kind of approximation the offered solution may have to the fact itself is a metaphysical question which we shall deliberately set aside in order to follow the scientific aspects of our problem. As a scientific problem, it is one of the most significant and provocative that has ever challenged the human imagination.

Part 1

THE PROBLEM IN ANTIQUITY

Pre-Socratic Solutions: Part One
Introduction

Our story begins with the emergence of the life-matter problem as a part of pre-Socratic Greek science (ca. 600 B.C. – 400 B.C.).

We have to ask first, however, whether the problem in fact existed for thinkers of this early period. The answer, we shall see, is that it did exist for the Greeks but had special meanings for them that it largely lacks for the biology of our own era. One of the challenges—and pleasures—of the historian is to escape the modes of thought of his own period and to place himself as far as possible in the intellectual position of those whose ideas he wishes to understand. To do this in the present case is not easy, but we may gain some help toward this objective by noting certain Greek words often translated by the English "life" and some of the ways in which the Greeks used these words.

Greek Concepts of Life

Three terms—*bios, zoe, psyche*—especially attract our attention. The semantic status of these terms is a knotty affair that the thesaurists have not completely untangled. The three words overlapped considerably in their content; yet each had also a distinctive cluster of meanings. The following generalizations are based on a preliminary study of the contexts in which these words appear in Greek literature from Homer through Aristotle.[1] Especially helpful are cases, of which many may be found, in which two of these words are juxtaposed in a single statement.

Bios, first of all, was a term of very general application, used in many of the ways in which we use "life" in English. It, rather than either of the other two, was used to express *mode-* or *duration-of-life,* serving the Greeks as the English term serves us in such expressions

---

[1] A fuller, documented study of these and other expressions is in progress.

as "the good life," or "life-span," or "a living" (livelihood), or even a written "life" (a biography).

The other terms, *zoe* and *psyche,* are more helpful to our quest than is *bios* because although they too were used somewhat interchangeably, they suggest two different ways in which the Greeks often thought about the things that *we* think of as "living," (a class to which the Greeks sometimes admitted things that we should exclude). *Zoe* came closer than either of the other two terms to meaning life as an *ensemble of actions* (such as generation, nutrition, and locomotion) whereas *psyche* meant life, in some sense, as *soul,* a designation we shall presently examine.

Zoe was relatively infrequent in earlier Greek usage but the various inflections of the verb-form *zen* or *zaien* (to live) were common, as were those of the substantive *zoon* (a living being). Life or living, in the sense conveyed by these terms, was often explicitly opposed, in Greek writings, to death. In the case of mortal creatures (plants, animals, men) life in this sense was the distinctive act, or ensemble of acts, that they perform *qua* plants, animals, and men and that they cease performing at a certain moment when they are said to die. If the verb *zen* came into general use earlier then the substantive *zoe,* that could have been due to the fact that the sort of life implied was not life as concrete entity or agent but, rather, life as action (suitably denoted by a verb).

"Soul" is the usual choice in translating *psyche* (and its Latin equivalent *anima*), yet the semantic role of *psyche* was far from identical to that of "soul" in modern English. For example, *psyche* carried more than a suggestion of what we signify by "breath." Some authorities render *psyche* by "life-soul" or "breath-soul," and we shall see presently that the suggested equivalence, or at least association, of life and breath is scientifically important.

By Homer (850 B.C.) *psyche* was used partly interchangeably with, and much less frequently than, another substantive, namely *thymos* (life, soul, spirit, courage, heart—according to context). Homer on "life" can best be studied by following his heroes on the field of battle where that commodity was constantly in danger. When a commander put "life" into his men, *thymos* was the term Homer used, and it was *thymos* that was in danger of being lost in combat. But when Posidonius pulled the spear out of the fallen Sarpedon, the latter's *psyche* came with it. And it was Posidonius' own *psyche,* in turn, that Achilles addressed later in Hades. *Psyche* as life in the physiological sense was uncommon in Homer; nevertheless, he and Hesiod used it occasionally in that sense, that is, for the sort of life that men share with animals. Much later, the Ionians made *psyche* the life that men and animals have in common with plants (see chapters 3–5).

The meanings of *psyche* came to be many and varied. It seems to have connoted, among other things, an ability to set things in motion and to keep them moving. As a motor, the life-soul was sometimes made responsible for such activities as walking and breathing. Some users made it the cause also of the detailed movements that underlie all bodily processes (such as assimilation of food or the flow of the body fluids). *Psyche* was used, too, to suggest various modes and degrees of perception, ego-awareness, emotion, and reason, but what is of special interest to us here is that while using *psyche* in all these and other ways, the Greeks also employed it where we should not use "soul" but "life."

This brings us to an important aspect of the Greek view of life-matter relations. Greek biology builds in effect upon a dual concept of life; life appears (*a*) as *action* and (*b*) as *soul*. When the Greek biologists became interested in a scientific analysis of *life-as-action* (i.e. in what we call physiology) they did not have to give up simultaneously their belief in *life-as-soul*. Both sorts could be present in the organism simultaneously. Psychic life could be thought of as inducing or permitting active life.

The foregoing way of looking at life differs markedly from our view. Twentieth-century biology has focused almost entirely on life-as-action whereas for many Greeks and, as we shall see, for many later thinkers organisms possessed two "lives"—soul *and* action. It is this duality that gave the life-matter problem a meaning for them that it no longer has for the biology of our day. Part of our longer story will concern the steps by which Greek ideas were replaced by a different concept from which life-as-soul was eliminated, with Descartes playing a pivotal role in the transition. But to understand this point more exactly we need to note briefly some of the Greek ways of looking at matter.

Greek Ideas of Matter

Scientific thinkers have always been curious about two aspects of the world as revealed by the senses: its nonhomogeneity in space (its diversity) and its nonhomogeneity in time (its transformations). Physics, viewed historically, is a continuing effort to understand correctly these two sorts of nonhomogeneity.

From the very beginning, physicists have been convinced that the world's diversity was not so great, nor its transformations so complex, as sensory evidence seemed to indicate. Behind the deceptive sensory world there lay a real world that was more orderly and often simpler. To understand the sensory world meant, for Greek physics, to recreate it in terms of the real or "physical" world that, in Heraclitus' phrase, lay "hidden."

The variety and mutability of sensible substances appeared, when thus recreated, as differing conditions of a real substance or substances that lay concealed. Concerning this hidden, natural world, physicists began to ask certain questions: how many differentiations, or real substances, are contained in it? have there always been that many (what is their origin and history)? how do these substances differ? to what sorts of change are they subject?

To the first of these questions—how many sorts of substances have a real existence?—the answers preferred by Greek theorists were: either *very many* (Anaxagoras said an infinity) or *very few* (one, two, and four substances were posited by various pre-Socratic theorists, we shall discover, and Aristotle added a fifth one).

Strikingly, the first theorists to think in a seriously scientific way about the whole question were the very ones who attempted the greatest simplification. Thus Thales, for whom *water* was reputedly the first and perhaps the only real existence, was already active at the beginning of the sixth century B.C.; he was the first person, in Burnet's view, who can rightly be called a man of science. It was only a few decades later, moreover, that two other Ionians chose, respectively, *air* (Anaximenes) and from a somewhat different point of view *fire* (Heraclitus).

The fact that monism came earlier than pluralism reflects the tendency of early Greek science to make use of prescientific materials of intuitive or mythological origins. Scientific monism, in other words, was a formal expression of the primal need of thoughtful men to find unity in nature—thus, Musaeus, the religious poet: "All things came into being from, and ultimately become, one thing."

Having chosen a primary substance, the Greek physicist posed the same question as before with only a slight modification. He asked not how many substances were to be acknowledged but, rather, *how* many states or conditions of the one and only substance. And again, the favored answers in Greek physics were: either very few, or very many. Thus, Heraclitus appears to have permitted fire and two derivatives of fire to compose his cosmos, whereas Democritus later postulated a great many sorts of atoms all made of the same single substance.

The postulation of first principles was linked in Greek thought with two different ideas about the nature of change. Changes were: either (*a*) transformations or conversions; or (*b*) mere spatial redistributions occurring without transformation. These two sorts of change figured differently, as we shall discover, in the physical systems of different theorists and had correspondingly different effects on their ideas of matter and of life. The problem of the nature of change is especially acute where more than one principle (e.g. the familiar four) are postulated, for here physics asks whether it is possible for one element to turn into another, fire into air, for example. We shall see that to Empedocles, the founder of the four-element theory, such a conversion seemed quite unthinkable, but to Aristotle, later on, it presented no difficulties. Between Empedocles and Aristotle stood Plato whose theory permitted interconversions between fire, air, and water. But on closer view, Plato's prime substances turn out to be composed of a common ingredient present in different amounts in fire, air, and water. Thus, in Plato's theory, conversion entailed, after all, a spatial redistribution of the ingredients that he supposed his first principles to have in common (see chapter 7).

Life and Matter

Life had for Greek biology, as noted earlier, a double signification in that the organism was viewed by many thinkers as possessing both *life-as-soul* and *life-as-action*. This dualism concerns us directly because many Greek thinkers imputed life-as-soul, *psyche,* to a particular substance—to a first principle or element, for example, or to several elements in combination. For such thinkers the life-matter problem became one of identifying the substantive vehicle of soul and of specifying the nature of soul's relation thereto.

Some Greek theorists took the further logical step of stipulating that only things containing this animated substance in proper apposition with other substances were able to move about, feed themselves, multiply, and, generally, display life-as-action. For such thinkers the solution of the life-matter problem depended on the posited relation between the animated principle or substance and other substances in the body.

Of the primary substances, three figured prominently in the Greek effort to identify the animate ingredient—namely, water, air, and fire. In tracing the interest of Greek thinkers in them we must, however, move with a certain caution. Evidence on the pre-Socratics is partial and circumstantial. We cannot always be certain who first made a particular statement or exactly what the statement meant to its author. To be sure, our present interest is not in whether this or that concept is attributable in a hard-and-fast fashion to this or that individual. Our concern is, rather, with the ideas themselves. But here we encounter the problem of meaning.

A modern student of Greek science has compared the earliest Greek scientific ideas, mixed up as they often were with poetry, myth, and religion, to those sculptures that Michelangelo and Rodin only half-carved from the rough stone out of which they are fashioned. It is true that some of these early ideas were only vague intimations or half-statements of later scientific concepts. But there were others, as we shall see in the succeeding chapters, that stand forth with considerable distinctness. In either case, vague or clear, they represent the first, often pregnant efforts to solve the most central of biological problems.

Possible Formulations

Life-matter relations appear susceptible, as judged by both Greek and later thought, to five distinct formulations. The differences will begin to be apparent in the immediately ensuing chapters, but it may be helpful to suggest them, at least in outline, in advance.

Life (1) as identical and (2) as immanent. First, some Greek and later thinkers appear to have thought of life as an actual variety of matter (as a material or corporeal differentiation of the cosmos), whereas others seem to have viewed it not as matter in the strict sense but as somehow immanent in it (related to it in the way that extension or mass is). These two putative sorts of life-matter relations—relations *(1)* of identity and *(2)* of immanence—are not always easily distinguished, and, in the case of certain Greek thinkers, we cannot be sure to which of the two they subscribed. Aristotle (384–322 B.C.) thought that Anaximenes (ca. 550 B.C.) meant actually to identify life-as-soul with air; that Heraclitus meant to identify it with fire; Hippo, with water; Critias, with blood; and so on. But from the existing fragments it is hard to be sure whether these thinkers regarded the materials in question merely as immanently "alive" or as somehow identical with "life" itself. In the eighteenth and early nineteenth centuries A.D., with the idea of life-as-soul very largely repudiated, we still find life equated, occasionally, with electricity or some other "subtle" or "imponderable" matter (see chapter 43). The idea of life as immanent also experienced a revival in the mid-eighteenth century (see Maupertuis, Buffon, Diderot) and even occasionally in the nineteenth (see E. H. Haeckel).

(3) Life as imposed. Plato viewed life-as-soul as a distinct but nonmaterial entity, an entity "bonded" to certain objects and inducing in them a characteristic ensemble of behaviors (life-as-action). The relation of life-as-soul to matter in this case may be termed a relation of *imposition.* We shall encounter a number of variants, with Plato as the pioneering proponent, of the idea of life-as-soul as imposed on matter and hence as causing it to display life-as-action.

(4) Life as organization. We shall occasionally encounter the idea that life is not a kind or immanent property but a special arrangement of matter that permits it to behave in a "lively" fashion (to eat, move about, grow, reproduce, and otherwise evince what we think of as "signs of life"). The most influential spokesman for this point of view was Aristotle who equated life-as-soul with *form.* By form he meant, among other things, a dynamic condition of matter permitting it to impart that same condition to other suitable matter. On the whole, this notion—that life *is* organization—has not been especially common in the history of biological theory. A later (and in comparison with Aristotle lesser) biologist to espouse this idea was Lamarck who equated life with *arrangement* and defined it as a special "state of affairs" (*état de choses*) existing in plant and animal bodies.

(5) Life as an emergent consequence of organization. Fifth and finally, life may be considered as an ensemble of distinctive activities that certain objects (organisms) engage in solely because of their material constitution. To this sort of supposed life-matter relation the term *emergence* has been aptly applied. The term emergence has

been employed by theoretical biologists in our own century in several overlapping ways. In evolutionary thought, it has been used to suggest that the increasingly complex organization of higher life-forms permits the appearance (the emergence) in them of new modes of life, new functions or behaviors, impossible in less organized forms. The term has also been used to suggest that the behavior and character of an organized system, living or other, cannot be predicted, at least in all its aspects, from a knowledge of its parts. The fundamental point in both these uses is that systems have characteristics that are absent from their isolated components, characteristics that only appear (that emerge) when the parts are assembled. In antiquity, Epicurus used life, explicitly, as an example of emergence, insisting that it was absent from the body's atoms considered singly. It is this idea of life, as an ensemble of activities produced by organization, to which the term *emergence* is here applied.

The idea of life-as-emergent may seem no different, at first, from the immediately preceding conception of life-as-arrangement, but the distinction between them is important. Lamarck equated life with *organization itself,* whereas emergentist thinkers make it, rather, action occurring as the *result of organization.* The latter view has won an immeasurably larger number of adherents than the former. Indeed, in modern biology (e.g. of the Renaissance and later periods up to the present) we shall find that the idea of life-as-emergent, that is as the product of organization, has tended gradually to displace all the others.[2]

Theoretically, all serious biological thinking should assume one or another of the five foregoing formulations.[3] Actually, we shall dis-

[2] The view that makes life organization itself, rather than an activity emerging therefrom, has been associated, especially by Renaissance theorists, with the observation that, under certain conditions, all visible action can be reversibly suspended in an organism, which is presumed, during its inactive period, to retain the organization indispensable to its recovery.

[3] The first two of these relations—those of identity and immanence—have been brought together in the past under the designation *hylozoic,* while the view of life-as-imposed corresponds roughly to that often termed *vitalistic,* and the emergent relation has been called either *materialistic* or *mechanistic.* These terms are acceptable enough, and there is no wish here to displace them. But, historically each has acquired a mixed constellation of meanings that make them, for our purpose, not particularly useful. Thus *hylozoic,* unless qualified, fails to specify whether life is a *sort* or only a *property* of matter. *Vitalistic* sometimes, but not always, means *psychic* and is in this sense ambiguous. Moreover, hylozoic and vitalistic are sometimes used interchangeably. *Mechanistic* is variously employed to mean "machine-like," "soul-less," and a number of other things that render its use equivocal.

cover that some thinkers remained rather vague with respect to the question; in the case of others, syntactical problems make it hard to pinpoint their opinions. But most of the biologists whose ideas we shall touch on have viewed life as either immanent, imposed, or emergent. Only a few have seen it as a peculiar kind of matter, or special arrangement thereof.

Some historians object to the expressions "pre-Socratic biologists," "Greek biologists," and the like, partly because the terms *biology* and *biologists* were only introduced about A.D. 1900. This objection is legitimate but, to our taste, too narrow. Many Greek thinkers were biologists in the sense of being interested in life *in general* as manifested in various ways in everything that lives. We shall not deny these thinkers the designation biologists merely because their interest in life was interwoven with even broader interests in man and the cosmos.

· 3 ·

Pre-Socratic Solutions: Part Two
Water, Air, and Fire

THALES

If one could credit Thales (fl. ca. 580 B.C.) with even half the ideas sometimes attributed to him, he would appear as a remarkably sophisticated theorist. The Thales presented to us by various ancient and some modern interpreters chose water as a first principle,[1] brought it to life by endowing it with a motive *psyche,*[2] and made living water a causal agent in generation and nutrition.[3] But how much of all this was put forth by the actual Thales?

The question is a vexed one. A scholar of our own time has called Thales a "fabulous figure."[4] Indeed, in following ancient sources, one seems to see a fable grow up around him. His image as a systematic thinker was questioned long ago by Paul Tannery in a short essay that was a masterpiece of urbane demolition.[5] Since then, Thales' image has alternately brightened and faded as one commentator succeeded another.[6]

The important point in any case is that behind the legend there was a real Thales who, in the early sixth century, pioneered the Greek effort to found a serious new science. This he did by substituting physical concepts for folk ideas, myths, and vague empirical apprehensions. On life and matter the Thalesian legacy may be summarized briefly as follows.

Aristotle, on evidence that is largely circumstantial, said that

---

[1] Arist. *Metaph.* 983b.
[2] D-K, A 23.
[3] Arist. *Metaph.* 983b5.
[4] Kahn, p. 6.
[5] In *La Grande Encyclopédie* (Paris, 1886–1902), 30:1149.
[6] K-R, pp. 74–98; WKCG, 1:45–71. For a skeptical view, see D. R. Dirks, "Thales," *The Classical Quarterly,* n.s., 9 (1959): 294–309.

Thales made life-as-soul (*psyche*) immanent in water and water a necessary ingredient for life-as-action (*zoe*).[7] It is believable that Thales did think something of the sort—though he would perhaps not have recognized his ideas in the account that Aristotle later gave of them. Aristotle had a habit of restructuring his predecessors' thoughts in order to fit them to a common frame for comparative and critical treatment.

Water as arche. Two considerations, though speculative, have inclined scholars to the view that Thales made water a psychically endowed first principle. First, long before Thales, men expressed, through myth and religion, a strong sense of the unity of the cosmos; Thales would almost surely have sought to express this unity scientifically, and perhaps substantively. Second, Thales would naturally have chosen as his unifying substance something mobile, transformable and, not improbably, vital. To consider the *arche* alive would not have required of Thales an act of conscious attribution. In early thought, life, variously conceived, had a wider distribution than in the thought of our times: life was often assumed in objects which today we consider nonliving. The idea of a living *arche* might well have arisen as an almost automatic supposition.

But—why water?

Water appeared repeatedly as a life-giver in the creation myths of Greece, Egypt, Babylon, the Near East, and Persia. The twentieth-century religious historian, Mircea Eliade, mentions scores of ways in which water appeared as a mythical or religious symbol of generation and regeneration, of formless preexistence and ultimate dissolution, of divinity, of fertility, of youth and health, of life both mortal and eternal. Eliade's examples are drawn from virtually all the major mythologies of Eurasia, America, and Africa.

When Greek science began, the substance of myth was not suddenly abandoned. It formed, rather, an important source of raw material for scientific speculation. So, moreover, did everyday observation. The connections between life and water were dramatized by such commonplaces as the narrow dependency of all organisms on moisture, the moist nature of food and semen, the excesses and deficiencies of body fluids in various diseases, the insensitivity of dry parts of the body (hair and nails), and such ecological phenomena

---

[7] See notes 5 and 6 above.

as the bursts of vegetation that occur in rainy seasons or moist micro-climates.[8] In prescientific civilizations, such empirical commonplaces must have furnished working materials for the elaboration of myth; when formal science began, the same materials were used in elaborating scientific doctrine.

All of which shows nothing more than that the ideas ascribed to Thales would have been received into an intellectually friendly climate. Perhaps the nearest thing to a suggestion that they actually were introduced by him is the fact that related ideas appeared in the same community, Miletus, only slightly later, when Anaximenes ascribed a life-soul, as we shall see, to air.

At the very least we may assume the establishment of a tradition associating life with a moist component of the living system; and we shall discover that this was to prove a very hardy tradition. We need to note, however, that Thales did not leave us a clearcut way of distinguishing things we ordinarily think of as living (plants, animals, men) from other things. We have already noted that primitive men often treated nonliving objects (nonliving by our standards) as if these objects were very much alive, and this way of looking at nature was carried over into Greek science in its incipient stage. Thus Thales, even while transcending primitive thought-modes, extended life-as-soul to magnetic stones,[9] amber, and perhaps to natural objects (especially mobile ones) generally. Moreover, when Thales spoke in this vein (as Aristotle makes him do) he gave the life-souls of things a *divine* character. According to Aristotle, "some say that it [*psyche*] is intermingled in the universe, for which reason, perhaps, Thales also thought that all things are full of gods."[10] Only gradually and imperfectly did pre-Socratic science move toward the sort of distinction between living and lifeless that today we take for granted.

Later developments. On the whole, the idea of water as the animate element earned only a modest success in antiquity. Although certain everyday data favored the view, others opposed it (too much water can suppress life). Its most explicit exponent was Hippo of Samos (fl. 450 B.C.), a Pythagorean physiologist coming long after

[8] WKCG, 1:54–72; see also William K. C. Guthrie, *In the Beginning* (Ithaca, 1957), p. 30, and M. Eliade, *Patterns in Comparative Religion* (Cleveland and New York, 1963), pp. 188–215.

[9] Arist. *De An.* 405a; DL 1. 24 (cited by volume and page).

[10] Arist. *De An.* 411a (trans. K-R). The same idea is imputed to Thales by Plato *Laws* x 899 B, and D-K, A 23.

Thales. There was also an interesting *Urwasserbegriff* in the pseudo-Hippocratic treatise called *Sevens* (date uncertain) where we are told in a tantalizingly brief paragraph that there is in man and the world one same fluid of which all apparent fluids are modifications due to heat and cold; however, in this work, water seems not to have been a first principle; the first principles were rather the hot and the cold, and even these were adapted to the author's aim of showing how the properties of the number seven apply to the cosmos and the living microcosm.[11]

As to Hippo, he seems to have been influenced by his observation that the sperm is moist, and he looked on sperm as endowing the future organism with soul. From Aristotle's pupil Meno we hear that "Hippo of Croton believes that there is in us a natural moisture whereby we perceive and by which we live."[12] But Aristotle himself dismissed Hippo from the company of significant theorists with the statement "Hippo no one would think fit to include among these thinkers, because of the paltriness of his thought."[13]

In assessing Hippo, or at any rate the attitude toward water which he represents, we should note that this view will have an elaborate and important future development. We shall hear about such a natural moisture (usually under the designation "radical humor") from certain Arabian authors, from the alchemical physician Arnald of Villanova, from Renaissance medical writers including Fernel and Paré, and in the seventeenth century from William Harvey, who equates the radical humor with the homogeneous substance ("crystalline colliquament") with which life starts. The radical humor will be different things to different authors, a substance determining the idiosyncrasies of individuals (or their parts), an innate vehicle of the innate heat, a formative substance capable of differentiation into tissue, etc. In the eighteenth century, the radical humor idea will die out, but Harvey's colliquament will produce a progeny that can be traced from C. F. Wolff, via Treviranus and others, to Dujardin whose "sarcode" will lead on to the protoplasm or "living matter" of nineteenth-century biology.

[11] *Peri Hebdomadon,* known only from a Latin text and Greek fragments, ed. É. Littré, *Oeuvres complètes d'Hippocrate* (Paris, 1853), 8:642.

[12] W. H. S. Jones, *The Medical Writings of Anonymus Londinensis* (Cambridge, 1947), pp. 52 (Gr.) and 53 (Eng.).

[13] Arist. *De An.* 405b and *Metaph.* 984a3 (trans. Ross); see also KF (PSP), p. 210.

ANAXIMANDER

Meanwhile in Thales' own approximate era the biological interest in water was continued, but from a radically modified point of view, by his younger fellow townsman, Anaximander (fl. ca. 565 B.C.). Anaximander thought that living things had their origin in water, but water was no longer for him a first principle nor was it preeminently soulful. Historically, Anaximander stands out more distinctly than Thales. Doxographic evidence makes it clear that he was a powerful pioneer in geometry, astronomy, meteorology, and geography—disciplines which he treated in a connected, synthetic fashion. Several important, durable cosmological notions are attributable to him, including the notion of a drum-shaped earth at the center of a geometrically orderly, concentrically differentiated cosmos.[14]

When we study the intellectual systems of serious Greek thinkers, they seem to be built on certain common assumptions and to reveal a common concern with certain cardinal questions. They *assume* the existence, "behind" the complex sensible world, of a simpler inferable world which it is their business to uncover. Concerning this inferable and somehow more real world, they ask what are the organization and modes of interaction of its differentiations (the cosmological question) and how it began and through what stages it developed to its present condition (the cosmogonic question). Moreover, their analytical interest in cosmology and cosmogony is involved at every step with a concomitant interest in the life-soul and in life-as-action. Indeed, our study of the life-matter problem reveals that, throughout the history of science, cosmology and biology develop not separately but through a complex pattern of interactions, as suggested in chapter 1. It is from this viewpoint that we may examine Anaximander's ideas.

Cosmogony: the Boundless

Anaximander taught that the cosmos commenced as an infinite, indefinite entity, "the boundless (*apeiron*)"[15] of which a portion, perhaps germlike, resolved itself into a ring or sphere of hot sur-

---

[14] D-K, A 10, 11, 21, 22; Kahn, pp. 46–58; WKCG, 1:89–101.
[15] D-K, A 1, 9, 9a, 11, 14–16, and B 1; Kahn, pp. 32–33, 231–39; K-R, pp. 104–17.

rounding a core of cold.[16] The hot was ripped off and was subdivided into many circumterrestrial rings like the rims of cartwheels. These rims were built of flame, each flamy ring being encased in a (sleeve-like?) coating of mist. The flame was thus hidden from view except where the mist was perforated by spokeholes. Through each hole the flame shone forth, and still shines forth, toward earth as a heavenly body.[17] The cold, which was presumably also dark and damp, differentiated into air, water, and earth, but it is debatable how Anaximander viewed the differentiative process as occurring.

Probably he believed that (a) fire, as represented by the sun, induced an evaporative resolution of the cold moist core of the cosmos into (b) aerial and (c) earthy components, (d) sea being left over as an unevaporated or partially evaporated residue.[18] This would give something like fire, air, water, and earth as four components of the cosmos, a fact which made Anaximander seem, to certain ancient commentators, to have anticipated Empedocles in stipulating these four as elementary substances. But our most careful modern student of Anaximander reminds us that there is no evidence that the famous Milesian saw these as elements in the usual senses of the term or in fact as limited to four.[19]

A suggestive but not readily understandable part of Anaximander's physics treats the effects of opposites on one another. Anaximander was aware, from daily experience, that fire is destructive of water and water of fire, and that hot and cold together yield a homogeneous warmth. Authorities are not in agreement on the matter, but some think that Anaximander meant to make a point of the native tendency of the cosmos to undergo dedifferentiation. A similar idea was to be formulated, in the physics of a much later and scientifically more sophisticated era, as the second law of thermodynamics (though we must not think that Anaximander anticipated in any serious sense the energy theorists of the early nineteenth century). Some classical scholars think it possible that Anaximander invented his *apeiron* as an inexhaustible source on which the world might draw to offset its tendency toward self-annihilation through a reduction of contrariety.[20]

[16] D-K, A 10; KF (PSP), p. 59, note a 1.
[17] D-K, A 11, 21, 22; JB (EGP), pp. 68–71 (4th ed., pp. 62–66).
[18] Arist. *Mete.* 355a, b; D-K, A 10.
[19] Kahn, p. 163.
[20] D-K, A 14, but see WKCG, 1:84.

A final and important point must be made about *apeiron*. We should form a very imperfect picture of it if we thought of it merely as an undifferentiated substance liable to differentiation. Its nature is, like that of Thales' water, in some sense divine. It can exert a directive or controlling influence.[21] The nature of this control was not specified (Anaximander supposed that it could "steer" things). Perhaps he believed that *apeiron* merely initially set the world to differentiating in a particular fashion, or perhaps he saw it as continuing to direct the world by some sort of sustained intervention.

Biology

The Anaximandrian legacy includes two items that are relevant to the life-matter problem. One concerns the origin of animals; the other, the locus of soul. We are told that animals arose from moisture undergoing evaporation or (which may be a less precise statement of the same idea) from water and earth when heated (*aqua terraque calefactis*). Man seemed to Anaximander to represent a special case, since the first newly generated men would have starved had they been created as infants. Anaximander's theory forestalls this eventuality by having the first men spend their embryonic and infant lives as aquatic—perhaps fish-like—creatures. These had a sort of bark to protect them against evaporation when they came out on land. Eventually, cracks appeared in the bark and through these the first self-dependent men and women were delivered. Other animals shared with man an aqueous origin.[22]

Soul and air. The doxography for Anaximander on soul is too slight to be reassuring. The boundless itself in its "steering" capacity was perhaps thought of by Anaximander as, in some sense, sentient or conscious. Beyond this, a single commentator, Theodoretus, would have Anaximander (with Anaximines, Anaxagoras, and Archelaus) give soul an airy nature.[23] That Anaximander held such an opinion, Theodoretus aside, is quite credible on general considerations.[24] The recognition of the connections between life and air did not await the rise of formal science. Thus, according to the (probably eighth-century B.C.) author of Genesis, after man's first formation it remained

21 Arist. *Ph.* 203b; K-R, p. 105.
22 D-K, A 10, 11, 30; K-R, pp. 141–42.
23 D-K, A 29.
24 Kahn, pp. 71, 114.

for God to "breathe into his nostrils the breath of life."[25] And there was Isaiah's resuscitation of the Shunamite's son by a similar procedure, after which, we remember, the boy "opened his eyes and sneezed seven times." Guthrie points to Orphic origins for the idea that we inhale the world soul and reminds us that air, especially the west wind, was also able in legend to impregnate (it was thus Achilles' horses were conceived). Such examples from the ancient literatures could easily be multiplied. Even more suggestive—but beyond the possibility of analysis in this context—was the intricate pattern of etymological and semantic cross-connections that evolved among various Greek words which are translated "life," "breath," "air," "spirit," and "soul." The overlappings among these are indicative of associations occurring in men's minds at the level of what Guthrie calls unconscious presupposition.[26]

In sum, if we cannot be sure that air was soul, or a vehicle for soul, in Anaximander's system, we can at least be sure that—in ancient thought generally—connections between air and soul evolved informally. We have already noted that many early scientific ideas were arrived at through a rationally ordered and explicit formulation of existing and sometimes implicit prescientific intuitions. Thus, ideas linking air with soul, acting explicitly or implicitly, influenced the concept of life from prehistory through the closing days of classical science (see below, the Stoics and Galen). With Anaximenes, the next Milesian scientist after Anaximander, we shall find that the connection between air and soul becomes explicit.

Importance of Anaximander. If Anaximander's theories seem at points strained or inelegantly developed, they were nevertheless destined to have a substantial scientific impact. They contained a wealth of ideas for Greek science to ponder, argue, modify, reject, build upon. Anaximander was the first of the Ionians to set down his views on nature in writing, and was thus presumably the first Greek to "publish" a prose work. He was the chief founder of both cosmogony and cosmology considered as serious, continuing, naturalistic traditions. His cosmogony was sophisticated even when compared, for example, with the much later and not much better cosmogony of Plato's *Timaeus.* In his physics, Anaximander raised the major issues

[25] Gen. 2:7; WKCG, 1:127–32.
[26] 2 Kings 4; see also William K. C. Guthrie, *The Greeks and Their Gods* (Boston, 1951), chap. 5.

about actually existent substances (later, elements), their nature and number, their origin, and the possibility of their transmutation. His thinking was specially penetrant on the varieties of physical changes. Of these the most important was the supposed process of separating out, or resolution, by which, according to the theory, the *apeiron* gives rise to a configured cosmos. We see in this idea the prototype of Aristotle's potential-to-real transformation and of a line of related conceptions—cosmological and biological—lasting through the Middle Ages (see Paracelsus, for example, and his theory of the Ilyaster) into seventeenth-century embryology (see William Harvey and his espousal of the theory of epigenesis).

But Anaximander's thought was important in still other ways. It seems likely that he viewed certain changes as two-way phenomena involving intertransformations between opposites, a concept of central importance in later Greek science. And he raised the persistent problem of cosmic catastrophism, another scientific question with an active future.

In regard, finally, to the origin of life, Anaximander strongly suggests that the crucial step was the evolution of something living out of something moist and semi-earthy—a theme which was to be repeated with many variations. This idea, which we may irreverently term the "wet mud" theory of the origin of life, was perhaps suggested by the belief that spontaneous generation seemed to occur in oozy places, or perhaps by the discovery of fossilized sea animals in dry rock at high altitudes. It is not even impossible that the potter's art emphasized the plasticity of mixtures of earth and water. (Pandora, we remember, was created by Hephaestus, at Zeus' command, out of the same two elements).

ANAXIMENES AND DIOGENES

Anaximenes

Anaximenes (ca. 550 B.C.) was possibly a friend, perhaps even a student, of Anaximander. His conceptual style appears, as Guthrie observes, to have been less metaphorical in a poetic sense than that of his slightly older fellow townsman. His thought illustrates, as had that of his predecessors, that what we think of today as partly separate disciplines, namely cosmology and biology, were then indistinguishable aspects of a newly emerging world outlook, the outlook we term scientific.

Of the views attributed to Anaximenes, two bear importantly on the problem that interests us: first, air is the primary substance;[27] second, air is soul.[28] Air perhaps appealed to Anaximenes because it seemed to combine the best features of the first principles specified by his two major predecessors at Miletus:[29] the substantiality of Thales' water and the multipotentiality of Anaximander's *apeiron.* Like Thales' water and Anaximander's *apeiron,* Anaximenes' air was divine: it acted as an agent of control and organization; the gods arose out of it.[30]

Anaximenes' physics looked forward, likewise, in being more naively micromechanistic than earlier physics. All objects in the world envisioned by Anaximenes comprised one basic substance, air, in varying degrees of compression. The result is a series of sorts of air that includes fire, air in its familiar invisible configuration, wind, cloud, water, earth, and stone.[31] We do not hear, however, that these are elements in any strict sense.

From biology to physics. We may draw a roughly valid contrast between ancient and modern science with respect to what may be termed the flow of analogy between biology and physics. Greek science tended at first to construct the cosmos along lines suggested by the better known, or at least more directly familiar, living microcosm. Modern science proceeds in the opposite direction—*to* biology *from* physics—and would model the microcosm as far as possible after the cosmos. Anaximenes' cosmology, his world theory, was as much an extension of his biology as vice versa. "As our soul is air and thereby holds us together," he said, "so wind and air gird up the world." Air, then, does not merely exist between things; it is responsible for their cohesiveness; it sustains their organization.[32] The

---

[27] D-K, A 5, 7; Arist. *Metaph.* 984a.

[28] D-K, B 2.

[29] Eduard Zeller, *Outline of the History of Greek Philosophy,* 13th ed. rev., trans. L. R. Palmer (London, 1963), p. 45.

[30] K-R, pp. 150–51.

[31] D-K, A 5, 7; see also Giorgio de Santillana, *The Origins of Scientific Thought* (New York, 1961), p. 42.

[32] D-K, B 2 (trans. ours). For a full consideration of the possible meanings of this, see K-R, pp. 158–62. For a discussion of *pneuma* and coherence, see S. Sambursky, *Physics of the Stoics* (New York, 1959), pp. 1–7.

idea of air (or something associated therewith) as responsible for the continuity in and among things was to be revived occasionally—most conspicuously, though in a more highly elaborated formulation, in Stoic medical and physiological thought.

Organization and Disorganization

Greek science here concerns itself with what will become a major subject for later physiological speculation. Organized matter would be believed, in all future scientific eras, to have a spontaneous tendency to lose its organization. After death, disorganization is unimpeded. In the living organism, it must be combatted if life is to continue. Disorganization must be either *prevented* (through a sustentive or cohesive agent, such as Anaximenes' air) or *counteracted* (through organizing activities that compensate for the simultaneously occurring processes of disorganization). Strikingly, nearly every major theorist whose ideas we shall consider will repeat, in his own way, the idea that life maintains itself in the face of an intrinsic tendency of its substantive substrate to destroy itself. In chapter 1, it was suggested that biology has always had a thematic character. The above idea—that life resists or counteracts disorganization—has been one of its dominant themes.

About two hundred years after the time of Anaximenes, Aristotle was to affirm that the essence of vitality is form and that vital form has a distinctive capacity for self-renewal. Two pressing challenges to biology, from then on, were to discover (*a*) how the form of the organism is initially established (through genetically guided development) and (*b*) how it is sustained in the face of wear and tear (through nutrition). Our point here is that these problems were not entirely new with Aristotle. Anaximenes appears to have identified them in at least an implicit and inchoate way and to have assigned to air a central role in both the establishment of vital organization (here the air directs and controls) and its preservation (here it acts as a sustentive agent). It is reasonable to suppose that Anaximenes expatiated further on biological subjects—but as to what subjects, and how construed, the evidence is wanting.

Diogenes of Apollonia

More than a century after Anaximenes, many of his ideas were revived and reformulated by Diogenes of Apollonia (ca. 435 B.C.).

The surviving fragments and references compensate, in the case of Diogenes, to some extent for the paucity of evidence on his great predecessor, Anaximenes. Diogenes wished to return, after the pluralism of Empedocles and Anaxagoras (see chapter 4), to the monism of the Milesians. Air appealed to him as the common substrate, or first principle, because of its multiformity and because of the readiness with which it changes its character (its taste, shape, position, temperature, moisture, etc.). Ordinary observation seemed to him to show that when air is present in the organism, so is life, which Diogenes associated with mind. The unique feature of organic bodies was to him their admixture with air in its primary form.[33] This theory he used to explain numerous biological and psychobiological details.[34] Not new with him was the association of moisture with lesser grades of perceptiveness or intellection, with drunkenness, with sleep, and so forth. The intellectual maturation of the individual he attributed to a decrease in body moisture. Pleasure he considered due to an adequate, pain to an inadequate, admixture of air with blood. A long fragment survives on the arrangement of the veins, and there are disconnected fragments on such matters as reproduction (he considered the offspring a derivative of the father only) and nutrition. Some of these interpretive attempts have a far-fetched quality and illustrate the dangers that lie, where understanding is the goal, in a too speculative and deductive habit in scientific thinking.

It seems fairly safe to believe that the indispensable requirement for life-as-action was, for Diogenes, a specific degree of warmth of the air the organism contains. Thus ". . . in all living things the [Life-] Soul is the same thing, namely Air, warmer than that outside in which we are, but colder than that on the sun."[35] He also believed this configuration of air to be similar, but not identical, in all men, and in animals; that is, it explains their basic similarities and superficial dissimilarities. If Zeller's interpretation is correct, Diogenes meant further to distinguish between merely "animate" beings in which air merely exists (entering by diffusion), and fully intelligent ones in whom it exists in much greater amounts (forced in by breathing).[36]

[33] D-K, B 2, 4, 5.
[34] KF (PSP), p. 279; and EZ (HGP), 1:285–305.
[35] D-K, B 5 (trans. Freeman).
[36] EZ (HGP), 1:297–98.

Lest Diogenes appear, as presented thus far, an exclusively materialistic thinker, it should be added that he does not appear so when we consider the full range of his theories. Like the first principles of the Milesians, Diogenes' air partook of the transcendental. Air was, for him, not only immanently intelligent and directive, but also divine.[37]

HERACLITUS

Problems of life and matter were a primary preoccupation of Heraclitus (in his prime around 500 B.C.), though his way of expressing himself makes it difficult at times to be completely sure of his meaning. He earned the epithet "dark philosopher" even among ancient commentators and, as Gomperz remarks, was "prone to satiate himself in a debauch of metaphors."

In spite of his obscurity, Heraclitus' style attracts us. The mood is enigmatic but evocative. The fragments remaining to us have a sharp and at times witty quality that entices us to try to understand them. By proceeding cautiously, moreover, we can piece together at least the skeleton of Heraclitus' physiological system. His scientific ideas have been the subject of increasingly good interpretive studies of which the last and in many ways the best is Guthrie's. We shall confine ourselves here to certain cardinal points of doctrine in which what he had to say is relevant to our interest in life and matter.

Fire as arche. We can be sure that Heraclitus saw the human body as comprising, like all bodies, earth, water, and fire.[38] Of the three substances, he viewed fire as fundamental in the sense of being that out of which all arises and into which all is ultimately reconverted. Fire was fundamental for Heraclitus, likewise, in being in some sense the basis of life-as-soul. "This ordered universe which is the same for all was not created by any one of the gods or of mankind, but it was ever and is and shall be ever-living Fire. . . ."[39]

The Way Up and the Way Down

Heraclitus is perhaps best known for his insistence on change as the primary feature of the cosmos. The world, as he depicts it, is

[37] D-K, B 3, 5; K-R, p. 407.
[38] JB (EGP), pp. 172–73 (4th ed., p. 151).
[39] D-K, B 30.

continually involved in a two-way transformation from earth to water to fire (Heraclitus calls this "the way up") and from fire to water to earth ("the way down").[40] Our interest here is in the way in which this idea bears on life-as-action.

The two transformations occur not episodically or alternately but continuously and simultaneously in everything, we are to believe, and the dynamic equilibrium thus postulated is central to Heraclitus' whole physical system. Apparent permanence generally conceals, he believes, an actual hidden mutation (we can never step twice into the same river). Apparent changes—birth and death and the vicissitudes of the organism and the cosmos—are, ultimately, only dislocations of the balanced flux that is everywhere basic.[41]

The flux in the light of earlier ideas. Heraclitus lived a half-century later than Anaximenes and at no great geographic distance from Miletus. To what extent was his theory a derivative of earlier Milesian thought? As with so many points in early science, it is the experts who are in least agreement on this question. Assuming continuity, the flux would have given Heraclitus a chance to synthesize two major traditions initiated by the Milesians: the dichotomic world order of Anaximander (whose cosmos was produced through a differentiation of the original boundless into pairs of opposites) and the serial world order of Anaximenes (for whom the sorts of matter differed in the degree of their condensation). Heraclitus developed a system that was both serial and dichotomic (i.e. it contained two series in opposition to one another). Judgment must be suspended as to whether Heraclitus' theory was or was not a synthetic derivative of earlier theories. In later periods, in any case, we shall find such fusions of thought a recurrent stimulus to new ideas.

Harmony. In Heraclitus' own view, everything is at all times pulled simultaneously in opposite directions (like the bow and the bowstring), yet this very tension of opposites is the sine qua non of the harmony, the integrity, of physical objects. Harmony entails an adjustment or balance, but by no means an abolition, of the strife that is everywhere present. To give a single biological illustration: the modern biologist will not deny Heraclitus a certain insight in his

---

[40] D-K, B 30, 31, 90; Plato *Cratylus* 402a; K-R, p. 195, 199.

[41] D-K, B 51, 53, 54, 60, 67, 80, 88; Plato *Cratylus* 402a; Arist. *Ph.* 253b. See also Philip Wheelwright, *Heraclitus* (Princeton, 1959), pp. 29–36.

insistence, which at first seems rather surprising, that things live partly by dying.[42] Claude Bernard's interest will not be entirely dissimilar when, in the nineteenth century, he will say *"La vie, c'est la mort."*[43] For Heraclitus, the underlying opposition is universal and fundamental (his metaphorical way of putting it is to assert that "war" is "father and king," by which he may mean that the *logos* expresses itself through a tension of opposites).

Logos. Heraclitus' forerunners at Miletus, Anaximander and Anaximenes, had endowed their first principles with a directive or "steering" facility whose origins were considered divine. Heraclitus emphasized this point even further. Fire and the flux of things to and from fire are material expressions of *logos,* he said, that is, of that which, governing all, causes natural events to occur according to measure or order.[44] Thus "the principle which orders the whole by gradually changing makes the world out of itself and again itself out of the world. . . ." In several metaphors, Heraclitus suggests that such transformations occur in a quantitatively regular way, with fire as a standard of reference.[45]

The life-soul. On the subject of *psyche* we should remember, first, that for Heraclitus cosmic fire was animate. He saw it as the locus (or the equivalent) of the soul of the cosmos, of a controlling intelligence, a *logos.* Where and what, then, was the microcosmic *psyche,* the soul of the individual, in Heraclitus' philosophy?

On this point the evidence is not decisive. Perhaps Heraclitus wanted us to think of a man's soul as his individual allotment of fire.[46] More probably, as Wheelwright and Kahn suggest, he intended to make the *psyche* hover somewhere in the zone between fire and water.[47] He may have considered it a process—an upward shift in the equilibrium or an event associated with the transformation of water to fire, "an exhalation" (*anathymiasis*) reaching upward to

[42] D-K, B 23, 51, 76, 80, 88; Plato *Sophist,* 242d; JB (EGP), pp. 163–68; WKCG, 1:435; see also Abel Jeannière, *La pensée de Héraclite d'Ephèse* (Paris, 1959), pp. 77–82, and Geoffrey S. Kirk, *Heraclitus: The Cosmic Fragments* (Cambridge, 1954), pp. 145–48.

[43] See chap. 46.

[44] D-K, B 1, 2, 30–32, 45, 50, 72.

[45] Kirk, *Heraclitus,* B 22.

[46] D-K, B 84 a; JB (EGP), p. 172 (2d ed.).

[47] D-K, B 12, where "Souls . . . are vaporized from what is wet," and B 36, where "To souls it is death to become water; to water it is death to become earth. From earth comes water, and from water, soul" (trans. Freeman). See also Wheelwright, *Heraclitus, pp.* 58–67.

the Fire from which it has come.[48] In that case its position would correspond roughly to that of the aerial soul of Anaximenes (and, perhaps, of Anaximander). This would bring Heraclitus conceptually close to his predecessors at nearby Miletus in relating soul to air. But, again, we are not justified in feeling sure that he owed his ideas to them. It was his way to condemn derivative thinking and strike out on his own.

Life-as-action; perception. One of biology's "classic questions" is the question of perception, of its roots in the material constitution of the organism, of its links with life in general, and of its relations with reason and intellection. Heraclitus raised and supplied provocative answers to this question (or cluster of questions). Perception is the one biological (or psychobiological) topic on which his thought was adequately documented by early commentators and critics.

Whether he viewed the individual soul (*a*) as fire or (*b*) as a movement toward fire from a point a little lower on the two-way path, he made clear his belief that downward displacement serves to depress the soul. A minor downward shift makes the individual less conscious, as in a drunken stupor (the wetter state)[49] or in slumber. A radical or major downward displacement makes the whole system unfit as a habitation for the individual soul which departs leaving the cold, physical body to proceed along the downward path through water to earth.[50] The soul itself may then unite with the cosmic eternal fire.

Upward shifts are generally desirable, in this scheme, because they enhance that awareness (*psyche*) which was for the Greeks the essence of living. The best soul for Heraclitus was a dry soul. Indeed, he viewed the whole process of perception as a transmutation in the hot, bright, dry, fiery direction.[51]

We here detect a special quality—an apparently intentional ambiguity—in Heraclitus' style of conceptualization that has something to say about the history of scientific theory-building. He makes soul seem now something substantial, now a property of a substance, now the transformation of substance from one condition to another. He

[48] K-R, pp. 205–7; see also Wheelwright, p. 61.

[49] D-K, B 117, where "A drunken man has to be led by a young boy, whom he follows stumbling and not knowing whither he goes, for his soul is moist" (trans. P. Wheelwright, p. 18).

[50] D-K, B 36, 77.

[51] D-K, B 118.

makes it seem now immanent, now imposed, now emergent. The science of our own era (which, 2,500 years later, is still trying to solve the "mind-body" problem) generally expects a man to take a stand, to choose one among such a group of alternative interpretations. But not all science in every era has assumed the necessity or even the desirability of arriving at the "one right explanation." Heraclitus seemed to delight in assigning to such a term as soul a multivalent meaning. We understand some things more deeply, he seems to counsel, by adopting complementary or, indeed, contradictory ideas of their nature. This makes his procedure quite distinct from that which we think of today as scientific (witness the impatience of modern physics with having had, for a time, to consider light as both corpuscular and wavelike).

Through the senses, according to Heraclitus, the individual soul may be kindled by the outer and universal soul, thus achieving some identity with the *logos,* the divine rationale, the fire that governs the cosmos.[52] And here is one of the grand themes of Greek theory of knowledge: viz. that the highest and most excellent activity of the individual soul is to identify in some manner with the world soul. This idea was to culminate in Plato's teaching that the ultimate good is a movement of the individual soul in harmony with the transcendent soul of the cosmos.

EARTH AS *Arche*

Earth as a first principle or vehicle for soul appears to have found no especially effective advocate. Aristotle in his role as historian of scientific thought suggested that fire, air, and water were selected as first principles by various monists, but never earth. Plato, in his *Timaeus,* made earth appear as an element but made it differ from the other elements in that he considered intertransformations possible among fire, air, and water—earth being unable to engage in these (see chapter 7).[53]

To be sure, the early fourth-century Hippocratic treatise *On the Nature of Man* mentions earth among the substances selected by earlier thinkers as *arche.*[54] The reference is perhaps to Xenophanes

---

[52] D-K, B 41, 101a, 107; KF (PSP), p. 118; K-R, p. 188; WKCG, 1:429.

[53] Arist. *Metaph.* 989a and *De An.* 405b. Plato, see below, chap. 7. Not all ancient philosophers, however, made earth nontransmutable (see below, Anaximenes, and Aristotle, chap. 6). See also WKCG, 1:64, 386.

[54] *On the Nature of Man,* chap. vi, 32.

(fl. ca. 500 B.C.) who is reported to have said that all things come from, and end by becoming, earth. The earliest person to report Xenophanes as saying that, however, was Aëtius who wrote after A.D. 100.

Xenophanes said also that "we are all produced from a combination of earth and water," and that "all creatures that come into being and exist are earth and water"[55] (compare with Anaximander). Such an earth-water formula would not contradict the possibility of an earthy *arche,* an earthy origin for living things, if we assume that living things can only "come from" earth that has been properly wetted—a recurrent idea, we have noted, in both early scientific and prescientific speculation about the origins of living beings.

To say that all things come from earth—if Xenophanes did say it—could have had several meanings. As science, it could have meant that earth was first principle or *arche.* As myth, it perhaps pointed to Gaia, the earth mother who generated things when fertilized by Ouranos the sky god. As one of those vague but cogent perceptions that men form out of everyday experience, it perhaps pointed to the cycle of nutrition and death. Life was often characterized in antiquity as cycling from dust to dust.

There is more than a possibility, unfortunately, that Xenophanes never really said that everything comes from earth. Ancient commentators (like modern commentators) were not always careful to avoid making their predecessors say what they wished them to have said. We must be content, then, to suppose that the poet-philosopher Xenophanes appreciated and in his own way posed the problem of the physical preconditions of life. Insofar as he suggested a solution, it was as probably dualistic (entailing earth and water) as monistic (entailing earth only). We must remember too that the life-matter problem presented itself in a special way to the Greeks who assumed life in many places where a later science would exclude it, that is, in objects we consider inanimate, in matter itself, in the cosmos as a whole. A problem can crystallize no more rapidly than do the terms in which it is stated.

SUMMARY

In the foregoing outlines, I have intended not to throw new light on the often analyzed theories of the Ionian thinkers, but to suggest how through them a path was marked out along which a serious

[55] D-K, B 29, 33.

science of life could begin to move forward. The life-matter problem—a subject formerly of vague intuition or, in myth and poetry, of vivid imagination—now began to become a scientific problem: it was now posed in terms of the new Greek interest in the nature (*physis*) of things.

The available evidence on these early theories is partial and circumstantial; much—most—of each theory is irretrievably missing. Certain critical questions of interpretation may prove unsolvable on the evidence we possess (for example, it is often difficult to be sure whether our "authors" conceived the life-soul as identical with the *arche* or as an immanent property of it). Fortunately, however, the surviving fragments of their thought tend to deal with the central and significant rather than the peripheral and trivial. What is most important is that a science of biology—though not yet a fully developed or separate science—had been born.

Pre-Socratic Solutions: Part Three
Dualistic, Pluralistic,
and Mixed Models

PARMENIDES

The foregoing world models were, with the possible exception of Xenophanes', monistic in their essentials. Each in its way used a single principle—water, air, fire, *apeiron*—whose differentiations were supposed to account for the complexity that greets the senses. Our next theorist, the poet-philosopher Parmenides (fl. ca. 475) posited, for purposes of comparison, two models, one monistic (though in a new sense, we shall see) and one dualistic, each developed in a separate part of his poem.

The Way of Truth

The first method Parmenides called, with a certain prejudice, the Way of Truth. To follow this method, one must reject the evidence of sensation, he argued, and follow the path of pure reason. Traveling this way, Parmenides arrived at a number of drastic conclusions. He convinced himself of two fundamental and embracing ideas: first, it is impossible that there should be any Not Being and, second, Being itself is completely uniform and absolutely unchanging. For the complex, illusory world of the senses Parmenides thus substituted a different but, to him, real world that was homogeneous, indivisible, unchangeable, motionless, and eternal. He envisioned it also as "complete on every side like the mass of a well-rounded sphere" which some scholars have taken to mean that Parmenides thought of the cosmos as spherical.[1]

Parmenides' cosmos seems, at first, a monotonous subject for scientific investigation. What was important about it historically was

---

[1] D-K, B 2, 6–9; DL 9.22; JB (EGP), p. 187 (4th ed., p. 175); K-R, p. 269. For a discussion of sphericity in Parmenides, see WKCG, 2:43–49.

that it raised (and answered negatively) two fundamental scientific questions: whether Not Being can be and whether Being can be many and mutable. Thus stated, these appear to be metaphysical rather than scientific questions. But when we translate them into scientific terms they ask: whether there is such a thing as a vacuum and whether there is more than one sort of matter (and if there are several sorts whether they are intertransmutable). From the time of Parmenides onward, these were central questions for physics.

Parmenides' relentless affirmation that reality was homogeneous in space and time turned out to be one of the most evocative of early physical concepts:[2] Its antagonists generally adopted one or the other of two possible responses to the challenge it posed. Some thinkers (see below, Empedocles and Anaxagoras) assumed a pluralistic position, acknowledging several principles rather than one. This step was crucial because up to this time Greek cosmologies had been largely monistic. Not all successors of Parmenides were pluralists, however, and for those who continued the monistic tradition the problem was to derive a complexly nonuniform but orderly visible world from an underlying reality regarded as uniform or single. This problem was solved by the invention of atoms, indivisible bits of internally uniform matter that were nonuniform—dissimilar—in their shapes, sizes, and movements. To make the atoms serve adequately in explaining the complexities of the cosmos, it seemed necessary to their inventors (Leucippus and Democritus) to invoke the void (the Not-Being which Parmenidian Truth had excluded).

The Way of Seeming

Parmenides' other "way," that of Seeming, or Opinion, was that followed, we are told, by "unthinking mortals" (including, he seems to imply, certain scientific thinkers of the period). By way of illustration, Parmenides took the trouble to build a fairly detailed cosmology showing what may happen when one proceeds in what he regarded as the wrong way.

Although he would have us follow the path of Truth rather than the path of Opinion, the results obtained in following the latter path have a certain interest; and, at the risk of troubling the ghost of

---

[2] See, e.g., Francis M. Cornford, *Plato and Parmenides* (London, 1939, 1958), p. 53; and Émile Meyerson, "The Unity of Matter" (chap. 7) *Identity and Reality* (New York, 1962).

Parmenides, we may glance briefly at the views he would have us repudiate.

Cosmology. Parmenides made use of two first principles in building the world model that was supposed to illustrate the wrong way to proceed. These two were: Fire (related to light and low density), and Night (related to darkness and high density).[3] Out of these two ingredients, he said, Opinion wrongly structures the cosmos. It is difficult—according to some scholars, impossible—for the modern reader to visualize this cosmos precisely from the remaining fragments of Parmenides' poem and from available second-hand information. The universe we are asked to envision is, in its general arrangement, concentric—the concentric "crowns" being either light, hot and fiery; or dark, dense, and cold; or mixtures of these. The earth is depicted as at the center. The course of all things is steered by a certain "goddess," located at the "middle" (halfway between the earth and the heavens?). Further details are presented, but in terms too ambiguous to merit analysis here. The theory sounds rather like a Pythagorean version of Anaximander's cosmology augmented by features borrowed from other important predecessors of Parmenides.[4]

Biology. Biology as such is absent from the surviving fragments of the first part of Parmenides' poem and is less adequately represented than physics in the extant verses of the second part. On partly circumstantial evidence, we may say that Parmenides probably set forth, under the heading of misguided Opinion, the following biological ideas. They deal with classic questions—about life and matter, about the origins of man, about the nature of perception—if not adequately, at least scientifically, that is, in terms of a physical theory linking the microcosm (man) with the cosmos at large.

First, Opinion teaches—deceptively—that organisms contain the same two ingredients, Fire and Night, that make up all other things. Man originated as a hot-cold mixture not unreminiscent of the wet earth with which other leading theorists have made us familiar.[5] Second, the vitality of living things is due to the fire they contain, perhaps because fire activates the other, passive component.[6] Consonant

---

[3] D-K, B 8–9.
[4] D-K, B 10–15; K-R, p. 284; JB (EGP), pp. 197–205 (4th ed., pp. 185–92).
[5] JB (EGP), p. 205 (4th ed., p. 192).
[6] EZ (PG), p. 719; KF (PSP), p. 151.

with this is the notion that old age is brought on by a diminution of warmth.[7] A preponderance of the cold component is found in the male and of warm in the female (in contrast to the view expressed at about the same time by Empedocles and later by the author of the Hippocratic treatise *On Regimen*).[8]

Perception is a property of everything; like perceives like; warm perceives warm; cold, cold. The mind is responsive to the physical composition of the body. Thus "according to the mixture of [the two basic principles in] the much-wandering limbs which each man has, so is the mind which is associated with mankind. . . ."[9] Intelligence varies as the prevalence of the warm—an idea at least distantly reminiscent of Heraclitus. Separation of the warm from the cold component leads naturally to death after which the cold is still perceptive but only of that which is, like itself, cold and dark.[10]

To summarize, the cosmos comprises a pair of contrary elements, one active and one passive, each endowed with primitive psychic powers (perception); the dynamic and/or configurational relations between these two permit the origin and continued expression of life-as-action, especially perception and mind; missing from the picture is anything definite on other aspects of life-as-action—nutrition, generation, and overt motion.

Of importance for the subsequent development of the life-matter problem and other cardinal problems is the distinction Parmenides insists upon between the sensible and empirical vs. the rational and intelligible approaches to "knowing." We have already heard from Heraclitus the warning that reason must be schooled to interpret rightly the potentially misleading testimony of sensation. We shall find, as we proceed, that the difference between these two ways of knowing and the proper balance and relations to be observed between them are subjects of explicit interest and varied preference to all further developers of Greek science (see especially Empedocles and Plato but also Aristotle, Epicurus, and Galen). Parmenides, more than anyone else, was responsible for dramatizing the difference between rationalism and empiricism and for compelling science from then on to be self-conscious in its choice between them—or

[7] D-K, A 46 a.
[8] D-K, A 52–53; EZ (HGP), 1:601; JB (EGP) (4th ed., p. 192).
[9] D-K, B 16; JB (EGP), p. 188 (4th ed., pp. 177–78).
[10] D-K, A 46.

in its synthesis of both into an integrated rational-empirical attempt to grasp the "nature" of things.

EMPEDOCLES

One of the central endeavors of twentieth-century biology is the building of physicochemical models of biological activity. Such biological subsciences as physiology, embryology, genetics, ecology, psychobiology, and even taxonomy and anatomy increasingly reflect this emphasis, though we often hear the admonition that biological problems should be studied at all levels and not merely the physicochemical. Biochemistry and physiological chemistry have recently made available to the biological subsciences an increasingly sophisticated and well-tested store of conceptual materials to be used in the building of models. What we witness is not so much a physicochemical theory of life in general as detailed interpretations of specific aspects of generation, nutrition, perception, volition, and movement (the classic questions are still classic).

A major outcome of our studies of life and matter will be the realization that modern model-building is the contemporary expression of an approach that dates back to antiquity. That this approach is an old one is intimated in the ideas of the early pre-Socratics with whom we have already become acquainted. It is much more clearly apparent in the three major thinkers with whom we shall round out our study of this period, commencing with Empedocles' (fl. 465 B.C.), whose life-matter conception is a mixed one containing emergentist intimations that give it a rather modern ring.

Cosmology

Empedocles saw the cosmos as oscillating in time between two polar conditions—one of maximum mixedness or integration, the other of separateness or segregation. In its maximally mixed state, the cosmos is a homogeneous, spherical "god" held together by the integrative influence of Love (*philia*). In its maximally unmixed condition, the four elements—for the first clear postulation of which science is indebted to Empedocles—are all separate from one another, kept apart by the segregative influence of Strife (*neikos*). During transitions, Love and Strife both work at once; but one gradually prevails over the other (segregation progressively supplanting integration or vice versa). Love and Strife exert their effects on four internally homogeneous, nontransmutable, and presumably particulately organized elements or "roots," fire, air, water, and earth, which compose the corporeal world and whose rearrangements in voidless space are the cryptomena underlying the patent phenomena of change. This formulation proved one of the most cogent and durable paradigms of pre-Newtonian science.[11]

Origin of life. A world can arise in either of two ways, Empedocles supposed, namely through differentiation of an initial amalgam (when Strife is on the increase) or else through integration of initially separate components (when Love is increasing).[12] Empedocles contrasts the two different ways in which organisms arise (*a*) when worlds are built by differentiation or *ekkrisis* and (*b*) when they are built through synthesis or integration.[13] When Strife is on the increase, the forms that arise tend to be generalized or "whole-natured," that is without limbs and with both sexes combined in a single individual. Monoecious trees and hermaphroditic animals come first; dioecious trees and bisexual animals, later.[14] When Love is on the increase life arises in a different fashion. First, separate body parts are formed and these are then brought together. All this occurs under the mixing or integrative influence of Love. Since the parts come together in an at least partly random fashion, nonviable forms—chimaeras, monsters—may result. But there are likewise viable forms and these survive and reproduce.[15]

A moment's reflection will show that this idea was only superficially related to Darwin's rule that the fittest survive. In both theories nature makes experiments. But Empedocles' scheme said nothing about crowding or competition. Only in a competitive situation, such as that postulated by Darwin, does progress result from nature's trial-and-error methods. Neither Empedocles nor any other Greek thinker, as far as known, subscribed to evolution considered as the general derivation of new species from preexistent species.

Chance and direction. There was another way, however, in which Empedocles' account was pointed toward the future. He said

---

[11] D-K, A 28, B 6, 7, 17, 21, 27a, 28, 30, 31, 35; also WKCG, 2: 147–52.

[12] See Guthrie, *In the Beginning* (Ithaca, 1957), p. 43.

[13] D-K, B 9, 17, 20, 21, 71.

[14] The primordial hermaphrodite is a theme with many variations and cosmological cross-connections in ancient and primitive thought, both western and oriental; in Plato's *Symposium* we hear both Pausanius, 181c, and at more length Aristophanes, 189e, speak about this subject.

[15] D-K, B 17, 19, 21, 23, 57, 59–62.

that active life as we see it in plants, animals, and men emerges when the four elements are properly brought together under the influence of Love. Life-as-action then is, in one sense, emergent. But what did he see as the nature of Love's influence? The question is crucial because it seeks light on Empedocles' theory of the cause of organization.

Did he view Love's role as blind and mechanical or rather as somehow directive? The latter, presumably, since to Empedocles life implied a harmony of elements, Love bringing about the necessary proportion. As an organizing agency, Love had had approximate antecedents in Anaximander's *apeiron,* in Anaximenes' air, in Heraclitus' fiery *logos,* and in Parmenides' goddess, all of which were concerned in the establishment or maintenance or both of organization. This idea was likewise to have many successors. Intermittently until the nineteenth century, theorists were to assign a variety of organizing functions to an assumed *anima, vis creatrix, principe vitale, nisus, Lebenskraft,* or *entelechy.* By no means would these organizing agents all be identically defined. Rather, each would be offered as in its way responsible for organization. But each would, at least implicitly, reject the possibility that organization emerges from the native, unaided properties of the elements. All would reject the view of life as an emergent product of mere configuration. We shall have much to say about these agents as our story grows.

This brings us to another and equally prescient part of Empedocles' theory. Love's role was for him directive, yet never fully decisive. What Love created had to meet a pragmatic test. The created product survived only if it was viable. Something in the creative process, then, was left to circumstance. It is not easy to be sure how Empedocles saw this as happening. To Aristotle, Empedocles' theory seemed to raise the question "whether or not it is possible that with their disorderly (*ataktos*) motion some of the elements might have united in the combinations that constitute natural bodies like bones or flesh . . ."—a pregnant question, surely, for the future of biology.

Had Empedocles been alive to defend himself, he might have replied that the question thus posed exaggerates the role of chance beyond anything he, at least, ever intended. Love would not have permitted such disorder as Aristotle proposes. All the same, Empedocles opened the door to the possibility, at least, of a partly non-directed creation. He did not open the door very far—but he opened it far enough to give future thinkers a foothold. We shall hear Empedocles himself make the degree of each individual's intelligence depend on the composition of the blood. He thus points the way to the much franker materialism of Democritus and Epicurus. For Empedocles Love was imposed; life was emergent—that is, Love is at least partly responsible for the organization that is responsible, in turn, for life-as-action.

Empedocles' Theory Tested in Application

We shall learn in our study of the evolution of life-matter theories to expect two things of such a theory. First, it should erect a conceptual micromodel of the living world and, second, it should test this model by applying it to familiar phenomena including nutrition, reproduction, locomotion, perception. Indeed, a central feature in the history of biology is the changing way in which in successive eras these four phenomena have been interpreted microdynamically. Empedocles was a pioneer in this sort of scientific interpretation.

Composition of the body. The different tissues differ in their mixture, he asserts, and the elementary components must be present in each in the mathematically proper proportion. This theory is accepted as linking Empedocles with the Pythagoreans. Blood and flesh combine the four elements in equal parts, if Empedocles is correctly reported by Aëtius, whereas bone has two parts each of water and earth to four parts of fire.[16]

TABLE 1

CONSTITUTION OF TISSUES ACCORDING TO EMPEDOCLES

	Fire	Air	Water	Earth
Blood	2	2	2	2
Flesh	2	2	2	2
Bone	4		2	2
Sinew	2		4	2

The key tissue, that in which thought and mind reside, is the blood. Those men are most intelligent and accurately perceptive in whose blood "the elements are equally mixed and not too widely spaced and in particles neither too small or too large. . . ." The further a person's

[16] D-K, B 96, 98.

blood components depart from this condition the less intelligent and perspicacious that person will be. "And those in whom the elements are present in a loose and rarefied arrangement are heavy and given to drudgery while those in whom they are close-packed and broken up into small particles move quickly, attempt much and accomplish little, etc. . . ."[17]

Soul. Students of Empedocles have been troubled by the difficulty that the life-soul seems in his earlier poem *On Nature* emergent and hence mortal; whereas in the presumably later *Purifications* this soul is made transmigratory—hence, immortal.[18] A happy solution has been offered by Kahn and in somewhat different terms by Kirk and Raven. They suggest that Empedocles associated the immortal soul in some manner with love. Love is as much a substance as fire, air, water, and earth. As substance it could well retain its immanent *psyche* even after its dissociation from the four elements at death. Kahn would "distinguish between the deathless soul which transmigrates and the conscious mind or thought which is the function of a particular compound."[19] Kirk and Raven remind us that a duality in the nature of soul (represented by *psyche* and *thymos*) is already intimated in Homer (see above, chapter 2) and that the tradition survives in the soul theories of both Plato and Aristotle.

In each department of physiology, Empedocles seems to acknowledge one or more ruling concepts and then develops these ingeniously in an effort to explain detailed factual phenomena. The surviving references to his views on these topics are copious, fragmentary, and enigmatic.

Sensation. Empedocles' theory of sensation assumed effluences emanating from objects.[20] To arouse sensation the effluences must fit properly into the interstices among the particles composing the sense organs.[21] The eye has a different *modus operandi* because here the object does not come to the sense organ as it does in the case of smell, taste, and sound. The key ingredient here is fire which per-

---

[17] The translation of Theophrastus is that of Friedrich Solmsen, "Tissues and the Soul," *The Philosophical Review,* 59 (1950):438. His paper contains a helpful discussion of Empedocles' ideas, and those of other pre-Socratics, on this topic.

[18] D-K, B 115, 117, 137.

[19] Charles H. Kahn, "Religion and Natural Philosophy in Empedocles' doctrine of the Soul," *Archiv für Geschichte der Philosophie,* 42 (1960):3–35.

[20] D-K, B 89.

[21] D-K, A 86.

ceives by moving out from the eye to meet the object.[22] All the senses illustrate a fundamental rule of the Empedoclean system, viz., the rule that like tends, if permitted, to unite with like.[23]

Nutrition. The same rule is conspicuous in Empedocles' theory of nutrition, where the central requirement is to supply to each body part with just those substances that it needs for sustenance and repletion.[24] The body parts appear to attract corresponding components of the absorbed nutrient.

Air enters not only through the nostrils but through the pores, into which it is sucked when the blood retires from the body surface to the heart. It is forced out again when the blood returns from the heart to the surface.[25] Since blood is (*vide supra*) the seat of thought and perception and is especially abundant in the heart, that region is the chief center of these activities.[26]

Reproduction. The central principle in reproduction is a corruptive interaction of fire and moisture, and Empedocles uses this to explain a variety of sexual phenomena. Both father and mother contribute materially to conception. Sex depends on the side of the womb the seed encounters; on the warm side a male develops; on the cold side a female.[27] By a curious deduction, hot paternal seed gives a male child with maternal dominance; hot female seed, a female child with paternal dominance. If maternal and paternal seed are equally hot, children most resemble the parent of their own sex.[28]

Concluding Note

The above notes suggest inadequately Empedocles' energetic efforts to utilize his physical system in the interpretation of a range of physiological data. If his efforts seem inept at many points, their impact was to be formidable. The long-term productivity of the approach far outweighed the understandable defects in its execution when first attempted. What we witness here are inchoate but in their way prophetic gestures in the direction of a serious partly materialistic biology. And Empedocles' four elements were to remain a fixture

[22] D-K, A 90, B 84–86; JB (EGP), p. 267 (4th ed., p. 248).
[23] D-K, B 109.
[24] D-K, A 77.
[25] D-K, B 100.
[26] D-K, B 105.
[27] D-K, B 63, 65–66.
[28] KF (PSP), p. 194.

of science—though not universally or continuously accepted—until after A.D. 1700.

ANAXAGORAS

The solution which we shall consider next posits a universal psychic entity, Mind, imposed in a special way on a special variety of matter which is thereby induced, or at least enabled, to display life-as-action. The theory is that of Anaxagoras (about 460 B.C.) whom we remember among other reasons for having brought science from Ionia to Athens.[29]

Cosmogony. Anaxagoras' cosmos comprised an infinity of corpuscles termed by him "seeds" (*spermata*).[30] In sharp contrast to these seeds was another entity invoked by Anaxagoras, and characterized as "infinite, pure, and unmixed, all-controlling." This latter entity he viewed as a comprehending and motive intelligence which he termed *nous* (mind). Originally the seeds existed in a monotonous condition of maximum intermixture, he supposed, with air and ether in some sense predominant. When the world began, *nous* gave the seeds an initial twirl and set them to sorting out in orderly arrangements,[31] a process in which, cosmogonically speaking, they persisted from then on. Sorting out gave rise, among other things, to human beings, not only our ancestors but other men "elsewhere."[32]

Anaxagoras supposed that groups of similar seeds constitute certain "homogeneous substances" that differ in important respects from the primary substances of the Ionians. First, he regarded the number of sorts of seeds (and hence of homogeneous substances) as indefinitely great.[33] He saw these as represented in living things by such things as hair, bone, and flesh, which we should designate tissues.[34] Second, he supposed that each sort of seed and substance contained "portions" of all the others, its distinctive character being due to the dominant portion.[35] This second point offered the possibility, fundamental in Anaxagoras' system, of endless transforma-

---

[29] Theodor Gomperz, *Griechische Denker* (Leipsig, 1896), trans. Laurie Magnus, *Greek Thinkers* (London, 1901), 1:208.

[30] D-K, B 1.

[31] D-K, B 1, 12, 13.

[32] D-K, B 4.

[33] Arist. *Ph.* 187a23.

[34] Arist. *Cael.* 303a28.

[35] D-K, B 12.

tions. Anything was, potentially at least, transformable into anything else.

Precisely in what sense he saw all substances as containing parts of all others, and how this fact was supposed to permit transmutation, are subjects of debate among interpreters of Anaxagoras' physics.[36] Their accounts come together, however, on a point that bears on our present concern, namely Anaxagoras' ideas about the nature of nutrition.

Nutrition. One of the persistent preoccupations of biologists has been the effort to explain physically how *food* (of a certain sort, e.g. grain) can turn into *tissue* (of a quite different sort, e.g. bone). The fact that food does this convinced Anaxagoras that the final tissue was in some sense already contained in the food.[37] This conviction squared well with his notion that "there is a portion of everything in everything." Nutritive transformation was, in his scheme, a special case of transformation in general. Although we cannot be sure which to accept among the rival scholarly interpretations of Anaxagoras' theory of transformation, we can feel safe in believing that, in his analysis of nutrition, his physiology and his physics came together, since transformation was basic to both. Indeed, one of the major conclusions of our study of the life-matter problem will be that, from the very first, cosmology and biology evolve not separately but hand in hand. In Greece, especially, conceptual changes in either field of inquiry tended to induce changes in the other.

Nonliving vs. Living

Anaxagoras appears, from available evidence, to have had a particular interest in nutrition. But there was another point at which his thought bore even more crucially on the life-matter problem. He said that "in everything there is a portion of everything but Mind (*nous*); and there are some things in which Mind is present as well."[38] It seems probable that Anaxagoras meant in this way to distinguish living from nonliving things. Such was the view of Burnet (1892)

---

[36] See, e.g., Gregory Vlastos, "The Physical Theory of Anaxagoras," *The Philosophical Review,* 59 (1950):31–37; Francis M. Cornford, 24: 14–30, 83–95, and A. L. Peck, 25: 27–37, 112–20, both in *The Classical Quarterly* (1930–31). Also Felix M. Cleve, *The Philosophy of Anaxagoras* (New York, 1949), pp. 3–18; and K-R, p. 386.
[37] D-K, B 10; K-R, p. 380.
[38] D-K, B 11.

recently reaffirmed by Kirk and Raven (1957) and Guthrie (1962).[39] These scholars think that Anaxagoras believed that *nous* stopped intervening, at an early stage in the creation of the cosmos, except in bodies that we ordinarily think of as living. Only in living bodies has *nous* continued to interfere and to sustain life-as-action in the face of metabolic turnover and change.

Another interpretation of Anaxagoras' ideas of life and matter is offered by Gershenson and Greenberg (1964) who emphasize his choice of "all the different tissues of living organisms" as elementary. To these authors Anaxagoras seems to say that an immanent sort of "life . . . is inherent in all things, but the *Life processes* can only come to view when the elements are suitably arranged. Inorganic matter is a mixture of organic molecules which does not display the articulated structure characteristic of living organism." In addition to immanent life and emergent life, Gershenson and Greenberg identify in Anaxagoras' image of the organism an individual Mind analogous to the greater Mind of the Cosmos. Anaxagoras' solution of the life-matter problem thus combines the three ideas of (*a*) an individual ordering entity, *nous,* which in each organism arranges (*b*) seeds possessed of latent life in such a manner as to generate (*c*) patent life or life-as-action. How close Anaxagoras came to this formulation is uncertain. It seems consistent with, but not conclusively evinced in, the documents (which Gershenson and Greenberg present in English, and in detail).[40]

DEMOCRITUS

With Democritus (fl. 420 B.C.) we come to a view that is, in the form in which it has come down to us, more forthrightly presented and more naively materialistic than those of Empedocles and Anaxagoras. His theory is an elaborated version of one put forth slightly earlier by Leucippus. A single matter—chopped up into a huge variety of indestructible, impenetrable, internally homogeneous atoms—underlies all physical existence. Upon the sizes and shapes of the atoms, infinitely varied, and upon their spatial distribution and position (inclination or slope) depends the variety of the world as known through the senses.

[39] JB (EGP), p. 297 (4th ed., p. 272); K-R, pp. 376–77; WKCG, 2:316.
[40] D. E. Gershenson and D. A. Greenberg, *Anaxagoras and the Birth of Physics* (New York, 1964), pp. 14–30.

Size and shape were thus for Democritus primary properties; all other aspects of things derivative from these two. He regarded atoms as separated by emptiness, by a void whose exact nature and ontological status has been much discussed by students of Democritus. These authorities agree, however, on Democritus' emphatic insistence upon the reality of the void. "Naught exists just as much as Aught," (fr. 156) was his answer to the challenge of Parmenides. He saw the atoms as subject to rearrangement but never to creation, alteration, or destruction. Democritus thus restates in atomic language the conservation law already proposed by Melissus (who, in turn, was paraphrasing Parmenides). "That which is," said Melissus, "was always and always will be."[41]

Atomic movement was seen by Democritus as both effecting and affected by mutual bombardment.[42] In the beginning, this movement was engaged in, as an intrinsic property, by all sorts of atoms existing in homogeneous intermixture. Later, however, atomic aggregates formed and the atoms within an aggregate displayed both individual and joint movements.[43] In time, the atoms in some aggregates began to whirl round and thus produced both our own cosmos[44] and an infinity of others,[45] all of them doomed to ultimate dissolution.[46]

Reason and Necessity

As to the genesis of worlds and their contents, Greek thinkers generally offered one or the other (or a mixture) of two explanations. Either organization required a directive agency of some sort, a fiery *logos,* a steering *apeiron,* a guiding *nous,* a goddess or synthesizing *philia.* Or else it resulted automatically from the native properties of whatever was organized (for Democritus, atoms). The latter interpretation is paradoxical, in a way, since the emergent pattern is at once *inevitable* (it is the necessary and unavoidable outcome of specific atomic events) yet *fortuitous* (the atomic events are unguided). The Greeks sometimes spoke of such an outcome as oc-

---

[41] D-K, A 1, 14, 37, 40, 49, 57, 61. Arist. *GC* 325a; *Metaph.* 985b; see also K-R, pp. 401–16, 418–21 and Bailey, p. 118. For Melissus, see D-K, B 1.
[42] D-K, A 47, 58, 66.
[43] D-K, A 60.
[44] D-K, A 1.
[45] D-K, A 1, 40.
[46] D-K, A 84.

curring of "necessity" (*anangke*) and sometimes as occurring by "chance" (*tyche, automatos*).

When Democritus says, as he does, that worlds are born accidentally, he means not in a haphazard way but as the inevitable consequence of the undirected activities of the constitutive atoms.[47] Scholars disagree, to be sure, as to the precise nature and extent of Democritus' commitment to a frankly emergentist view, but he certainly went farther in that direction than any earlier thinker. His commitment was strong enough, in any case, to evoke a prompt reaction from others. In his *Timaeus,* Plato was to distinguish *logos* (reason) from *anangke* (necessity) and assign the former a partly directive role over the latter. In the *Laws,* Plato would decry the notion than man's essential nature emerges from his mere physical existence. Plato acknowledged, all the same, that necessity had a certain autonomy. He admitted that physical factors may cause a man to behave in a wicked way, in opposition to his moral disposition. Later, in his *Physics,* Aristotle reviewed the history of ideas about chance—which meant, to him, the absence of immanent purpose. Aristotle insisted, sometimes rather lamely, that final causes operate even when they seem not to do so: a man who "accidentally" meets a debtor in the *agora* does so because he went there for a purpose (albeit a different one).

The important point for our story is that Democritus more definitely than Empedocles attracted attention to an issue which would never thereafter be completely absent from biological thought, and which would frequently occupy a central position in it: Does organization entail direction or does it result from a chance association of elementary components? We shall not hear the last of the debate on this question until the latter part of the nineteenth century.

Living Things

Among the atoms that segregated in the cosmogonic process, according to Democritus, were certain small, round, rapidly moving ones which in sufficiently pure aggregation appeared as fire and when properly intermixed with others as soul.[48] Given the proper configuration, such an intermixture was able to display life-as-action,

47 K-R, pp. 412–13; Arist. *Ph.* 2.
48 Arist. *Cael.* 303a12; *Resp.* 472a3; *De An.* 404a5, 405a8; Bailey, pp. 154, 156.

Democritus supposed, such configurations sometimes arising spontaneously by an intermixture of fiery soul atoms with the familiar duo of earth and water.[49] But what did Democritus regard as the right configuration? On this question, if we may accept early commentators, he gave a quite specific formulation. He viewed the living body as in reality two bodies—one psychic (composed of fire), so to speak; the other, somatic (mostly earth and water). He regarded the two sets of atoms, psychic and somatic, as everywhere alternately interposed, perhaps in a one-to-one relation. Epicurus would later take over this idea of psychic and somatic atomic systems (insisting, however, that the psychic atoms are fewer in number than somatic ones).[50] Epicurus would likewise relinquish the idea that certain atoms are individually and immanently sentient or soulful, a point which is the keystone of Democritus' biology. To Epicurus, life seemed entirely epiphenomenal, or emergent (see chapter 9).

Respiration. An indispensable requirement for life, in Democritus' system, was some mechanism to keep an adequate population of fiery soul atoms at home, in order to combat their tendency to go away (in sleep, in death). According to Aristotle, Democritus saw the surrounding air as tending, perhaps by pressing on the body, to squeeze out the psychic atoms. While life lasts, according to Democritus, their escape is counteracted by breathing. What we inhale contains, among other things, fire atoms whose intake opposes the pressure tending to extrusion.[51]

Democritus here joins those theorists who, in one metaphor or another, make breath responsible for the structural integrity of the body (see Anaximenes, the Stoics, and in later times Francis Bacon). It was Democritus' view that through breathing, organization is maintained during life. At death, the psychic atoms depart from the body, but death need not come all at once or abruptly. If the individual is killed by a blow or wound, certain parts may retain their soul atoms and thus remain sensitive—but only temporarily since such parts do not breathe and so do not replace the soul atoms that escape from them.[52]

Soul and mind. In a tradition that would later be adopted by

[49] D-K, A 139.
[50] Lucretius *De Natura Rerum* 3.370 ff.
[51] Arist. *De An.* 404a10; *Resp.* 471b30–472a16.
[52] D-K, A 117, 160.

Epicurus, Democritus associated soul materially with mind but also drew a distinction between them. He considered soul and mind as alike comprising the same sort of small, round, fiery atoms,[53] but as different in distribution. It was his idea that soul atoms are interposed throughout the body among somatic atoms; mind atoms are concentrated in the breast region, the difference between them being primarily a difference in density.[54]

Sensation. Democritus used his psychic-somatic theory to explain a host of biological phenomena, especially the phenomena of perception. He thought the most fundamental sense was that of touch, the mechanism of which is not detailed in the sources that survive to us. The central supposition in his theory was that sensory perception and the thoughts it gives rise to generally involve dislocations of soul atoms by atoms entering from without.[55] In the sense of taste— nearest to touch in nature—the shape of the impinging atoms is crucial (a sharp taste is caused by small angular atoms, a sweet taste by smooth ones, etc.).[56] How atoms of a particular shape affect the organism depends to some extent, however, upon the bodily constitution ("there is nothing to prevent what is sweet to us being bitter to some other animals").[57]

Democritus' story of vision has been variously reported. Did he postulate effluent atomic films, or "idols," moving from the object to and into the eye to stamp an image there? Or did he (as Theophrastus suggests) envision a meeting in space of two streams of atoms—one from the eye and one from the object? Later we shall encounter again the idea that vision involves an emanation from the eye.[58]

Generation. Unhappily, from the historians' point of view, Democritus' presumably extensive biological writings not only have been lost but are inadequately reported in ancient sources. The doxography does yield, however, some salient points about Democritus' theory of generation. He thought of the seminal substance as collected from the whole bodies of male and female parents. Sex he viewed as determined by a presumably quantitative predominance

[53] Arist. *De An.* 405a8.
[54] KF (PSP), p. 314; K-R, p. 422.
[55] D-K, A 119–20; Arist. *Sens.* 442a29; K-R, p. 422.
[56] D-K, A 128–29, 135.
[57] D-K, A 130, 135.
[58] D-K, A 135; Bailey, pp. 165–70.

of paternal or maternal semen. He considered the umbilicus to be the first organ formed.[59] These and a few other details make it only the more regrettable that most of what Democritus believed is irretrievably lost. How extremely interesting it would be, especially, to have access to his ideas of nutrition.

Conclusion. Living things differ from nonliving ones in being composed of a network of immanently animate fire atoms interwoven with and producing life-as-action in another network of immanently inanimate atoms. The interactions of these two atomic networks are posited as the latent equivalents (cryptomena) of the patent processes (phenomena) of life.

[59] D-K, A 140–41; Arist. *GA* 764a6; WKCG, 2:467.

· 5 ·

Pre-Socratic Solutions: Part Four
Conclusion

The lasting influence of the pre-Socratic thinkers lay less in the factual biological knowledge they amassed or the theoretical concepts they evolved than in the intellectual procedures they laid down. As biologists (and they were all more than that), they created a strategy of interpretation which was to be used with certain gradually added improvements in every subsequent period and was eventually to pay enormous returns. We mentioned these procedures earlier, and are in a position now to outline them more fully. Seven features of the strategy may be identified whose importance will become apparent as we follow them in use not only in later antiquity but during succeeding centuries up to and including our own.

1. *Classic Questions*

Both pre-Socratic and later biologists have addressed themselves to certain "classic questions." What is reproduction? nutrition? respiration? locomotion? volition? sensation? reason? What happens when these activities occur in an abnormal fashion? How is the organization of individual living things established? dissipated? temporarily sustained in the face of an apparently inherent tendency toward dissipation? How did plants, animals, and men first come into being? How, in general, is life related to the physical condition of the material in which it appears?

Our sources of information are too fragmentary to permit us to be sure that every major pre-Socratic scientific thinker individually identified and systematically investigated all of these questions, though the later, pluralistic theorists clearly studied many of them. The important point in any case is that all these and other crucial questions were in fact raised at this early period and that, once identified, they were to remain the central questions of all later biological inquiry, the precise connotation of a given question naturally changing with the passage of time.

2. *Phenomenon and Cryptomenon*

In answering the foregoing classic questions, both pre-Socratic and later biologists have reasoned simultaneously on two conceptual levels: they have studied both (a) the perceptible aspects of living things and (b) their imperceptible but inferable aspects. To "explain" or "state the nature of" something has implied for both pre-Socratic and later thinkers a reformulation of its perceptible aspects in terms of inferred aspects that have been deemed somehow more "natural" or "real." It would be proper to think of a distinction as running through the history of physiology, then, between the study of the apparent aspects of things, or *phenomena,* on the one hand and of their hidden aspects, or what we may term *cryptomena,* on the other.

The suggested duality is just as well illustrated by twentieth-century molecular biology as by, for example, the microhydrodynamic physiology of Boerhaave (1708 A.D.), or the corpuscular biomechanics of Descartes (1630 A.D.), or the "facultative" biomedical system of Galen (175 A.D.), or the atomic biology of Epicurus (290 D.C.) In subsequent chapters we shall turn to the cryptomenal schemes of these and many other thinkers. The goal of all these systems has been: a reformulation of observed phenomena in terms of the presumably more real or significant cryptomena supposed to underly them.

Cryptomenon, as we have used this designation, refers to things so subtle, so slight in effect, or so small as to lie beyond reach of sensation. We shall find it helpful, as we proceed, to give the term more precision; we shall also need to extend it. We shall need to ask, for example, *why* the putative "hidden" entity or agent is hidden, and whether an extension of the senses—a better microscope, a more *sensi*tive (!) chemical indicator—would give more direct and concrete evidence of its presence. We shall not even hesitate to ask whether the cryptomenon is "really" there at all. For we shall discover that certain of the cryptomena we encounter, considered ontologically, are only abstractions—"products of mind not of nature"—paradoxically useful in solving problems even though their nature, their very existence, is doubtful; philosophers of science have called such putative explanatory devices "useful (or heuristic) fictions." We shall likewise discover—and this will cause us to extend the term to entirely different sorts of "hidden" events or objects—that hidden-ness is not always due to subtlety or smallness; some cryptomena are (hypothetically) sensible phenomena that are hidden by their re-moteness in space or time (they may be not micro- but macro-entities that existed, or events that occurred, a long time ago or far away from the earth).

3. *The Reductivist Rule*

An important assumption that has operated in both pre-Socratic and later interpretive endeavors is what we may term the reductivist rule or canon, which directs the interpreter, in moving from phe-nomenon to cryptomenon, to reduce complexity to simplicity as far as possible and to interpret apparent disorder in terms of underlying order. It does this because of the conviction of scientists that the cryptomenal world is orderly in fact, and that it is simple by com-parison with the phenomenal world. The operation of the reductivist canon manifests itself in pre-Socratic and later thought especially well in the complementary doctrines of *stoicheia* (elements) and *archei* (first principles). Reductive interpretation is traceable historically to the prescientific and intuitive search for oneness in nature, to what Guthrie calls "myths of primal unity."

We shall be impressed, as we proceed, with the immense diffi-culty theorists have found in trying to follow the reductivist rule in biology. The difficulties are intrinsic in the extreme complexity of even the simplest living systems when compared with the most elab-orate naturally occurring nonliving systems of comparable dimen-sions. The resulting frustrations have given rise—at first in antiquity but later especially in eighteenth- and nineteenth-century European biology—to the evocation of a variety of supposedly irreducible entities ("principles," "properties," "powers," etc.) considered by their inventors to be peculiar to living as opposed to nonliving things. It is to the belief in such irreducible or inexplicable entities that "vitalism" was applied when the term was first introduced in the late eighteenth century.

4. *Life and Matter*

Both pre-Socratic and—as we shall see—a great many major biolo-gists in later periods have developed theories, varying in explicitness and in the importance assigned to them by their authors, concerning the material prerequisites or concomitants of life. The life-matter problem has been, then, a real and persistently challenging one. The immediately preceding chapters have suggested solutions to this problem advanced by a selected series of pre-Socratic authors. Later chapters will be concerned, again in a selective way, with solutions advanced in the centuries that follow.

5. *System-building*

Both pre-Socratic and later biologists have attempted, to varying extents, to assemble their interpretive ideas into consistent, integrated conceptual systems. In particular, a specific solution of the life-matter problem may offer its author a common basis for interpreting in a consistent and integrated way a full range of vital manifestations, so that his answers to the classic questions acquire a systematic relatedness to one another. We shall find the systematic habit of thought signally apparent in the "form"-dominated biology of Aristotle, for example, as well as in the consistently micro-mechanical physiology of Descartes and his followers, in the irritability-oriented interpretations of life that evolve in the eighteenth and nineteenth centuries, and in the systematically cellular interpretation that will form a common ground for physiology and pathology from about 1860 onwards. In these and many other cases, it will be clear that a thinker's convictions about the physical preconditions of *life in general* may impart an integrated (a systematic) character to his interpretations of *life's particular manifestations.* It is true that for the earlier, monistically-oriented Greek natural scientists, firm evidence of systematic physiological thought is wanting. But the roots of it were there, and these roots gave rise to the patently systematic theories of later, pluralistically oriented authors. Whether and, if so, under what circumstances system-building is scientifically wise, whether it is desirable or fruitful, is a separate question to which we shall return.

6. *Cosmology and Biology*

In both pre-Socratic and later science, cosmology and biology have evolved not independently but hand in hand. For the Greeks, the cosmos often assumed human or animal characteristics; or, conversely, living things were given a cosmological interpretation. In either case, there was a continuous feedback between cosmology and biology, changes in either inducing changes in the other. As our study develops, we shall note a gradual and unevenly realized reversal of thought in regard to this matter. In early antiquity, the tendency was to reason, or imagine, from biology to cosmology and to construct the universe in the human image. Later, commencing with the Greek atomists, but much more pronouncedly after the Renaissance, science tended to reverse this tendency; it became usual to construct the organism in the image of a lifeless cosmos. We shall return from time to time to the developments that mark this reversal in interpretive procedure.

7. *Conceptual Themes*

Both pre-Socratic and later biological theorizing have been dominated by certain organizing ideas that have given its development a thematic character. Such organizing ideas have varied pronouncedly in their origins, in their relative degrees of generality and explicitness, and in their ways of affecting interpretive endeavors. We may remind ourselves of a few of the themes that played an organizing role in the pre-Socratic period.

a) Among the most influential was the assumption mentioned above of a fundamental parallelism between the cosmos on the one hand and the living microcosm on the other; this assumption, varying almost limitlessly in its expression, was never subsequently completely absent from biological thought.

b) Another was the assumption, especially common in the period we have been studying, to which the term hylozoism (the idea that matter itself is immanently alive) has often been attached. Hylozoism had prescientific roots and was possibly grounded in the intuitively animistic thought modes of primitive peoples. It came under a partial and preliminary attack, we noted, by the atomists. The atomists thus raised a question which oriented a great deal of subsequent speculation, the question, namely, whether life is immanent, or emergent, or imposed. We shall hear Epicurus (ca. 290 B.C.) raise this question more insistently than the early atomists have done (see chapter 9).

c) Still another early orienting theme was the doctrine of opposites, or of contrariety, conspicuous in the thought systems of Anaximander, Anaximenes, and especially Heraclitus and achieving a particularly generalized expression with the Pythagoreans including Alcmaeon. This theme significantly influenced early medical opinion (see chapter 6) and the analytical schemes of Plato and especially Aristotle (see chapters 7 and 8). The doctrine of opposites was to remain influential until at least the sixteenth century when a countermovement away from Greek medicine and physics began; its repercussions were to be felt, however, for a long time after that.

d) Another major organizing conception was the identification or, at least, association of life with one particular substance (e.g. in pre-Socratic thought with fire, air, or water). The "special single substance" theme was to prove one of the most frequently recurrent in the effort to specify the material preconditions of life. We shall encounter it at every stage in the history of biological theory-building. We shall hear life attributed by various theorists to particular elementary substances, or particles, or unique combinations thereof (see, in subsequent chapters, Plato's *panspermia,* the Stoics' *pneuma,* the radical humor of certain medieval thinkers, Harvey's crystalline colliquament, blood specifically designated "living" by Harvey, Lower, John Hunter, and others, and, finally, the sequence of substances culminating in and succeeding "protoplasm"). The differences between all these hypothetical living substances are more marked than any physical similarity among them. They represent successive expressions, however, of a persistent assumption, namely that a particular kind or configuration of matter is, par excellence, the locus of life.

e) We may mention, finally, an assumption whose roots lie deep in everyday observation and which has influenced biological interpretation from pre-Socratic times onward, the supposition, namely, that life is, or entails, a reconciliation of simultaneously occurring processes of destruction and reconstruction. The empirical bases of this theme must have been the observable facts of material intake and output and, related to these, the fact that living things tend to decompose as soon as they die (in a manner that is not characteristic of things that have never been alive). The inference from the very first was that even during life destruction continually occurs but is counteracted by simultaneous reconstruction. Among the pre-Socratics, this idea received especially dramatic expression in Heraclitus. We shall encounter it in the Hippocratic corpus, in Plato and Aristotle, in Epicurus, in Galen, and, as suggested earlier, in the analytical systems of almost all major physiological theorists from the sixteenth to the twentieth centuries.

The foregoing is a representative and not at all an exhaustive enumeration of themes that molded early Greek biological thought. What we would emphasize here is not the full range but the active presence of these mobilizing ideas and their cardinal influence not only in Greek science but in the science of later periods up to the present. Certain early Greek themes persist (we shall find this to be true, for example, of the idea of life as opposed transformation); others continue for a time but ultimately cease to function (e.g. hylozoism). Each era sees the rise of new organizing ideas, the influence of which is more or less controlling, and more or less enduring.

SUMMARY

In this chapter we have suggested that the principal contribution of pre-Socratic science to later biology is an interpretive strategy which (1) identifies certain "classic questions," (2) answers them partly by reformulating perceptible phenomena in terms of imperceptible but inferable "cryptomena," (3) seeks for simplicity and order, (4) concerns itself with the material prerequisites (or concomitants) of life, (5) tends to reason in a generally patterned or systematic fashion, (6) attunes its biological to its cosmological speculations, and (7) is organized by certain orienting or directive ideas or themes.

6

Examples of Early Medical Opinion

The oldest Greek biological books that we still have in a reasonably intact form were written not by philosophers but by physicians. Three such works—two from around 400 B.C., one slightly before that time—will serve to suggest early medical approaches to the life-matter problem. All of the treatises are from the Hippocratic collection; the authorship of all is uncertain.

ANCIENT MEDICINE
(ca. 420-400 B.C.)

The first of our three treatises is notable for a certain stylistic purity and conceptual independence. Religious, occult, and sacerdotal elements are neither present nor in any way implied. Philosophy of the sort that builds on hypothetical assumptions is mentioned only to be repugned. The treatise is short, readable, and direct. It has been admirably analyzed in recent years by W. H. S. Jones (1946)[1] and A-J. Festugière (1948)[2] with fresh translations by Jones into English and Festugière into French. Jones says that *"Ancient Medicine,* one of the outstanding books of the Hippocratic *Corpus,* is little more than an essay in defense of Alcmaeon's theory," Alcmaeon having probably introduced the important idea that biological differences depend on the *krasis* (temperament or blend) of body constituents.[3] We shall confine ourselves here rather closely to what *Ancient Medicine* seems to say about life and matter (with the usual warning that we are not thus doing full justice to the variety of interesting ideas contained in the book as a whole).

Ancient Medicine is admirably illustrative of certain trends and

[1] W. H. S. Jones, "Philosophy and Medicine in Ancient Greece, with an edition of *On Ancient Medicine," Supplements to the Bulletin of the History of Medicine,* 8 (1946):1–100.

[2] A-J. Festugière, *Hippocrate, l'ancienne médecine, introduction, traduction et commentaire* (Paris, 1948).

[3] Jones, *Ancient Medicine,* p. 4.

issues of the day. Should medical biology derive its wisdom (*a*) from philosophy (from theoretical ideas about the nature of man) or (*b*) from technology (from experience with the things that make sick men well and vice versa)? Again, do we learn (*a*) about disease through the study of health or (*b*) about health through the study of disease? This may seem a specious question but to the author of *Ancient Medicine* it was a real one. Again, should the emphasis in biomedicine be placed (*a*) on therapeutics or (*b*) on prevention? In either case, since the physician is inescapably engaged in both, will procedures stress (*a*) regimen (exercise and diet) or (*b*) medication (there was a developed *materia medica* at this time, but note the infrequent mention of drugs in the three treatises we shall consider)? Again, pre-Socratic physics has presented medical science with two rationales for "explaining" the phenomena of both the cosmos and the living microcosm—one essentially dynamic (or microdynamic), the other essentially dimensional (or microstructural); on which of these explanatory procedures will the physician base his view of the body in health and disease? These are only some of the issues that worked, explicitly or implicitly, to mold biomedical thought at the end of the fifth century B.C.

In one sense, *Ancient Medicine* played down the importance of the life-matter problem. Its author disagreed with "certain physicians and scientists who say that it would be impossible for anyone to know medicine who does not know what man consists of, this knowledge being essential for him who is to give his patients correct medical treatment." He thought such thinking appropriate only to philosophers like Empedocles (the only author mentioned by name throughout the treatise). "For my part, I consider, first, that all that has been said and written by scientist or physician about natural science has less to do with medicine than it has to do with the art of writing. Next, I consider that clear knowledge of nature can be derived from no source except from medicine; . . ." nature meaning here what a man is, the causes of his being, and similar concerns.[4]

The position taken by our author is illustrative of a significant point about the evolution of physiology and pathology. The two sciences have developed in an interrelated way, but whereas some thinkers have tended to start with the well man (with physiology) and

---

[4] Jones, *Ancient Medicine,* chap. 20. Our quotations are from Jones's edition, with permission of the Johns Hopkins Press.

from that have derived their ideas about disease, others—including the author of *Ancient Medicine*—have tended to begin with the sick man (with pathology) and from that have learned about men's normal constitution. But how, more precisely, did our author view medicine as producing a knowledge of man and what did he believe that medicine has told us?

We find well-thought-out answers to the first part of this question and a nucleus of interesting ideas, or suggestions, about the second part. The very first thing we are told (chapter 1) is that medicine is not to be approached through the formation of hypotheses because it is in fact "an art (*techne*) that really exists, one used on the most important occasions by all men, who pay the highest honour to its good craftsmen and practitioners." The implication is that the art has been established through experience, and the author is explicit on the way the experience was obtained. He treats the subject in a somewhat anthropological manner.

Dietetics Reveals the Nature of Man

We hear that primitive men living on harsh foods learned that not all foods suited them equally well; they "endured much terrible suffering from strong and brutish regimen when they partook of foods crude, unblended, and possessing powerful qualities, just such, in fact, as they would endure through them today, falling into violent pains and diseases speedily followed by death." In this way men evolved a dietetic art. It became clear likewise, during the development of this art, that refined men require more refined food than other men do, and that sick men often require less, or milder, food than well men require. Perhaps at first men learned mostly to reduce the quantitative intake of foods that did not agree with them, but they learned presently—and this was especially important for the sick— to moderate its quality by blending and coction. Thus physicians, when there began to be such, were following in the path of those who discovered what was good for men in general. They discovered and removed from the diet of the sick what was harmful to the sick.[5]

On the practical side, there is danger not only when food is too much or too strong but also when it is too little or too weak. Hence, "it is necessary to aim at some measure." The real test of the physician is his accuracy in gauging such matters not in mild but in serious

[5] Ibid., 5–8 (cited by chapter).

affections, just as the test of those who steer is their accuracy when the ship is in a gale. The proper measure is not a mathematical matter, however; it must be determined by how a man feels. The physician must take into account, too, the patient's habits. A person accustomed to two meals a day cannot with impunity omit one meal (or vice versa). "If a man is in the habit of taking lunch, that meal being beneficial to him, refrains from doing so, he experiences at once, as soon as the lunch hour is past, severe prostration, trembling, and faintness." A strongly constituted person will, however, be less affected by a departure from habit than a weakly constituted one.[6]

Discoveries concerning diet are still being made, our author acknowledges, for example by athletic trainers who do research to discover by what food and drink a man can best increase his strength. In Jones's commentary, he notes how much wiser the Greeks were than we in making their medicine more preventive and less merely remedial. "The Greeks were not so stupid; they recognized, in theory and in practice, that the art of healing must be combined with the art of health."[7] One senses here, too, a Greek interest not merely in making the sick man well, but in making the well man even more so.

The Constituents of the Body

As to constituents, the author of *Ancient Medicine* argued against making these merely the hot and the cold—the least potent pair of powers[8]—and rather saw the constituents (here seeming to follow Alcmaeon) as "a great many other things" (endowed with other sorts of opposite powers). As to what things exhibit these powers, he speaks of them, rather infrequently and without specifying what he means in much detail, as humors (*chymoi*). By inference, one judges that he visualized many humors, and not just the four that we shall find in the slightly later Hippocratic treatise *On the Nature of Man*.[9] He thought that what makes the difference in prevention and therapeusis is not whether food be hot or cold, or moist or dry but whether it be, if hot, hot and stringent or hot and insipid.[10]

As had Alcmaeon, our author supposed that absence of pain was due to the constituents being mixed and blended so as to prevent the

---

[6] Ibid., 10–12.
[7] Ibid., Jones's running commentary on chapter 4.
[8] Ibid., 16.
[9] Ibid., 22, 24.
[10] Ibid., 15.

existence of extremes. Both inadequately blended food and inadequately blended body constituents were regarded as injurious; pathological exudates were likewise considered to be, as it were, unblended. The therapeutic deduction from this theory was not that one should treat an extreme by applying its opposite; rather, one should temper the extreme expression of a particular power by supplying well-tempered foods. For the healthy man to eat blended food (e.g., bread and barley cake) was the proper way to stay healthy. "When these are taken in great quantity by a man, no disorder arises in the least degree, nor isolation of the powers in the body, but a very high degree of strength, growth and nourishment, simply because the food is well blended, with nothing unblended or strong, forming a single, simple whole."[11]

Homeostasis

In his attack on the hot and the cold as rather impotent opposites, our author hints at a doctrine of compensatory adjustment or thermal homeostasis. When hot and cold are blended they naturally cause no pain. But they may cause pain if isolated, so that if something cools the body unduly "for this very reason speedily before anything else heat appears internally out of the man himself, needing neither reinforcement or preparation; these effects are produced both in healthy men and in sick." If having thus adjusted internally to a cold situation, a man were to move from the cold outer air to a normally heated room without removing some clothes, he would feel uncomfortably hot for a while. If it be objected that "cold does not come to the rescue against the heat," our author answers that a feverish person is feverish not primarily through heat but through some other quality combined with it, such as acid or salt, which cold would not be expected to overcome.[12]

Dynamic and Schematic Aspects of Body Components

We discovered in our study of Greek biology, and this was already clear in the pre-Socratics, that different authors place different emphases on two partly separate interpretive traditions. Some are primarily concerned with the qualities or powers of the components of the system they are describing, whereas others pay more attention

[11] Ibid., 14.
[12] Ibid., 16, 17.

to the dimensional and structural aspects of these components. The difference among authors on this point is partly one of emphasis, since most interpretive systems pay some attention at least to both dynamic and dimensional considerations and relate them to each other in various ways. Thus far, we have listened to the author of *Ancient Medicine* primarily in connection with the powers (*dynameis*) of the body constituents, but he was also interested in their dimensions, their shapes or structures (*schemata*).

His interest in spatial organization extended from macro- to microstructural considerations. He regarded hollow organs with narrow apertures as best able to attract and draw moisture (the mouth does that better when almost closed than when wide open). Man's bladder, his head, and, in the female, the womb are examples. Or at the microlevel "spongy and porous organs, like the spleen, the lungs and the breasts, will drink up most readily what is in touch with them, and these will harden most and increase in size on the addition of fluid. They will not be emptied each day, as is the belly, etc."

Summary

The central doctrine of the author of *Ancient Medicine* was that what we know about the man we know largely through medical experience, acquired by physicians through generations of experiment with the diets of sick and well human beings. Our author had a secondary and derivative interest in man's constitution which seemed to him to have two essential aspects, dynamic and schematic. From Alcmaeon he derived the idea of the normal condition as a blend of many opposite powers possessed, the writer probably believed, by humors that constitute the body.

On the Nature of Man
(ca. 400 B.C.)[13]

Physical theories of whatever period typically specify both (a) the number of elements that exist and (b) their differing properties. We have just noted that, as to properties, Greek physicists tended to enlist under one or the other of two opinions, one school (whose chief exponent, ultimately, was Aristotle) emphasizing the "qualities" or "powers" (*dynameis*) of the several elements whereas the

[13] Quotations except where noted from "Nature of Man," *Hippocrates,* trans. W. H. S. Jones, Loeb Classical Library (London and New York, 1931).

other school (the atomists) saw the dimensional characteristics of the elementary particles as fundamental. The author[14] of our next treatise, *Nature of Man,* belonged to the first of these two persuasions, depending more upon qualitative or dynamic than on dimensional or schematic assumptions in explaining the appearances of things.

Physics and Physiology

He did not discuss explicitly the physics of the cosmos, but limited himself to that of the living microcosm. His biophysics paralleled the physics of Empedocles in specifying four elements in the living body. These physiological elements he termed humors (*chymoi*).[15] He surveyed the evidence for—and to his own satisfaction rebutted— the view that would form the body out of a single multimodal *arche.*[16] The humors were for him distinct and nontransmutable.[17]

His way of developing the humoral doctrine was in emphasis biomedical rather than generally physiological, the distinction being drawn in the very first sentence of the treatise where we are warned not to look for information about the nature of man in general. Nevertheless, the treatise conveys by implication—and to some degree explicitly—much general knowledge. Its author sought to say what health is and how to achieve it. The four humors seemed to him to make it possible to do this, it being assumed that together they form a properly balanced and thoroughly compounded intermixture (*krasis*). Variations in this mixture were also part of the theory, accounting for many phenomena, especially diseases, and suggesting methods of prevention and treatment.[18]

There is an important point to be noted about the historical role of the doctrine of humors. It did two things that a good theory is supposed to do, and it did them fatefully well. First, it imposed order upon—it organized—phenomena that seemed otherwise largely disordered. And second, it posited principles (humors) that were simpler than the phenomena (the data of health and disease) that they were meant to explain. The humoral doctrine was in these respects

[14] Or, authors? Jones, *Hippocrates,* Introduction, p. xxvii, surmises that this along with other Hippocratic "books" is "a chance collection of fragments varying in size and competence, and perhaps put together by a librarian or book dealer."

[15] Ibid., "Nature of Man," 4–5 (cited by chapter).

[16] Ibid., 1–3, 6.

[17] Ibid., 5.

[18] Ibid., 4–6, 9.

a highly successful pioneering application of sound scientific procedure. The theory was ultimately abandoned, but only after 2,000 years and after many changes and reformulations. What impresses us, then, is not its ultimate wrongness, but the basic rightness of what it tried to do and of the way in which the attempt was made. Few scientific theories have remained so persuasive for so long.

Applications of Theory

Temperament. The variations in humoral balance (*krasis*) are in some cases normal and healthful, in others irregular and harmful, our author informs us. Each of us has his humoral individuality and to preserve his health must follow an individual regimen. The temperament (blend) of each of us changes normally in a seasonal way, each humor becoming dominant in its proper season. Phlegm is, like its season winter, wet and cold; blood, like spring, wet and warm; yellow bile, like summer, dry and hot; black bile, like fall, dry and cold. "All these elements, then, are always comprised in the body of a man, but as the year goes round they become now greater and now less, each in turn according to its nature. . . ."[19] The cause of these seasonal changes is the weather, and each season has its corresponding diseases.[20] At a later period in medical history the idea of temperament was to be used as a basis for classifying men into four somatopsychic types (sanguinary, phlegmatic, choleric, melancholic). Contrary to a rather widespread opinion, this typological grouping is not found in any of the Hippocratic treatises. It was probably introduced by a contemporary of Galen and transmitted to Europe through Arabia (see below Table 3).

Cause and cure of disease. Disease itself was viewed by our author as a general or local dyscrasia, or imbalance, typically involving a translocation of a particular humor with resultant exhaustion of one locality and flooding of another. It was his view that when this happens, both localities are painful. He recommended that a local excess be treated by evacuation induced by drugs or by regimen (bleeding seems not to have been used for the relief of excess by the early Hippocratics).[21]

As to the causes of disease, he thought the two principal ones were

[19] Ibid., 7.
[20] Ibid., 8.
[21] Ibid., 4, 9; bloodletting has a long if not honorable history stretching back to the Neolithic; it was widely practised by Egyptians, Babylonians, Hindus.

bad regimen and bad air, regimen-caused diseases occurring sporadically or endemically, atmospheric ones often epidemically. To treat the former called for an altered regimen, in his view, whereas the latter were to be treated by reduced inhalation of the suspected air, as well as by weight-reducing diets (because reduction minimizes the need for air).[22]

Therapy. Humoralism supplied the theoretical basis for erecting a system of general therapy, our author setting this forth succinctly in directing that "diseases due to repletion are to be cured by evacuation and those due to evacuation by repletion; those due to exercise by rest and those due to idleness by exercise. To make the whole matter plain, the physician should set himself up in opposition to [the untoward effects of] (*a*) diseases, (*b*) idiosyncrasies, (*c*) seasons, and (*d*) ages. . . ."[23]

Conclusion

There are aspects of humoralism that we could wish, for present purposes, to have had explained more fully. The theory did stipulate the physical preconditions of life—but almost exclusively of *human* life; it would have been helpful to have had the discussion extended to animals and plants. But then, the author's interests were primarily in human hygiene; and we do hear, in chapter 3, that generation is composition and that death is decomposition not only in man but in "animals, and all other things."

ON REGIMEN
(ca. 400 B.C.)

Life as Opposed Transformation

The third of the treatises which we choose as illustrative of early medical opinion on life and matter uses a dualistic physics.[24] The four elements are reduced to two, fire (which causes motion and at times erosion) and water (which causes nourishment and repletion). The author of *On Regimen* held the view—contrary to that developed in *Ancient Medicine*—"that he who aspires to treat correctly of

[22] Ibid., 9.
[23] Ibid., 9.
[24] "On Regimen," *Hippocrates,* trans. W. H. S. Jones, Loeb Classical Library (London and New York, 1931). Our quotations are from this edition except where noted.

human regimen must first acquire knowledge and discernment of the nature of man in general—knowledge of its proper constituents and discernment of the components by which it [man's nature] is controlled." He adds that the physician must know further the power possessed by all the foods and drinks of our regimen.[25]

It was the central thesis of the author of *On Regimen* that exercise consumes the body while food replenishes what exercise has consumed, and that "indeed . . . (if) it were possible to discover for the constitution of each individual a due proportion of food to exercise, with no inaccuracy of excess or defect, an exact discovery of health for men would have been made."[26]

This system was not as simple in application as in theory, since it taught that the balance of food to exercise must be variously adjusted to take account of constitutional idiosyncrasy, time of life, season of the year, direction of the wind, geographical locale, and other factors. Moreover, food was, for our author, not just food; nor was exercise just exercise. Kneaded barley cake was different from unkneaded. Wrestling in the dust, different from wrestling with the body oiled. The physician must not depend on general canons. He must study the patient's physique in the gymnasium to see what should be added or taken away through proper food and exercise.[27]

The Physics of On Regimen

Turning to the book's physical principles, we find that its author regarded fire as hot and dry (but with a certain moistness); water as cold and wet (but with a certain dryness). He posited an interplay between fire and water, fire tending to volatilize water and water tending to quench fire, but never, in either case, completely. As fire threatens to annihilate water, the fire itself flags (because the water it is annihilating is its food); and as water threatens to quench fire, the water is immobilized (because the fire is the source of the water's movement). We have here an early appreciation of the phenomenon that modern technology terms negative feedback.

Students of Greek medicine are not in complete agreement as to the sources from which the author of *On Regimen* derived his physical ideas. The theme of contrariety, which is a conspicuous part of the

[25] Ibid., 1.2, 3 (cited by book and chapter).
[26] Ibid., 1.7, where we hear that spring, like blood, is moist and warm.
[27] Ibid., 1.2.

theory, we have already encountered in Anaximander, Heraclitus, Alcmaeon, and Empedocles. Contrariety was also fundamental to Pythagoreanism, where, however, the list of (usually) ten pairs of opposites generally omitted the hot and the cold, the wet and the dry. But the early fifth-century physician Alcmaeon, who may have been associated with the Pythagoreans, had "taught that what preserves health is equality between the powers—moist and dry, cold and hot, bitter and sweet and the rest—the prevalence of one of them produces disease, for the prevalence of either is destructive."

The notion that fire could be a little wet and water a little dry could have stemmed, and the modern commentator Joly thinks it did,[28] from Anaxagoras (who had accounted for change on the assumption that everything contains a part of everything else). Joly and others suppose that an immediate source of our present author's physics may have been the relatively minor thinker Archelaus of Athens (445 B.C.?) who was a student of Anaxagoras. The idea of mutual encroachment had appeared even earlier, however, in Anaximander.

Whatever its sources, the supposed interaction of fire and water seemed to the author of *On Regimen* to explain many familiar phenomena. One of its most remarkable consequences was the spontaneous generation of seeds and even of fullfledged animals[29]—though the reader gets no detailed ideas as to how the oppositive interaction of fire and water would produce such results.

Obscurity. Such views warn us about the spirit in which we must approach a book like *On Regimen*. It has not always been, as it is today, a goal of science to express its meaning clearly. Many medieval and ancient writings, including parts of this one, attempted a deliberate mystification. In such cases the modern reader must guard against an initial feeling of repugnance. An obscure passage may show us the author hovering very close to but just missing a profound intuition. Or the author may be playing a kind of game; he may be playing the oracle, or even hoodwinking the reader gently. Such games can be quite rewarding for the reader prepared to enjoy them.[30]

---

[28] Robert Joly, "Recherches sur le traité-Hippocratique Du Régime," *Bibliothèque de la faculté de philosophie et lettres de l'université de Liège,* fasc. 156 (Paris, 1960), p. 19. For Alcmaeon, see D-K, B 4; the translation is Guthrie's.

[29] "On Regimen," 1.3–4.

[30] See the article by W. H. S. Jones, "Intentional Obscurity in Ancient Writings," in *Hippocrates,* 4:ix ff.

Applications of the Theory

If we cannot always understand, we can at least admire the super-structure of explanations which our author reared on his assumption that everything is fire and water. However gratuitous, these explanations illustrate the basic determination of science to find actual order in the midst of apparent disorder and to understand the human body as only another aspect of nature.

Sexual differences. "The males of all species are," we learn, "warmer and drier, and the females moister and colder. . . ."[31] Each sex can, however, produce a male and a female secretion (an activity partly responsive to diet). "If a man would beget a girl, he must use a regimen inclining to water. If he wants a boy, . . . inclining to fire."[32] The author of *On Regimen* considered conjunctions of these differing seeds responsible for sex-determination and even for the production of supersexes and intersexes[33] not totally unsuggestive of those which twentieth-century biologists (Goldschmidt, Bridges, Dobzhansky) later discovered in insects.[34]

In A.D. 1902, C. E. McClung showed that sex is often determined through the production of two sorts of sex cells within the same parent (typically the male).[35] One can admire the shrewdness of the hunch about sex-determination suggested in the following table which summarizes our author's ideas on this subject.

There were conflicting ideas in early Greek medicine as to the

TABLE 2

DEGREES OF SEXUAL DIFFERENTIATION

If the contribu-tion of the FATHER is:	And of the MOTHER is:	The RESULTING INDIVIDUAL will be:
Male	Male	A strong and brilliant man
Male	Female	A brave man
Female	*Male*	An hermaphroditic male
Female	Female	A brazen or mannish woman
Male	*Female*	A bold but still modest woman
Female	Male	A fair woman

NOTE: Italics indicate partial dominance, or "mastery."

[31] "On Regimen," 1. 34.

[32] Ibid., 1.27.

[33] Ibid., 1.38–39.

[34] See especially C. B. Bridges' reviews in the first and second editions of *Sex and Internal Secretions* (Baltimore, 1934, 1939).

[35] "The Accessory Chromosome—Sex Determinant?" *Biological Bulletin* III, nos. 1 and 2 (May and June, 1902), pp. 43–84.

source within the parents' bodies of their respective contributions to the offspring. A careful modern student of these views, E. Lesky, sees opinions on this subject as divisible into three categories. First, according to Alcmaeon, soul stuff and seed stuff are one, and both are identified or associated with the contents of the brain and spinal chord (*myelos*); another proponent of this scheme was Plato (see chapter 7). Second, the atomists, who may have obtained the idea from Anaxagoras, derived the genetic material from the body in general. Third, Parmenides and Diogenes of Apollonia derived it from the blood. Among these theories our present author seems to incline toward the second, or pangenetic, view, though the passage dealing with this subject is not easy to understand.[36] Materials—presumably pangenetic in origin (i.e. from various parts of the parents' bodies)—can only form an offspring, he said, if they achieve a certain harmony or attunement of the tonic with the fourth, the fifth, and the octave.

Morphogenesis. The meeting of the two secretions leads on, in our account, to embryonic differentiation in which fire acts as a molding agent to produce a small copy of the whole. Among the effects of fire are: activation of the moisture, a compaction of part of the moisture to give rise to sinew and bone, and a carving of passages through the residually moist materials; of these passages, the innermost and greatest is the belly, from which the fire leaps forth and makes other passageways for the breath and for the food. Other details of morphogenesis are given which, although they are too filled with poetry and astrology to be more than semicomprehensible, reflect an important theme in the history of biological ideas, viz. the parallelism of the cosmos and the (human) microcosm. The belly is analogized to the sea, and the surrounding firmer tissues to the earth. In the latter, three circuits are provided by fire for itself—one circumferential, another intermediate, another central. These in some way correspond to the power or powers of the stars, the sun, and the moon. The nature of this correspondence and the character of the circuits (*periodoi*)—much discussed by the students of this treatise—are obscure.[37]

Less perplexing are certain other assertions, such as that conception can occur only on one day of the month,[38] that an amalgamation

[36] "On Regimen," 1.8. See, also, Joly, "Recherches sur le traité-Hippocratique. ..." pp. 26–35.
[37] "On Regimen," 1.9, 10, 26.
[38] Ibid., 1.27.

of the male and female genetic matter is accompanied by an amalgamation of souls,[39] that although we may see larger parts sooner all are formed simultaneously,[40] that the newborn is relatively warm and moist, a young man dry and warm, and an old man dry and cold.[41]

Nutrition. We have followed the idea of nutrition as a summation of opposed transformations from sources at least as early as Heraclitus. From our present author, Heraclitizing for the moment, we hear that "all things both human and divine are in a state of flux upwards and downwards by exchanges."[42] In a more Anaxagorean vein, he informs us that "what enters [as food] must contain all the parts [of what is already there]." This must be so, because "anything [already existing in the body] for which [a corresponding part], was not present [in the food], would fail to grow, whether much or little food were available, because it would not have the wherewithal [i.e. the necessary parts wherewith] to do so." But, our author continues, "each [existing part], having all [that it requires], grows where it is, thanks to the addition of dry water and moist fire, part being driven *in* [in nutrition] and part being driven out [in elimination]. Similarly, when carpenters saw a log and one pulls and the other pushes, they accomplish the same thing—without which the saw will not move. The one who exerts pressure below pulls the one up above and if force is applied all is lost. So with the nutriment of man; this pulls, that pushes; what is forced in presently comes out; but if an untimely violence accedes, nothing happens." This passage is not easy, but neither is it impossibly occult. The idea it evokes—of life as opposed transformation—is one which will have a long and active later history.[43]

The Life-soul

Our author's allusions to this subject—and he does little more than allude to it—are far from clear, and, after the efforts of qualified interpreters, the lack of clarity remains. The following outline seems consistent with the texts, translations, and authoritative interpretations, but is tentative. Our author saw the soul as made, like the body, of water and fire and as being capable, like the body, of nourish-

[39] Ibid., 1.29.
[40] Ibid., 1.5.
[41] Ibid., 1.33.
[42] Ibid., 1.5, retranslated with additions.
[43] Ibid., 1.7. See also T. S. Hall, "Life as Opposed Transformation," *Journal of the History of Medicine and Allied Sciences,* 20 (1965):262–75.

ment[44] and consumption. Soul seemed to him to be uniform in itself but yet able to express itself differently according to the different bodies or body parts. A man's soul, for example, develops in a man and in no other creature. In another and rather spiny passage, we hear that the soul has its parts, which the author probably saw as corresponding to parts of the body.[45]

On Regimen depicted the soul, further, as "entering" the body (not metempsychotically, presumably; perhaps by passing from parent to offspring when life begins). We hear that at the beginning of life there is an amalgamation of male and female souls. Also that soul is burned up in early life to permit growth, and diminishes through cooling with the mitigating influence of old age, but can be added to in adulthood.[46] Perception is a shaking of the soul by impinging fire, and we are about to see that the blending of fire and water is the basis of the soul's intelligence or want of it.[47]

Individual Differences

A major challenge to any physiological system is to account, if it can, for the fact of individual, constitutional difference. The present theory meets this assignment by permitting certain differences in condition of the two elements fire and water:

TABLE 3

CONSTITUTIONAL TYPOLOGY

THUS IF, IN AN INDIVIDUAL		
WATER IS:	AND FIRE IS:	THAT PERSON'S CONSTITUTION WILL BE:
Fine	Rare	Healthy
Dense	Strong	Physically robust but susceptible to illness
Dense	Fine	Cold and moist; unhealthy in winter and spring
Dense	Moist	Warm and moist; unhealthy in spring, especially during childhood
Fine	Strong	Warm and dry; unhealthy in summer, especially in prime of life
Dry	Rare	Cold and dry; unhealthy in autumn

[44] "On Regimen," 1.4. In one passage he alters the picture and composes the soul of water and fire mixed with the parts of the body. See chap. 25.

[45] Ibid., 1.5. In the passage in question, the author speaks about changes in the size of the parts (perhaps in the embryo, perhaps in the developing child) insofar as these are affected by nutrition; and he seems to say that no new parts are added nor are any taken away but spatial conditions must permit the requisite change in size of those already present.

[46] Ibid., 1.25.

[47] Ibid., 1.35.

Note the similarity of the last four constitutions to the phlegmatic, sanguineous, bilious, and atribilious constitutions specified in the later (medieval) doctrine of four temperamental types.[48]

In the second book of *On Regimen* we learn further that "wild beasts are drier than tame; small eaters than great eaters; hay eaters than grass eaters; small drinkers than great drinkers; . . . males than females; entire than gelded; the black than the white; the hairy than those which have little or no hair."[49]

In still another and equally gratuitous application of the fire-water hypothesis, our author accounts at length for differences in human intelligence (*phronesis*). Very moist fire blended with very dry water produces outstanding intellective faculties including acute discrimination and strong retention. Other blendings yield: the less intelligent; the merely attentive (but not ineducable); the silly; the grossly stupid (they weep for no reason and fear what is not fearful); the half-mad; the utterly mad.[50]

If the balance of fire and water be displaced in the fiery direction, the result, our author seems to suggest, is increased quickness and perception tending even to madness; if in the watery direction, the result is slowness tending to stupidity. In an interesting analysis of this part of *On Regimen,* the modern commentator Siegel suggests that the author may have grasped in a preliminary and intuitive way those psychic variations which we recognize today as resulting from hyper- and hypothyroid activity.[51] The debt to Heraclitus is, in any case, apparent.

For each of these psychological variants an appropriate regimen is recommended. The diet for a half-mad man should "consist of un-kneaded barley bread, boiled vegetables (except those that purge), and sardines, while to drink water only is best, should that be possible, otherwise the next best thing is a soft white wine." The highly intelligent are to be cherished through careful attention to food, drink, gymnastics, and the amount and timing of intercourse. In sum, it "is this blending, then, that is, as I have now explained, the cause of the soul's intelligence or want of it; regimen can make this blending either better or worse." Such characteristics as irascibility, indolence, craftiness, simplicity, quarrelsomeness, and benevolence, however, depend

[48] Ibid., 1.32.
[49] Ibid., 1.49.
[50] Ibid., 1.35.
[51] R. E. Siegel, "Hippocratic Description of Metabolic Diseases in Relation to Modern Concepts," *Bulletin of the History of Medicine,* 34 (1960):355–64.

not upon the fire-water mixture but upon the character of the body passages through which soul flows. Regimen does not affect these traits.[52]

Prodiagnosis. We shall resist the temptation to explore further the many biological, psychological, and pathological questions developed ingeniously, if often elusively, in *On Regimen* except to mention finally that the work was partly developed around a central theme, an approach to disease through "prodiagnosis." This theory, according to its author, "reflects glory on myself, its discoverer." Prodiagnosis is superior as a system, we are told, to mere passive prognosis. The prodiagnostician interprets the disease in its early stages as due to a particular imbalance of food and exercise and can both predict and modify its further development. W. H. S. Jones calls the author of *On Regimen* the "father of preventive medicine."[53]

Conclusion

On Regimen is a rich source of information about the daily life of the Greek as he proceeds from his bed (there is a section on dreams)[54] to the table, market, shop, gymnasium and theater. The work is highly derivative. Anaximander, Heraclitus, Alcmaeon, Anaxagoras, Empedocles, the Pythagoreans are its authors' lineal ancestors. But the synthesis is new and if the book looks backward, it looks forward, too, both in the way it asks the classic questions, and in its endeavor to bring physiology and pathology together in an ordered conceptual system.

[52] "On Regimen," 1.35–36.
[53] Ibid., Introduction. Prodiagnosis is mentioned at 1.2.
[54] The use of dreams is diagnostic; the topic of the dream reveals the nature of the disorder. Most of the fourth and final book of "On Regimen" is concerned with dreams.

7

The Life-soul as Imposed

PLATO [427? · 347 B.C.]

The Timaeus

Plato's *Timaeus* was, from two points of view, one of the most significant of early scientific writings. First, it contained what is the oldest extant complete physical theory of the universe—its origin, contents, evolution, and present condition. Second, it arrived at this theory through a sound formulation of sound scientific questions, among them, the life-matter question.

The answers to these questions are supplied in the dialogue by a certain Timaeus (not otherwise known from ancient sources) who is heard expounding cosmology to Socrates and two companions. We shall follow Plato's didactic device of ascribing the theory to the presumably fictitious Timaeus. The presentation is Plato's own synthesis of ideas partly borrowed (many of them will be familiar) and partly his own.

A special point needs to be noted by the modern reader with respect to the *Timaeus* as science. Plato did not offer it as dogma or certain knowledge but as probable and hypothetical (Timaeus calls it a "likely tale"). Its subject matter—the world and man in their physical aspects—could not be known surely in Plato's epistemology, since for him the only things surely knowable were certain transcendent and eternal ideas, or Forms, of which the phenomena we see are merely transient copies. Only the gods and a very few mortals especially dear to the gods can have knowledge, he believed; the rest of us can only make guesses and form opinions.[1]

Otherwise unidentified numbers refer to corresponding passages in the *Timaeus.* Translations, unless otherwise identified, are those of Francis M. Cornford, *Plato's Cosmology* (London, 1956), and are published by permission of Routledge & Kegan Paul, Ltd. and Humanities Press.

[1] 27d–29a.

Timaeus' "likely tale" was different in still another way from what we think of as scientific exposition, namely, in incorporating mythical elements. As science its purpose was, like that of much modern science, reductive; it intended to make man and the world comprehensible in terms of biology and physics. Moreover, it proceeded, partly, by telling the story of man's and the world's creation. Yet—curiously, it seems at first—we are not meant to take the creation story literally. Most Platonists, at least, agree that Plato thought of the world as having always been in existence. The story of its creation in successive steps, as we hear it from Timaeus, is merely an explicative device to help us understand the world's permanent nature. Just as the geometer helps the student grasp the nature of a polyhedron by "constructing" one, so Timaeus shows how, if the cosmos had not always existed, one might have proceeded to build it. Although the cosmogonic part of the account is thus a myth and although, as Paul Friedländer notes, "nothing could be stranger to a modern reader than a myth on physics,"[2] that myth serves in the present case what Aristotle later termed pedagogical purposes.

So many accounts of the *Timaeus* have emphasized its mythological elements, in any case, that we are in danger of overlooking the large parts of it that were straightforwardly scientific. The cosmogony of the dialogue was mythical, but its cosmology, especially its physics and physiology, were offered as serious science. And on the subjects of our special interest, life and matter, it rewards us richly.

First, Timaeus gives us a sophisticated theory of "matter." According to the theory, four sorts of elementary particles account by their dynamic properties, shapes, and juxtapositions for the complexities of sensible reality. Second, living things are explicitly identified: they are things that have *soul* bonded to their constituent particles. Third, this bond can exist only where the elementary particles (or their constituent subparticles) are properly compounded. Soul, moreover, is not a loose concept. It is as precisely and specifically formulated as is matter, even more so. Furthermore the compound of elements to which soul is bonded is, in the human body, actually visible. It is present also in animals and confers vitality upon them, as it does upon plants. This whole way of looking at nature, finally, has important implications for the general way in which man must

[2] Paul Friedländer, *Structure and Destruction of the Atom According to Plato's Timaeus* (Berkeley, 1949), p. 227.

look at himself; ethical questions are raised in the *Timaeus* which have profound relevance to the human condition.

Precautions. In exploring Timaeus' ideas, we need to observe two precautions. First, from our acquaintance with the pre-Socratics, it will not surprise us that the terms "life" and "matter" had special meanings for Plato. Plato's physics was borrowed from, but also differed from, that of his most eminent predecessors, Democritus and Empedocles; it also differed from that of his student and intellectual successor, Aristotle (who, incidentally, would be the first to use a word, *hyle,* for matter as such). Thus, when we say that Plato attempted an answer to the life-"matter" question, we mean, more precisely, that he had a theory about (*a*) the physical differentiations of reality in general ("matter") and about (*b*) the special physical differentiations of reality that permit perceptible life-as-action.

Second, there is a point to be noted about the relations of the *Timaeus* to the whole corpus of Plato's writings. The life-matter problem did not have for Plato, when we consider the totality of his interests, the central significance that it was to have for scientists of two thousand years later. Plato was first of all a general philosopher for whom questions arranged themselves according to a certain hierarchy of significance. The significant questions for him were "what *is*?" "what is *being*?" "what is *good*?" "Protoplasm" occurred in his system as a ramification, albeit an important one, of these more fundamental questions. Knowledge of nature was not an end in itself for him, but part of an attempt to link human order and excellence with the order and excellence of the cosmos. He makes Socrates complain that Anaxagoras failed to use Mind as a way of showing why the physical world is the best of all possible worlds. But this fact does not in the least lessen our interest in what Plato says (or reports the probably fictitious Timaeus as saying) about what, for us at least, is a burning question, namely: What latent condition of the body permits it to exhibit the patent phenomena of life?

Matter

Primary bodies. Timaeus' (Plato's) cosmos was distinctive in being particulate yet not strictly material. Its primary constituents consisted not of differentiated matter but rather of dynamically differentiated space. A fire particle for him was a very small sample of ignified space.

The total number of basic differentiations in Plato's physics was, in imitation of Empedocles, four. It was his view that we become aware of these four and distinguish them from each other by our senses: all four are "visible bodies." Yet he saw the sensible attributes of things as stemming from properties of a more fundamental nature. And these more fundamental properties of things were two in kind. First, each of the four primary bodies possessed certain dynamic powers or abilities to effect changes in the others. Second, each differed from the rest in the geometrical characteristics of the particles composing it.[3]

Earlier Greek physics had presented Plato with two ideas about the properties with respect to which principles or elements might differ. According to one tradition, conspicuous in Heraclitus, the differentia were preeminently dynamic. According to the other, conspicuous in Democritus, they were preeminently (but not exclusively) dimensional. Plato combined these two traditions, the dynamic and dimensional, and developed them further. He went beyond Empedocles and Democritus to assign specific geometrical shapes to each of the four sorts of particles, and he linked these (cryptomenal) shapes to the (phenomenal) effects the four exert on the senses. Particles of fire, smallest of all, are regular tetrahedra, Timaeus tells us. Those of air, octahedra. Those of water, icosahedra. Those of earth, cubes.[4]

One's first impulse is perhaps to dismiss this conception as fanciful. In fact, it merits an important place in the history of conceptual model building. It represents a serious early effort to account for sensible phenomena by assigning hypothetical mathematical properties to a supposed insensible substrate. The selection of polyhedra was not strange. Interest was being focused on them in the Academy through the creative work of Theaetetus, and they were thus a logical choice. Among the sensible phenomena which we hear Timaeus "explain" in terms of the dynamics and/or dimensions of particles are: the transmutation of one fundamental substance into another, transition from state to state, volatility, solubility, viscosity, hardness, heaviness, color, sound (pitch, volume, timbre), and many other physical, physiological, pathological, and psychic manifestations.[5]

[3] 31b–32c; 48k–e; 49a–50a.
[4] 53c–56b.
[5] 56c–68b.

Transmutation. Timaeus' (Plato's) primary bodies were, as are the "atoms" of modern physics, unstable. Through a disassembly and reassembly of the individual particles, one element was supposed to be able to turn into another. Timaeus posits the transformation of earth into water, water into air, air into fire, and the reverse transformations. Elsewhere he retracts the part of the assertion that has to do with earth. Earth has to be left out of the transmutation cycle, for a definite reason which we now need to examine.[6]

To posit that a particle of element A can turn into one of element B assumes, or at least suggests, something about the makeup of these particles. It suggests that the particles themselves are divisible, permitting transmutation through rejuxtaposition of something smaller (as twentieth-century physics interprets transmutation through radioactive decay). It did not escape Plato that he could account for transmutation by assuming common components in the particles stipulated for each of the four elements.

Triangles. We discover such common components when we pay attention to the flat surfaces by which the particles were supposed to be bounded. In the case of fire, air, and water these surfaces are equilateral triangles, Timaeus tells us, and in the case of earth they are squares. Timaeus presents these surfaces as built up of smaller triangles properly fitted together. Not just any triangles will do; the "fairest" must be chosen. For fire, air, and water, one must use a right scalene triangle whose sides are in the ratio $1:2:\sqrt{3}$, fitting six of them together to form the required equilateral. For earth, one must use the isoceles $1:1:\sqrt{2}$, fitting four of them together to make up each side of earth's cubic particle.

Transmutation involves a fragmentation of the initial particles into component triangles and their reassembly to form particles of another sort. Two four-sided fire particles may fragment and reunite to form one eight-sided air particle. Water, with twenty sides, is formed from one four-sided fire and two eight-sided air particles. But in the case of earth with its square side, built of isoceles rather than equilateral triangles, transmutation is impossible.[7]

Timaeus presents the elements as, in a sense, isotopic, not every

---

[6] 49b–c; 56c–57c.

[7] 53c–57c. Note that when two fire particles unite, the volume of the water particle thus formed is greater than the sums of the volumes of the two uniting fire particles. For an analysis of the difficulties thus created, see Friedländer, *Structure and Destruction;* and Cornford, *Plato's Cosmology,* pp. 229–30.

fire (air, etc.) particle being precisely like every other. Their shapes are alike, but their sizes differ. Ice is a kind of water whose constituent particles are larger and more uniform than those of regular water. Gold is solid water comprising uniformly small particles.[8] Modern students of Plato's science think that Timaeus meant to build his isotopes by combining scalenes in different numbers, e.g. two, six, and eight:

(Note: each of the four equilaterals of the third figure can be divided into two scalenes by dropping the perpendicular from any apex to the opposite side). A similar possibility would exist in the case of the isoceles triangles different numbers of which could make up squares of different sizes:

We follow, here, the interpretation agreed to by both Cornford and Friedländer in their studies of Plato's physics. Their interpretation may be correct—it gives the precision that seems to be needed for orderly transmutation, for example—but it is considerably more precise than anything Timaeus actually says.

Proportionality of the primal bodies. Another crucial innovation in Plato's view of elements, or primary bodies, was the proportionality he assigned them. When first compounded to form a cosmos, he thought, they bore the relation

$$\frac{F(ire)}{A(ir)} = \frac{A(ir)}{W(ater)} = \frac{W(ater)}{E(arth)}$$

Fire and earth (the visible and tangible elements respectively) were used first in the building process, and air and water were inserted between them, in proportionate amounts, to secure that "amity"

[8] 57c–59b.

which makes the cosmos indissoluble except by the Creator who composed it in the first place. Had fire and earth been two-dimensional, a single mean would have been adequate. Thus, if the average diameter of all the fire were a and of all the earth were b, the proportion could be written

$$\frac{a^2}{ab} = \frac{ab}{b^2}$$

(which holds good for any desired values of a and b).[9] But, since fire and earth are three-dimensional, two means must be utilized, and the proportion must be written

$$\frac{(F =\ )a^3}{(A =\ )a^2b} = \frac{(A =\ )a^2b}{(W =\ )ab^2} = \frac{(W =\ )ab^2}{(E =\ )b^3}$$

(which, once again, holds good for any desired values of a and b).[10]

This scheme was more than a whimsical exercise in numerology. It represents its author's effort to account for the stability of the cosmos—a stability produced, he believed, by the orderly quantitative relations existing among its constituents. A question that has been argued by students of Plato was: What specific parameter or quantity of each of the elements would have to be measured to reveal their harmonious and stabilizing proportionality? The answer to this question is not revealed in Plato's text; but Cornford is almost certainly correct in thinking that Plato viewed the proportionality as existing among the total volumes of fire, air, water, and earth used in the creation of the cosmos.[11]

Two questions remain: what were Timaeus' particles particles *of?* and, what were they *in*—what were they surrounded by? In discussing these points Timaeus is not consistent or lucid, and the experts have disagreed about Plato's meaning. Nevertheless, it seems that without undue violence either to Plato or his interpreters we can make the following sense of his story.

Idea, Image, Receptacle. In Plato's metaphysics a particle or anything else was only an Image, a copy of a real "original" existing in some realm of Forms or Ideas apprehensible only by gods and

---

[9] e.g. $\dfrac{2\times2}{2\times3} = \dfrac{2\times3}{3\times3}$

[10] e.g. $\dfrac{2\times2\times2}{2\times2\times3} = \dfrac{2\times2\times3}{2\times3\times3} = \dfrac{2\times3\times3}{3\times3\times3}$

[11] 31b–32c; see also Cornford, *Plato's Cosmology*, pp. 43–52.

by those whom the gods hold dear. Indeed, the entire cosmos was but an Image of an ultracosmic and transcendental Idea or constellation of Ideas. But, in addition to Images and Ideas there existed for Plato a third something to which he variously referred as the mother, the nurse, the place, or the receptacle of the Image. He viewed this something, whatever it was, as homogeneous in itself. But he thought it had the ability to assume certain appearances causing it to be called earth, water, air, or fire—though it might more precisely be designated "earth-like," "water-like," et cetera.

Timaeus' (Plato's) Receptacle seems to come close, at times, to what we mean by "space" when we use that word nontechnically. Yet his Receptacle, unlike our space, is never empty. It wears everywhere the aspect of one or a combination of the elements. There is thus nothing corresponding to a vacuum. The Receptacle "appears different at different times," is "made fiery," is "liquefied," is to an element as gold is to an object or as a mirror to an image. If the modern mind boggles at Plato's Receptacle (which seems to be at once both something and nothing, both substance and space, both differentiated and homogeneous) we can take comfort in the realization that to Plato himself the Receptacle was no easy problem. Timaeus says that "if we describe (the third something) as a Kind invisible and unshaped, all-receptive, and in some most perplexing and most baffling way partaking of the intelligible, we shall describe her truly."[12]

Soul

Timaeus advances a physicomathematical theory of soul which may seem, on first reading, gratuitously specific and elaborate. The theory can be partly understood in the light of Plato's personal acquaintance with the creative geometricians of the period. Plato's science, viewed as a whole, was highly eclectic. We have already seen how important was mathematics among the disciplines whose contents he borrowed, altered, and integrated into his own intellectual system.

Timaeus presents soul as an incorporeal agent separate from but coterminous with the cosmos and responsible for celestial motion. For the latter role, the soul needs two portions, a circling outer portion to sweep the fixed stars round the earth and a subdivided inner

[12] 29b–d; 48e–52d (trans. R. G. Bury).

part to move the sun and the planets. In Plato's cosmology, all this is duly provided.

Creation. But provided by what? by whom? To solve this problem, Timaeus postulates a provider, variously designated as Artificer, Maker-and-Father, Architect, Constructor, Modeller, Demiurge. Timaeus' Demiurge has been understandably but quite erroneously identified by certain commentators with the single Judaeo-Christian Creator-God. Plato, a polytheist and pagan, had something less personally deistic in mind. The Demiurge, as presented, is symbolic; it is useful heuristically but not to be taken as factual. The Demiurge is a symbol of the transcendentally intelligent, the reasonable, character of the cosmos. Divine reason, acting as a cause, determines cosmic events.

Reason, Necessity, Chance. Not every event, however, is determined solely by Reason. Sequences of events occur such that, given one, the others follow necessarily (that is, without intervention by Reason). Every event, then, is caused by Necessity, by Reason, or by Necessity partly ordered and restrained by Reason.[13] In the sense that Necessity left to itself wanders from the course set by Reason, it becomes, in Timaeus' phrase, the Errant Cause (Aristotle was soon to draw a comparable distinction between Purpose and Chance). Some foolish and dangerous persons, disbelieving in the divine, would leave too much to errant necessity, making out that stars are mere stones; gods, fiction; religion, a cooking up of words. Timaeus' cosmos is one in which Reason and Necessity exist in a productive condition of reciprocal restraint.[14]

The cosmic soul, according to Timaeus, should be both motive and perceptive. It should both move and be moved. This motive-perceptive cosmic soul was brought into existence, he says, as soon as or even before the material world was created by that divine Artificer created by Timaeus more or less in the human image. The steps of the creative process are recounted in lavish detail. The story need detain us only briefly.

To perform effectively, Timaeus implies, soul must be able to perceive (*a*) whether or not bodies have being, i.e. whether they exist, and (*b*) if they do exist whether in the case of several objects

---

[13] 29d–30c; 34a–b.

[14] 47e–48e; see, on this passage, Cornford's commentary (indispensable reading for anyone interested in the history of the idea of causes in physics).

they are alike or not alike. The Artificer accordingly builds the soul out of three primordial ingredients: "being," "sameness," and "difference." Each of these three primal components already existed in two different modifications, one indivisible (unchangeable, eternal) and the other divisible (changeable, ephemeral), corresponding to the two realms postulated in Plato's theory of knowledge—the realm of realities or Forms and that of appearances or Copies. Taking the two members (indivisible and divisible) of each pair, the Artificer makes a third out of them intermediate between the other two. Then from these three (intermediate existence, intermediate sameness, intermediate difference) he finally fashions the soul. At any rate, the present world picture is what it would be had a hypothetical Artificer proceeded in the indicated manner.

The Artificer now subdivides soul into a linear series of segments which represent a combination of the geometrical series, 1, 2, 2^2, 2^3 and 1, 3, 3^2, 3^3 which, dovetailed together, give 1, 2, 3, 4, 8, 9, 27. This sequence is meaningful geometrically but not musically—not, at least, until the Artificer in two further steps divides each segment into smaller segments and these into still smaller segments corresponding mathematically to the intervals of the diatonic scale. Quantitative values are assigned to each "note" by a method that is a model of sophistication.

We shall later hear Timaeus say that the human soul was framed less purely but otherwise in the same general manner as the soul of the cosmos. The resulting arrangement permits us, perhaps through a sort of resonance action, to distinguish the metrical properties of objects and makes possible individual recognition of numerical order in the cosmos. This part of Plato's theory of knowledge squares well with the Pythagorean idea that "number, fitting all things into the soul through sense-perception, makes them recognizable and comparable with one another. . . ."[15]

Returning to the soul of the cosmos, the Demiurge next splits the linear series lengthwise to form two parallel strands and brings each of these together at its ends to form a ring. The two rings are set in rotary motion. The outer of the two supplies the motive force for fixed stars. The inner and oppositely moving ring has to be further subdivided into seven circular bands, an arrangement roughly analogous to seven separate endless belts for five planets and the

---

[15] 35a–37c, the quotation is from Philolaus; see K. Freeman, *Ancilla to the Pre-Socratic Philosophers* (Oxford, 1948), p. 75.

sun and moon to move on. The two original bands coincide with the planes of the earth's equator and ecliptic.

Many further details are given. The whole scheme is an ingenious analysis of the relative apparent movements of the heavenly bodies. It is far too complex to be given here in detail but we may note, by way of illustration, that the animate god that is earth is made to rotate from east to west by the rotating world soul that pervades it. Why, then, do planets and stars seem to rotate round the earth? Because the earth god's own soul moves it, simultaneously, from west to east, thus permitting it to remain in motion relative to the heavens but at rest in an absolute sense.

LIFE AND MATTER

Using the ideas just outlined, Timaeus posits, in a manner of speaking, not one sort of living system but three; these are: first, the whole cosmos which he presents as a divine superorganism or animal;[16] second, the gods, of whom more in a moment; and finally, terrestrial beings, namely plants, animals, and men.

The Animate Cosmos

A living thing for Plato was matter properly configured to permit effectual intervention of soul. This point is made not only in the *Timaeus* but in other dialogues as well.[17] Certain students have considered Timaeus' use of the term *animal* (*zoon*) in connection with the cosmos to be a metaphorical designation. This is only partly true. The account of creation in the Timaeus is figurative, to be sure; we are not meant to think that a Demiurge literally used mixing bowls, took stitches, "glued" things together, etc. But the view of an animate cosmos is literally meant. For Timaeus, the universe lives. That it lacks the organs of locomotion of an ordinary animal is understandable, he explains. Such organs would have no adaptive value; there is no place for the animal to go; its only motion is rotation. Does the universal animal lack eyes and ears? There is nothing outside itself —nothing to be seen or heard. Does it lack organs of ingestion and egestion? It is nutritionally self-sufficient, nourishing itself by its wastes.[18]

16 30c–31a.
17 87d–88e.
18 33b–34a. For the relevant fragments from Empedocles, see D-K, B 28, 29, 34.

Timaeus' animate spherical cosmos had an antecedent in Emped-ocles' image of the divine cosmic Sphere as it exists when Love rules supreme. In possessing both *psyche* and *soma,* it is interesting historically as an example of the parallels drawn in every scientific generation between the cosmos on the one hand and the living microcosm on the other. Moreover, we hear Timaeus assign life-souls not only to the cosmos as a whole, but also, as we shall learn, to the heavenly bodies and the earth. For these, like the cosmos, are divine living things. The Artificer made their souls by the same mixing process he used when he made the soul of the cosmos. Then he furnished these souls with bodies using fire as the principal constituent material, and placed them in the outer shell of the cosmos where they could sweep round and round as fixed stars. In thinking of the stars as gods one has to get used to the fact that a being for Plato could live, feel, and reason even though it lacked the familiar animal form. The dualist who sees soul as separate from body does not require a nervous system as sine qua non for sensation, intellec-tion, or volition.[19]

Life on Earth

The gods' help was solicited, once they were created, in the next business—the creation of the mortals. Timaeus says that the Artificer kept for himself the responsibility for making the immortal part of men's souls. These immortal part-souls of future mortals the Artificer distributed one to each star where "as if mounted on chariots" they were equipped for later terrestrial or planetary life by being given a glimpse of the real nature of the universe. We come into the world "trailing clouds of glory," in Wordsworth's phrase (it was also in a chariot, we remember, that Parmenides pursued the way of truth).

To the gods the Artificer delegated the task of fabricating the mor-tal part of the soul and of equipping souls with bodies. This brings us to the heart of our problem since the crucial question for our inquiry is what mortal bodies were made of.

The soul is bonded, according to Timaeus, to a universal seed-stuff (*panspermia*) found in its purest form in the cranial and spinal cavities where it appears as "marrow." Timaeus thus posits "mar-row" where we see brain and spinal cord (which for Plato were not nervous structures in our modern sense). The marrow is the primary

[19] 39e–40b.

life-stuff in which "were fastened the bonds of life by which the soul is bound to the body."

The "marrow" does seem not to comprise the familiar four elementary substances, fire, air, water, and earth, but to be made directly out of specially well-formed examples of those triangles which are the common components of these elements. We may let Plato speak for himself about the way in which this was accomplished.

> Taking all those primary triangles which, being unwarped and smooth, were best able to produce with exactness fire and water and air and earth, God separated them, each apart from his own kind, and mixing them one with another in due proportion, He fashioned therefrom the marrow, devising it as a universal seed-stuff (*panspermia*) for every mortal kind. Next, He engendered therein the various kinds of Soul and bound them down; and He straightway divided the marrow itself, in His original division, into shapes corresponding in their number and their nature to the number and the nature of the shapes which should belong to the several kinds of Soul.[20]

Timaeus refrains from a quantitative spelling-out of the "due proportion" in which the triangles were mixed to form a bonding substrate for soul. The important point is the recognition that, in human and all other living beings, the essential components of matter must be mixed in definite proportions if the mixture is to manifest life. This idea we have already encountered in approximate form, in certain Hippocratic treatises and in Empedocles. We shall hear in a moment about the connection Plato establishes between *panspermia* as a vehicle for the life-soul and as the genetic link between parent and offspring.

Life-as-action

In chapters 1 and 5, the point was made that physiology has always pursued, though with differences in application, a strategy of interpretation that has remained on the whole remarkably stable. This science has dealt with classic questions, among which the life-matter question has been central in the sense that the answer to it has tended to control and give consistency to the answers to other questions, among them questions especially about nutrition, generation, and motion. The rest of this chapter will suggest that Plato followed this

[20] 73b–c, trans. R. G. Bury with permission of the Harvard University Press.

standard strategy within the limits of his own philosophical ideas and his peculiar intellectual style.

Heat and respiration. Heat, in Timaeus' account, has something to do with maintaining the proper proportions in the body. In men and mammals the speaker posits an inward fire that is generated in the blood and concentrated in the region round the heart. This fire naturally tends to escape and rise to the fiery outer sphere of the cosmos. In exhalation it drives the air along with it. Since there is no void, the escaping fire and air displace the fire and air of the body's environs. This fire and air enter the pores of the body to replace that which was exhaled. Inhaled fire and air, conversely, push body fire and air out through the pores. Breathing is a two-way reversible cycle.[21]

Inhalation draws fire and air into the stomach as well as into the lungs. In the stomach, fire is effective in breaking down the food; because of the sharpness of its particles, fire has a chopping action. From the intestine fire drives food into the veins and through the veins to all the parts of the body. The intestines permit continual absorption of meals eaten at widely spaced intervals, permitting time between meals (delightful suggestion!) for cultural and philosophic pursuits.[22]

Displacement-replacement. Modern biochemistry informs us that the materials we absorb are only temporarily embodied in the molecular architecture of the body. Each molecular component is subject to reconstruction or replacement.[23] An atom's sojourn in the body can be measured by making it radioactive. In this and other ways, the turnover rate can be calculated. Apparently static or stable conditions in the body are often the expression of a dynamic equilibrium or balanced summation of simultaneously occurring, opposed transformations. This idea of a turnover of materials appears repeatedly in ancient physiology. For Plato, life occurred in a system subject to simultaneous depletion (*anachoresis*) and repletion (*plerosis*). As the tissues are depleted by the dissolving impact of the elements,

---

[21] 77e–79e.

[22] 73a (trans. R. G. Bury)—but, then, why entrails in irrational animals? Timaeus is not without a sense of humor.

[23] See, e.g., Rudolf Schoenheimer, *The Dynamic State of Body Constituents* (Cambridge, 1942). See also T. S. Hall, "Life as Opposed Transformation," *Journal of the History of Medicine and Allied Sciences*, 20 (1965):262–75.

they are replenished from the blood.[24] Blood components take their proper position in the tissues because of their tendency to move toward their likes.[25]

Depletion and repletion figure, further, in Timaeus' accounts of pain and pleasure. Pain is occasioned in the soul by sudden depletion of the affected part; pleasure, by its sudden restoration or repletion. (Elsewhere, he gives rather different accounts of pain and pleasure.)[26] If that proper equilibrium of assimilated elements which is so necessary to life be disturbed, the results can be catastrophic. A disproportion of elements—and especially of the triangles in marrow—can bring on diseases and death. Such disproportion is caused specifically by irregularities in the rate of flow of substance into or out of the affected part of the body. Excess bile in a part, for example, influences it adversely. It can overpower the fibrine of the blood whose normal function is to give the blood a proper consistence. If bile inflames the marrow, the soul may be loosed from its moorings. Disease is also caused by a reversal of those synthetic processes by which food is readied for incorporation into the tissues, that is by an imbalanced occurrence of dissolution and decay.[27]

In our survey of the pre-Socratics, we were struck by the interdependence in their world schemes of biology and cosmology. The point is vividly illustrated in Plato's interpretation of the reciprocal flux (*anachoresis, plerosis*) of the living body. This flux occurs because of the fact that in the body as in the cosmos like tends to move towards like.

> The manner of this replenishment and wasting is like that movement of all things in the universe which carries each thing towards its own kind. For elements besetting us outside are always dissolving and distributing our substance, sending each kind of body on its way to join its fellows; while on the other hand the substances in the blood, when they are broken up small

---

[24] 42a; see, also, Harold W. Miller, "The Flux of the Body in Plato's *Timaeus*," *Transactions and Proceedings of the American Philological Association,* 88 (1957): 103–13; and Cornford, *Plato's Cosmology, pp.* 142–46. Also, T. S. Hall, "The Biology of the *Timaeus* in Historical Perspective," *Arion,* 4 (1965):109–22.

[25] 81a–b.

[26] 64c–65b; see also T. S. Hall, note 24 above.

[27] 81e–87b.

within us and find themselves comprehended by the individual living creature, framed like a heaven to include them, are constrained to reproduce the movement of the universe. Thus each substance within us that is reduced to fragments replenishes at once the part that has just been depleted, by moving towards its own kind.[28]

Life, old age, and death. One way of defining life is to distinguish it from its opposite condition. In the nineteenth century, we shall find Bichat defining life as sustained opposition to death. To be scientifically useful, such a statement requires that something relevant be known about death. In the twentieth century, for example, certain biologists will regard the organism as a material system which somehow evades the entropy law obeyed by all nonliving systems including those previously living.[29]

Timaeus is aware of the relevance of death to an understanding of its opposite, life. If life is (or is made possible by) the presence of soul in a body, then the death of the body implies its absence. (In the *Phaedo,* Socrates cheers himself with the thought that, when he drinks the hemlock, the part of him which is better will gladly relinquish the part which is worse.)[30]

The soul's separation from the body at death should, according to the theory, be caused by some change in the particles of the marrow to which soul is bonded in life. This, according to Timaeus, can happen in two ways. It can happen suddenly when, through violence, the equilibrium of particles is disrupted. Or it can happen gradually in case the particles are slowly abraded. Indeed, senescence is considered to be just such a slow abrasion, one consequence of which is that the body's deteriorated fire particles are less capable of cutting up the ingested food.[31]

But death permits the survival of only part of the soul. For the soul, while still with us, is tripartite. An immortal reasoning part (*nous*) resides and rotates in the head; a mortal, passionate part (*thymos*) resides in the breast; and an appetitive part (*epithymetikon*) resides in the liver. Only *nous* is immortal.[32] After death, the im-

[28] 81a–b.
[29] For a discussion of this problem, see Erwin Schrödinger, *What is Life?* (Cambridge, 1944, 1963) and chap. 51 below.
[30] 113d–114c.
[31] 81b–e.
[32] 69c–72d.

mortal soul is susceptible to reincarnation, Timaeus subscribing wholeheartedly to the notion of metempsychosis. The highest reward of the soul is to occupy the star to which that soul was assigned temporarily at the time of its creation. When this reward is not deserved, less desirable fates await the soul; to these we shall return.

Perception. Complex and subtle speculations surround the supposed movements of the soul while still in the body, in Timaeus' psychology. The soul of the cosmos rotates. So do the souls of the star gods. So does the human soul, at least its intellective part (*nous*), within the cavity of the cranium. Induced alteration of this spinning movement has something to do with conscious awareness. Perhaps we are to believe that the carefully proportioned soul segments can move and that their movement somehow contributes to sense perception. But our lecturer is not really clear on this subject. Plants' souls do not turn; hence, plants cannot think, or opine, or move around.[33]

Tissues

Timaeus explains that tissues other than marrow were fabricated by the gods out of proper amounts of the familiar elements loosely pegged together to permit dissolution at death. As for bones, they become "soaked in marrow" during the process of fabrication. Whether Timaeus would have marrow enter into the makeup of other tissues is obscure. Marrow must in any case give rise to them since it is the sole contribution of parent to offspring (and is probably for this reason termed seed stuff).[34]

Soul stuff and Seed stuff

Timaeus posits an open passage through which soul stuff or seed stuff moves from the vertebral canal to the sex organs. (Later the great Swiss physiologist, Haller, will observe in this connection that the late pre-Platonic medical thinker "Alcmaeon said that the semen was a dew that emanated from the brain, and according to Plato, it comes from the spinal marrow; I think they only said that," Haller notes, "because of the lively sensation it occasions at the time of its ejaculation.") The existence of a passageway from the spinal canal to the genitals was to have adherents in Renaissance times. The

[33] 76e–77c.
[34] 73e–76e.

famous coition figures of Leonardo show two passages in the penis, one connected to the spinal marrow, one to the testes.[35]

Women

Life is not confined to the marrow of male human beings. What about women? What about birds and terrestrial and aquatic animals? Timaeus has answers to these questions. Women were created by the gods as receptacles for souls which had not adequately mastered their passions in their first incarnation as males. They represent an inferior order of being. Biochemically they are presumably similar, as far as the marrow is concerned, to males.[36]

Plants

Modern evolution theory proposes that the earliest living organisms evolved from aggregations of very large molecules and that they continued for a time to use such large molecules or something like them as a food source. Ultimately, the original source of large nutrient molecules tended toward depletion, and the continuance of life was threatened. Extinction was averted by the appearance of new sorts of organisms that were able to manufacture large molecules de novo, using as raw materials the small molecules formed as an end product of all vital processes. Among these new organisms were green plants, now understood to exist with other forms of life in a condition of biochemical equilibrium.

Timaeus, too, sees animal life as impossible without adequate nutriment. "Wherefore," he hypothesizes, "the Gods contrived succour for mortals by engendering a substance kindred to that of man, so as to form another living creature: such are the cultivated trees and plants and seeds which have been trained by husbandry and are now domesticated amongst us; but formerly the wild kinds only existed, these being older than the cultivated kinds. . . . Certainly that creature which we are now describing partakes of the third kind of soul, which is seated, as we affirm, between the midriff and the

[35] Albrecht von Haller, *De la génération* (Paris, 1784), 1:38 (a French trans. of parts of the seventh and eighth vols. of Haller's *Elementa,* see chap. 27, note 2.) For an authoritative account of the relations of soul stuff to seed stuff, see Erna Lesky, 'Der enkephalomyogene Samenlehre,' "Die Zeugungs—und Vererbungslehren der Antike und ihr Nachwirken," *Abhandlungen der geistes—und sozialwissenschaftlichen Klasse, Akademie der Wissenschaften und der Literatur in Mainz* (Wiesbaden, 1950), pp. 1233–54.

[36] 42b–d.

navel, and which shares not at all in opinion or reasoning and mind but in sensation, pleasant and painful, together with desires."[37]

Note that plants are based on a "substance kindred to that of man"; kindred, that must be, to marrow (the substance to which soul is bonded).

Animals

Modern evolution theory posits, for the animal kingdom, an evolving intelligence. Along with intelligence goes the capacity of the individual, where stimulated, to select what response it will give. Stored information helps the animal to select its response according to its individual advantage. Rudiments of this faculty have been discovered in animals as behaviorally simple as flat worms. Man, with his complex behavior, stores a greater variety of information and is capable of an infinitely larger variety of choices.

The precise pattern of the evolution of intelligence, its distribution among different animals, is far from understood or established by modern biology. Behavioral complexity does not itself imply freedom nor the capacity to store information. Witness the elaborate but only limitedly flexible behavior of the social insects, for example. About all that may be safely assumed is that man's recent ancestors experienced a progressive increase in intelligence with a developmental spurt occurring during the early history of his species.

Timaeus correlates, as we do, the degree of development of mind with that of body. This assumption leads to some amusing ideas about what we think of today as vestigial structures. Men's fingernails reflect the forethought of the gods. "For those who were constructing us knew that out of men women should one day spring and all other animals; and they understood, moreover, that many of these creatures would need for many purposes the help of nails...."[38] Timaeus' account of these and related matters is delightfully set forth.

> And the tribe of birds are derived by transformation, growing feathers in place of hair, from men who are harmless but light-minded—men, too, who, being students of the worlds above, suppose in their simplicity that the most solid proofs about such matters are obtained by the sense of sight. And the wild species of animal that goes forth on foot is derived

[37] 77a–b.
[38] 76d–e; 91a.

from those men who have paid no attention at all to philosophy
nor studied at all the nature of the heavens, because they ceased
to make use of the revolutions within the head and followed
the lead of those parts of the soul which are in the breast.
Owing to these practices they have dragged their front limbs
and their head down to earth, and there planted them, because
of their kinship therewith; and they have acquired elongated
heads of every shape, according as their several revolutions have
been distorted by disuse. On this account also their race was
made four-footed and many-footed, since God set more supports
under the more foolish ones, so that they might be dragged
down still further to the earth. And inasmuch as there was no
longer any need of feet for the most foolish of these same creatures,
which stretched with their whole body along the earth, the
gods generated these footless and wriggling upon the earth.
And the fourth kind, which lives in the water, came from the
most utterly thoughtless and stupid of men, whom those that
remoulded them deemed no longer worthy even of pure respiration, seeing that they were unclean of soul through
utter wickedness; . . .[39]

It perhaps needs no emphasis that this account does not suggest
anything like an evolutionary origin of species. In the first place,
Timaeus posits the creation by the gods of animals to receive
directly the transmigrating souls of various grades of human beings.
In the second place, it is probable that Plato, while accepting the
idea of transmigration, would have rejected the idea of creation at
least in anything like the form in which for pedagogical purposes he
makes Timaeus depict it.

Physics and ethics. The author of the ideas set forth in the
Timaeus could scarcely have concerned himself as exhaustively as
he did with the implications of his theory of matter without extending
that theory to an analysis of human behavior. About a third of the
text of the *Timaeus* is an application of physical theory to physiological and pathological functions, and to normal and abnormal behavior. Outrightly irrational behavior is ascribed to excess pain or
pleasure accompanied by inordinate desires to avoid the former or
gratify the latter and by distorted perception and incapacitation of the

---

[39] 91e–92c, trans. R. G. Bury, with permission of the Harvard University Press.

reasoning part of the soul. Misbehavior is viewed in the *Timaeus* as partly educational but also partly biochemical in origins. Sexual incontinence, for example, should not be construed as voluntary wickedness. It is, in reality, a physiopathological manifestation due to particle imbalance—specifically, a superabundance of seed stuff. So too with rashness, cowardice, ignorance and stupidity.[40]

Conclusion

In keeping with the focus of this book, we have chiefly shown Plato's science as a self-contained and integrated system. It would be totally wrong to view that system as isolated from or outside of the broad flow of scientific thought. The *Timaeus* was a masterpiece of synthetic reasoning that brought together pre-Socratic particle theory, the Empedoclean four-element doctrine, and geometrical discoveries of Plato's own mathematical colleagues. To mention these strands of his thought is to be struck by his deep direct and indirect indebtedness to the Pythagoreans. Indeed the late Pythagorean, Philolaus, may even have invented the polyhedral elementary particles that were central to Plato's physics. Pythagorean too, though much modified by Plato, was the notion of the body as a temporary receptacle for an ontologically separate transmigratory soul. More generally, Plato belonged to the ruling tradition, in Greek science, that was simultaneously animistic and reductive. Plato's animism and reductivism were challenged, successfully for a time, by Aristotle. But they were by no means demolished. They were to have a major influence—over adherents and opponents, we shall see—up to, and beyond, the time of Descartes.

---

[40] 86b–87b.

8

Life as Self-replicative Form

ARISTOTLE [384 · 322 B.C.]

We found Plato's biology to be broad in its concerns, compact in presentation. The biology of his student, Aristotle, is, by contrast, of oceanic proportions. In four major and a half dozen minor works he invites inquiry into almost every biological question of importance. His interests range from ants to zoophytes and from acoustics to zephyria (wind eggs). Bees, bats, blood, bile, chickens, crocodiles, crabs, crows, dwarfs, dreams, digestion, decay—no accessible biological phenomenon appears to escape his attention.

It goes without saying that Aristotle's investigations will include a study of life-matter relations. He has, in fact, much to say on this topic, but we shall confine ourselves to two pregnant ideas that are central, in a way, to all that he says on this subject. The first is Aristotle's identification of life with form—form which has, in plants and animals (including the rational human animal) by contrast with other things, a capacity for self-replication. The second is his way of expressing the correspondence, as he sees it, between the cosmos and the living microcosm with special reference to their material components and to the way in which cause works in both worlds.

Matter

Elements. To the four elements of Empedocles, or rather to his own version of these elements, Aristotle adds a fifth which he terms *aether*.[1] Aether constitutes the entire translunar sphere of the cosmos including the stars that abound there. It possesses neither levity nor gravity, is ungenerated, indestructible, eternal. Unlike the naturally centrifugal movement of fire and air and the naturally centripetal movement of water and earth, the movement of *aether* is in a circle. The cause of its movement is the Unmoved Mover.

---

[1] *Cael.* 269a19–270b26 (see list of Abbreviations).

Just as *aether* completely fills the outer cosmos, so the four elements account for everything, or almost everything, sublunar. What, then, is an element?

Aristotle defines an element as "a body into which other bodies are divisible but which is not itself divisible into bodies unlike itself."[2] Although his elements are indivisible they are, nevertheless, transmutable. The transmutations of elements have nothing to do, however, with micrometric properties as specified by Plato. Plato's particles, along with the triangles on which they were founded, are specifically rejected by Aristotle.[3]

Rather, Aristotle sees the elements as having certain distinctive potencies (*dynameis*)[4] that are more fundamental, in a way, than are the elements themselves. These basic potencies—hot, cold, moist, and dry—are paired in the four elements in such a manner as to endow each element with its familiar characteristics. Fire is hot and dry; air, hot and moist; water, cold and moist; earth, cold and dry. These ideas are much more highly elaborated than the related ones we encountered in the Hippocratic treatise *On the Nature of Man,* a work with which Aristotle was familiar.[5]

Intertransmutation. The transmutation of fire into air involves an alteration of *dry* to its opposite potency, or quality, *moist.* So through the rest of the cycle:

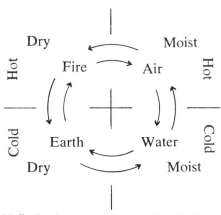

[2] *Metaph.* 1014a26 ff. (book 5, chap. 3) trans. W. D. Ross.

[3] *Cael.* 304a15–20, 305a30–34, 306a1–b15; *GC* 315 b 29–32.

[4] See A. L. Peck in the Introduction to his trans. of *GA* (Cambridge and London, 1943), pp. xlix–liii.

[5] *GC* 330a30–b6. For a helpful study of the four elements and the four humours in ancient thought, see Erich Schöner, "Das Viererschema in der antiken Humoral-pathologie," *Sudhoff's Archiv für Geschichte der Medizin, etc.,* Beiheft 4 (Wiesbaden, 1964), pp. vii–x, 1–114.

Note that fire and earth are directly intertransmutable. This is made possible through the reversible transformation hot ⇌ cold. Similarly, earth and water transmute via the transformation dry ⇌ moist. In Plato's scheme, we remember, earth was nontransmutable (because it was built up of isoceles triangles, all other elements of scalene triangles).

Contrariety. The intertransformations (hot ⇌ cold, moist ⇌ dry) exemplify Aristotle's conviction that *every* transformation is a change to, or at least toward, an opposite condition.[6] Death transforms living to nonliving; convalescence, unhealthy to healthy; coitus, nonfertile to fertile. When left to itself, a musical instrument changes from being in tune to being out of tune. For Aristotle, the shifts that occur back and forth between contrarieties are the most fundamental of natural phenomena.[7]

A common substrate? Are the four elements really four differentiations of a single more primitive substrate? Not, according to Aristotle, of any separately perceptible substance such as fire, air, or water. Yet Aristotle acknowledges that certain facts—the facts of change and transmutation—do argue for a common substrate of some sort. But this *prima materia* is never perceptible as such, since it always appears in the guise of an element.[8]

Composition of the Body

The living body is depicted by Aristotle as a hierarchy of structural components. The elements are differently compounded to form flesh, sinew, blood, bone, and the like—structures to which Aristotle refers as uniform or "homoeomerous" parts of the body. By homoeomerous he means like-parted or homogeneous in composition. The uniform parts are arrayed, in turn, to form such nonuniform or heterogeneous parts as a hand, the heart, or the liver.[9] In the *Timaeus,* Plato had recognized a not dissimilar arrangement of the body into primary and secondary body parts, and in A.D. 1801 Bichat will arrive at a comparable classification and call his uniform components "tissues." Bichat will repeat Aristotle's assertion that

[6] *Ph.* 188b23–26.

[7] *Ph.* 189b.

[8] *Ph.* 190a32–36, 191a3–6. The problem is also discussed at *GC* bk. 1, chaps. 1 and 2 and bk. 2, chap. 7, and by F. M. Cornford in the Introduction to his trans. of *Ph.* (London and New York, 1929), pp. xxii ff. See also E. J. Dijksterhuis, *The Mechanization of the World Picture* (Oxford, 1961), pp. 19–20.

[9] *PA* 646b12–24.

in each tissue a single function predominates, whereas organs, possessing several tissues, have several functions. But Bichat will not see his tissue as thoroughly blended or homoeomerous from point to point (see chapter 36).

Aristotle's histology and histochemistry (if we may use the phrase) differ considerably from Plato's. With Aristotle, flesh is the primary organ of touch by which animals are distinguished from plants. Flesh exists for the sake of the soul and is primary partly in the sense that marrow is primary for Plato.[10] The other parts exist for the sake of the flesh[11] out of which nature first fashions the animal.[12]

Blood, the final form assumed by nutriment before incorporation into the tissue, has a more watery component and a more earthy component.[13] The balance of these affects the individual's character and intelligence. Thin (watery) and cold blood disposes to intelligence; thick (earthy) and hot blood, to passion and courage. In the ideal situation, blood will be hot but also thin and clear, since the result will be a combination of courage with intelligence.[14] Aristotle's theory differs from that of Empedocles, however, in that Aristotle makes the heart rather than the blood the seat of perception and intelligence.

Nutrition and the Life-soul

Like Plato, Aristotle follows the standard strategy of physiological inquiry that evolved, if somewhat indefinitely, in Ionia and was crystallized by later pre-Socratic pluralistic thinkers. He treats classic questions; he deals with these on the levels of both percept and concept; and he builds an integrated system which derives its consistency in part from his general views of life and matter, views which reflect, in turn, the Greek tendency to consider the cosmos and the living microcosm as analogous. We may begin with his treatment of the most universal differentia of life, nutrition.

The problem. Aristotle correctly surmises that the parts of the body experience considerable erosion and replacement[15] and is acutely interested in the related phenomenon of growth. The intimate

---

[10] See F. Solmsen, "Tissues and the Soul," *The Philosophical Review*, 59 (1950): 463.
[11] *PA* 653b30.
[12] *PA* 654b29.
[13] *PA* 650a35, 651a12–17.
[14] *PA* 648a1–10, 650b13–651a17.
[15] *De Longitudine et Brevitate Vitae* 466b29 – 467a1, and *De An.* 412a14–21.

arrangement of elements is different, in his opinion, for each differ-ent part of the body.[16] Hence, normal vital action depends on the organism's ability to acquire the right kind of food and handle it in such a way that it will maintain each part properly.

When food reaches the tissue as blood, it must be ready for the final step of assimilation, a fact that brings us directly to grips with our problem—since the ability to effect assimilation is close to the essence of living. In a word, one of the most central of biological questions (and it is very much a question today) is: what agency governs the assimilative process? To this Aristotle answers: soul, specifically, the nutritive soul. We need to focus briefly, then, on Aristotle's ideas (*a*) of soul, (*b*) of nutrition, and (*c*) of their inter-connection.[17]

Soul in general. With Aristotle, as with Plato, it is soul that dis-tinguishes the living.[18] But soul is not understood in the same way by master and pupil. Indeed, our purpose, or one purpose, in studying the history of science is to gain an understanding of the way in which through time the solution of a problem develops. Conceptual themes (e.g. the soul theme) will recur, but with variations caused by the discovery of new data and the acceptance of new theoretical assump-tions. This point will be amply illustrated, as we proceed, by the developing interpretation of *psyche*.

Aristotle discusses the nature of soul from many viewpoints of which the following is especially germane to our narrative. We have to remember, first of all, that the word he uses (*psyche*) is very far from meaning what we mean when we use *soul*. We noted earlier that *psyche* has for Greeks a constellation of meanings partly cov-ered by the modern term *soul* and partly by *life*. *Psyche* (the life-soul) is then, for Aristotle, a single entity endowed with an ensemble of faculties or powers: first, a nutritive faculty (shared by everything living); second, a set of sensitive faculties (limited to animals in-cluding man); third, a locomotor faculty (limited to animals that can move around, which not all animals are able to do); and fourth, a rational faculty (in human beings only).

Of sensitive faculties some require contact (touch and taste) and others are effective at a distance (smell, hearing, and vision). Along

---

[16] *HA* 487a2–10.
[17] *De An.* 415a23, 416b19 ff.
[18] *De An.* 413a1, 415b9, 434a22; see also *De Plantis*, bk. 1, chap. 1.

with sensation goes desire; hence, the sensitive faculty carries with it a desiderative function.[19] The rational and intellective faculty—or the part of the soul possessing it—has, at least in its most active manifestation, an altogether different status from the others: it is immortal.[20]

TABLE 4
DISTRIBUTION OF SOUL FACULTIES

	PLANTS	ANIMALS		
		Stationary Animals	Animals that move about	
			Other than man	Man
Nutritive	+	+	+	+
Sensitive				
Immediate (touch, taste)	0	+	+	+
Mediate (smell, hearing, sight)	0	0	+	+
Locomotor	0	0	+	+
Rational	0	0	0	+

Nutrition and the nutritive faculty of soul. Of all these faculties we are most interested here in the nutritive, since it controls the material composition of living things. By an apt simile, Aristotle suggests what occurs in nutrition. Were one to try to make wine "grow" by adding water one would succeed only if the added water really turned into wine. But this is not what happens; all that happens is that wine is diluted. The body, however, really grows when food is added, since "the active principle of growth [the nutritive soul] lays hold of an acceding food which is potentially flesh and converts it into actual flesh."[21] The key problems, then, are to discover (*a*) what the conversion process really is and (*b*) how soul controls it.

The preparation of food occurs in a series of steps the first of which take place, under the influence of heat, in the stomach. Even before this "concoction" (*pepsis*) takes place, the mouth breaks up food mechanically. "But this operation does nothing whatever towards causing concoction: it merely enables the concoction to turn out successfully; because when the food has been broken up into small pieces the action of the heat upon it is rendered easier."[22] Semi-

[19] *De An.* 413b22 ff.
[20] *De An.* 413b25.
[21] *GC* 322a11–13.
[22] *PA* 650a9–15, trans. E. S. Foster.

concocted food is brought off from the intestines via the arborizations of mesenteric veins which are compared by Aristotle to roots.[23] These veins bring the food to the heart.

Here occurs a second concoction; in the heart and in the neighboring spleen and liver, food is changed into blood.[24] In these two processes, a key role is played by heat. We hear that "since it is by the force of heat that [this nutriment is] . . . concocted and changed, it follows that all living things, plants, and animals alike, must, on this account if no other, have a natural source of heat."[25]

Role of heat in nutrition. Heat, in Aristotle's scheme, is fundamental to life processes in general. Soul cannot even exist, much less function, in the absence of heat. From too little heat come decay and senescence. Too much heat is equally disastrous.[26] The brain, composed of earth and water, is a refrigerant organ.[27] Lungs and gills have a similar function.[28] In plants which generate less vital heat than do animals the circumambient air is an adequate refrigerant, especially when stirred up by food passing into the body.[29]

In nutrition, heat serves in certain ways as an instrument of soul. For "of all substances there is none so suitable for ministering to the operation of the soul as that which is possessed of heat. For nutrition and the imparting of motion are offices of the soul, and it is by heat that these are most readily effected."[30]

In agreement with the above notion, Aristotle postulates a certain parallelism between the distribution of heat and of soul. Both are everywhere in the body but their chief locale and origin is in the heart.[31] The heart is the source of the vital heat of the blood. This idea, that the heart is hot, is to have many future adherents, including Galen, Descartes, and William Harvey; in medieval and Renaissance biology it was the standard view until, as we shall see, it came under question in the latter part of the seventeenth century.

Final act of assimilation. Although blood is the "final form taken by nutriment as such," a further conversion occurs in chang-

---

[23] *Juv.* 468a10; *PA* 678a1–20; see also *De Longitudine et Brevitate Vitae* 467b1.
[24] *PA* 647b4–7, 670a19–22; *Juv.* 469a1, 469a27–b20.
[25] *PA* 650a4–7, trans. Ogle.
[26] *Resp.* 474b10, 478b31, 479a8; *Juv.* 469b7–470a5.
[27] *PA* 652b26, 653a21, 656a20.
[28] *PA* 688b35–670a14; *Resp.* 478a28–b22.
[29] *Juv.* 470a22.
[30] *PA* 652b10–14; *Resp.* 474a25.
[31] *Juv.* 468a1, 469b6; *Resp.* 474a25 ff.; *PA* 678b5; *GA* 735a23.

ing blood into actual tissue. This final act of assimilation stands at the very heart of Aristotle's physiological theory for he thinks it is at this point that "as fire lays hold of the inflammable [and converts it to fire] so the active principle of growth [nutritive soul] lays hold of an acceding food which is potentially flesh and converts it to actual flesh."[32] The conversion is occasioned in the case of some tissues (bone, sinew) with the help of heat and in the case of others (e.g. flesh) with the help of cold.[33]

This much information concerning the detailed steps in nutrition permits us a partial insight into the causal role of the nutritive soul. In his work on *Physics* and elsewhere, Aristotle notes the operation in nature of four different though importantly interrelated kinds of causes often referred to as the efficient, the final, the formal, and the material. He says that scientific knowledge is, in effect, a knowledge of these four kinds of causes. In biology, soul serves in the first three of these causal roles. To illustrate: (*a*) Soul instigates both muscular movement and the movements that constitute growth and organic change, and in that sense operates efficiently. (*b*) It operates in the final sense in that "all natural bodies are instruments of . . . and exist for the sake of the soul."[34] (*c*) It acts as formal cause especially, as we shall see, in nutrition. But first what, more precisely, does Aristotle mean by "form"?

Form not separate. Forms have for Aristotle as for Plato a very real existence. Plato's forms, we remember, were separate existences in a realm of reality higher than that known to the senses. Aristotle's forms on the other hand are inseparably bound up with sense objects. The form of a particular man exists in connection with that man, and not elsewhere. The same holds for living things in general.[35]

Form and soul one and the same. In this concept of immanent form lies the key to Aristotle's notion of life-matter relations, for form itself is equated by Aristotle with the life-soul. Soul is the "form of the living being." Or in a slightly different formulation "soul may therefore be defined as the first actuality of a natural body potentially possessing life. . . ."[36] To us this formulation may seem

---

[32] *GC* 322a10 ff., trans. H. H. Joachim (Oxford, 1922).
[33] *GA* 743a8 ff.
[34] *De An.* 415b9–27.
[35] *De An.* 403a3 ff., 412b5, 414a19 ff.
[36] *PA* 641a18; *De An.* 411a29, trans. W. S. Hett (Cambridge and London, 1935); see also John P. Anton, *Aristotle's Theory of Contrariety* (London, 1957), pp. 114–24.

at first obscure and unfamiliar, but in the following ways it is penetrating and useful.

Life and matter. If we equate "form" with organization (which is not far from Aristotle's intention)[37] his theory states, in effect, that living matter is matter whose organization can perpetuate and extend itself during growth and reproduction and throughout cycles of material renewal. It perpetuates and extends itself by imparting its organization to matter possessing the potentiality of being so organized. Such a statement finds acceptance even in the science of our times. It fulfills one, and a very fundamental, requirement of any adequate solution of the life-matter problem by grasping, more than vaguely, the notion that life entails a dynamically self-perpetuating arrangement of elements.

It will not offend modern ears to hear that the final *nutriment* must be *potentially* what the *tissue* is *actually,* and that it thus matters urgently how the final raw materials are composed.[38] "The heat, however, to produce flesh or bone does not work on some casual material in some casual place at some casual time . . . : that which is potentially will not be brought into being by a motive agent which lacks the appropriate *actuality;* so, equally, that which possesses the *actuality* will not produce the article out of any casual material. No more could a carpenter produce a chest out of anything but wood; and, equally, without the carpenter no chest will be produced out of the wood."[39]

We must be cautious, of course, to avoid crediting Aristotle with insights that belong in fact to later eras. His formulations sometimes sound more modern than they are. They illustrate a certain point about the development of scientific ideas. In the case of persistent problems, such as that which we are exploring, new solutions are offered by succeeding generations of scientists. Ideally, but not invariably, a new solution represents an improvement over that which preceded it—it is more useful; it squares with more data; it is more powerfully and richly predictive. Each solution, as suggested earlier, may be thought of as having a metaphorical relation to others in the sequence and to the objective fact.

Aristotle's notion of a self-perpetuating arrangement of elements

---

[37] *De An.* 412b. Soul is to body as shape is to wax.
[38] *GA* 743a20.
[39] *GA* 743a20–27, trans. Peck.

in tissues was, then, an early metaphor for the fact modern science formulates in the concept of enzymatically governed replication of macromolecules. The details of Aristotle's model were arrived at by his insertion of empirical data into the axiom systems he lived by. Aristotle's model conformed to Aristotle's doctrines of motion, matter, and form.

Generation

Aristotle brought his ideas of life and matter to bear on the problems of sex and of reproduction in general. He adduces, on these subjects, an impressive array of empirical data obtained from a broad spectrum of animal species, and he submits these data to his own brand of scientific interpretation. Typically he enumerates alternative hypotheses, lengthily refutes those he considers erroneous, and argues in behalf of the interpretation he considers correct. The argument is conducted according to his well-known method of inserting his data into an almost too well-oiled apparatus of syllogistic inference. Aristotle seems at times to be bemused by the beautifully spinning wheels of the syllogistic machinery. However this may be, the major premises of the syllogisms in question are his own broad views on the nature of life, soul, form, matter, movement, and causation.

The resulting inferences about reproduction are far from simple and are not easily abstracted. We may get a partial idea of them by comparing what Aristotle says about generation with what we have heard him say about nutrition. In nutrition, we remember, the building of tissue involves a final assimilative act by which highly concocted blood which is *potential* tissue is converted into *actual* tissue. Generation, too, involves a conversion of potential into actual.

The catamenia. In the female, an unused residue of highly concocted blood becomes menstrual fluid which is ready, because of its concocted condition, to be turned into all the parts of a new individual. Indeed, the catamenia really contain the parts—not actually, to be sure, but potentially. And this fact defines the essential feature of generation which is: the conversion of the potential animal (present in the menstrual fluid) into an actual animal (recognizable as such).[40]

In nutrition, the final conversion is a step in which matter already

[40] *GA* 727b30, 729a22 ff., 732a10.

possessing actual form or soul (i.e. organization) communicates that form to highly concocted blood that is ready to receive it. In generation, too, actual form (or soul) must somehow be communicated to the menstrual fluid which possesses it only potentially. The menstrual fluid must undergo those transforming movements (*alloiosis*) by means of which structure arises. It is induced to do this by semen.

Role of semen. One possibility, later to be adopted by Paracelsus, Buffon, Maupertuis, and even Darwin, is that every part of the male body supplies a small part of itself to be transferred to the female at coitus where they react in some fashion with similar particles derived from her body and so form the new individual, the offspring. Such particles were proposed in pre-Aristotelian antiquity by Democritus and, closer to Aristotle's own time, in the Hippocratic treatise *On Generation* (ca. 370 B.C.). Aristotle weighs and (as will Weismann two thousand years later) rejects them. As Weismann will, Aristotle questions the transmission of acquired characteristics (he says mutilations can, but need not, be transmitted).[41] In lampooning the pangenesists, he complains that they "talk as if even the shoes which the parents wear were included among the sources from which the semen is drawn!"[42]

He likewise considers and eliminates the possibility that a tiny preformed animal is the sexual product of the male. Here he anticipates those seventeenth-century observers who will "see" such creatures in the spermatozoa which their microscopes revealed to them.[43] For these erroneous ideas Aristotle substitutes what he considers to be the right interpretation.

He tells us that the effect of semen on the catamenia is like the carpenter's effect on wood, or the effect of the medical art on the patient. Aristotle says that the "female always provides the material, the male provides that which fashions the material into shape; this, in our view, is the specific characteristic of each of the sexes: that is what it means to be male or to be female."[44]

Soul. If we ask *how* semen shapes or "sets" the menstrual fluid, we hear that it "imparts to it the same movement with which it is itself endowed."[45] Semen is, in Aristotle's view, a residue of blood

41 *GA* 721b28 ff., 724a4 ff.
42 *GA* 723b30.
43 *GA* 734a24.
44 *GA* 729a22–b19, 738b20 ff.
45 *GA* 737a20.

carried to an even higher state of concoction than menstrual blood. Since the semen communicates to the menstrual blood its own organized pattern of movements, we may properly say that semen possesses soul, because soul *is* form (or organization).[46]

Indeed, even before their meeting, the more perfect male and less perfect female *semina* possess nutritive souls in a potential sense. When they meet to form a fetation, their former potential nutritive souls become a single actual nutritive soul. But sex has a deeper significance.[47] There is a crucial difference between catamenia and semen. The latter possesses not only a nutritive but also a sensitive soul (this being what distinguishes every animal from every plant). An infertile hen's egg leads temporarily a plant-like (merely nutritive) existence; to be actualized as an animal it requires the influence of a sensitive soul introduced by the semen.

Vital heat. Pneuma. In "setting" (imparting the proper movement to) the catamenia, semen acts, according to Aristotle, like rennet when that substance "sets" milk. "Rennet," in fact, "is milk which contains vital heat, as semen does. . . ."[48] We learn further that vital heat is neither fire nor a product of fire but is something sui generis which resides in a special physical substance, *pneuma.*

Pneuma *in Aristotle's biology.* How Aristotle viewed *pneuma* has occasioned considerable scholarly debate. We shall limit ourselves here to some of the things he actually says about it. He says that souls of various sorts seem to exist in relation to a physical substance different from and more divine than the four elements. In semen the substance thus associated with soul is termed by Aristotle *pneuma.*[49] Its nature is partly suggested by Aristotle's reference to it as "hot air" or simply a "hot substance"—but with the understanding that its heat is no ordinary fire-produced heat. Rather, this hot substance, *pneuma,* with which soul is associated in semen, is analogous to the substance *aether* out of which the stars are made.[50]

A. L. Peck supposes that Aristotle means to make *pneuma* a quite general intermediator between *psyche* and *soma,* a role which would give it great importance in life-matter relations.[51] In nutrition, we saw

[46] *GA* 741b5.
[47] *GA* 741a10 ff.
[48] *GA* 729a9 ff., 739b20 ff.
[49] *GA* 736b30.
[50] *GA* 736a1–2, 736a35.
[51] A. L. Peck, *"Symphyton pneuma"* (Appendix B), in his trans. of *GA* (London and Cambridge, 1943), pp. 576–93.

that soul's instrument is heat, but there soul heats and concocts the aliment without the intermediation of *pneuma,* as far as we are told.[52] In willed movement, and in sensation, *pneuma* is given an intermediary role not easy to grasp from Aristotle's presentation of it.[53] In all these activities we hear nothing of the relation of *pneuma* to the translunar "fifth element." Peck sees *pneuma* as "analogous" to *aether* in two ways. First, both *aether* and *pneuma* are to some extent generative substances. Second, both mediate motion: *aether* transmits the motor effect of the Unmoved Mover to the sublunary world; and *pneuma* transmits the motor effect of soul to the menstrual blood. Peck is perhaps correct in reading this meaning into what Aristotle says which is simply that "the parts are differentiated by means of *pneuma.*"[54] Aristotle gives no clear statement, but he gives a clue as to the way in which *pneuma* might cause differentiation (see below).

Once the catamenia has been "set," there is an orderly, stepwise conversion of potential parts into actual parts; this is what is meant when it is stated (as historians often do) that Aristotle is an epigenesist. He is an epigenesist, too, in his suggestion that the parts are first mapped out in broad outline and later filled in in detail. The first part to be actualized is the heart, then its blood vessels. This is necessary since the actualization of the other parts requires nourishment provided by blood which the fetal heart obtains from the mother via its umbilical vessels and distributes to the fetal body.[55]

Here we get an intimation of the role of *pneuma.* We are told that as blood oozes out of the fetal vessels like water out of an unbaked earthenware vessel, it is subjected to varying degrees of heating or cooling. Aristotle probably means to make *pneuma* responsible for both. If the *pneuma* evaporates, the result is cooling; in this way flesh is formed, for example, and, from the earthier parts of blood, such things as hair, nails, and horns. Or, on the other hand, the vital heat may be imparted by *pneuma* to the blood if that is what is needed to convert it into tissue; this happens, for example, in the case of sinew and bone.[56]

---

[52] *PA* 652b7–10; *Juv.* 470a19–23.

[53] *MA* 703a4–28; *GA* 744a2 ff., 788a23 ff.

[54] *GA* 742a5. The heat-life-*pneuma* problem has been helpfully interpreted by F. Solmsen, "The Vital Heat, the Inborn Pneuma and the Aether," *Journal of Hellenic Studies,* 72 (1957): 119–23.

[55] *GA* 739b34–740a24.

[56] *GA* 743a10–17.

Summary and Conclusion

Generation (in warm-blooded animals) is viewed by Aristotle partly as a complex analogue of nutrition. In nutrition an existing tissue is able to communicate its distinctive intimate movements to blood, if the blood is fully ready, and so convert that blood into tissue like itself. In generation, too, the proper movements must be communicated to fully prepared (catamenial) blood in order to convert it into tissue. The movements are communicated in this case by the semen. Semen is blood carried not only to but even beyond the point where it is ready for conversion into tissues. It is endowed with all the varied movements of the tissues themselves, and it is able to communicate these movements to the menstrual blood whose potentialities for tissue formation are thus actualized. The semen's way of communicating its movement to the menstrual blood is to heat or cool it, and Aristotle seems to suggest that in order to accomplish this end, the semen uses its contained *pneuma* as an intermediary agent.

Aristotle has a great deal to say that is relevant to the life-matter problem, and this includes both empirical information and conceptual interpretation. We have attempted to make a microtome slice through some of his central contributions. Of these, two stand out as especially relevant. The first is his equation of the life-soul with the form of the living system. Life *is* organization. But the organization is intrinsically dynamic. In nutrition—which is, for Aristotle, the most general of vital phenomena—this organization acts to extend itself through a series of transformations in matter whose potentiality for organization is in this manner actualized.

The other and equally pregnant insight of Aristotle, the teleological insight, is troublesome to modern readers. But it makes sense when we consider the empirical observation that occasions it: separate events—steps in nutrition, generation, development, movement, sensation—have only limited meaning when considered in isolation. Their true significances emerge only when we consider these isolated events in relation to the larger functions they make possible. A single step in development is remarkable to some extent in itself but in a much more dramatic way when considered in the light of the end product which it makes possible. The same holds equally true for an isolated step in nutrition. And as with function so with organization. Any structural component of the system is significant only secondarily in itself and primarily in its relation of the whole to which it contributes. Ultimately, but only with the utmost difficulty, science will show that material and efficient causes are better instruments than formal or final causes for interpreting the organism even when it is viewed as a spatiotemporal whole.

Plato and Aristotle. Plato and Aristotle follow the established procedure of portraying the organism simultaneously on two different levels. They present us with (*a*) an organism whose features can be seen and (*b*) a conceptual model whose features are inferred. The character of the model is complexly determined. It depends partly on the overt biological phenomena to which it corresponds and partly on the model-builder's ideas of life and the cosmos.

In Plato's system, *psyche* (life) may exist independently of *soma* (matter); in Aristotle's, this is impossible. But the fact that both stipulated *life-as-soul,* however differently envisioned by them, as the primary condition of *life-as-action* is one of the most fateful of Greek scientific ideas. The Arabs were to transmit this idea to medieval Europe where it would ultimately come under attack. This attack, with Descartes as its most articulate spokesman, heralds the birth of a new biological science—the science we think of as biology today.

Both Plato and Aristotle developed versions of Empedocles' theory of four primary bodies. Plato distinguished these bodies in terms of dimensional, Aristotle in terms of dynamic, properties. According to Plato, the four sorts of elementary bodies (or triangles that compose them) must be properly configured to permit the bonding to them of soul, and only such a bonded soul-body complex can manifest life-as-action. According to Aristotle, too, a proper configuration of elements is requisite. But in his model that configuration, or form, is itself life. The configuration or form is responsible for physiological function (life-as-action), but its most remarkable property is its capacity for self-communication or replication.

Plato and Aristotle both concerned themselves, then, with the causes of natural order. Order arises in Plato's system out of the interplay of (*a*) reason and (*b*) necessity. In Aristotle's system, it arises out of the interplay of material, efficient, formal, and final causes. Occasionally Aristotle borrows the terminology of his teacher and refers to material and efficient causation, taken together, as necessity. And frequently, he acknowledges that his formal and final causes are almost indistinguishable from one another. From this point of view, the major difference between the two thinkers is that, in addition to necessity, Plato assigns the major causal role to *logos* (reason) whereas Aristotle assigns it to *telos* (purpose). The two thinkers apply their respective systems of causal analysis to nature as a whole, to both the cosmos and the living microcosm.

9

Life and the Atom

EPICURUS [342? · 270 B.C.]
LUCRETIUS [96? · 55 B.C.]

As our story grows, we become increasingly aware that science has traditionally viewed nature from a basically dual perspective. It sees an open aspect of things (directly revealed to sensation) as well as a hidden aspect (known only through inference and reason). It is significant that the inferred hidden aspect of things has generally seemed the closer of the two to their real nature. Any phenomenon as directly observed is thus considered in need of reformulation. It needs restatement in terms of those hidden phenomena—cryptomena, we have termed them—of which it is a mere sensory surrogate or summation. Epicurus says that "it is time to consider generally things that are obscure," meaning specifically things not perceivable through the senses; and he adds that it is reason that "attempts to infer the unknown [i.e. the imperceptible] from the known [the perceptible]."[1]

We noted earlier that Democritus, in endeavoring to "grasp imperceptible things" had assigned a major role to the sizes, shapes, and changing spatial relations of atoms, and that Plato, too, had interested himself in size, shape, and motion, with the difference that the particles in Plato's theory were of four basic sorts distinguished by their geometrical forms and lacking empty space between particle and particle. Aristotle was, by contrast, little interested in microdimensional questions; to him, qualitative and kinetic cryptomena were crucial. Though microarrangements were occasionally mentioned in Aristotle's physics, he explicitly rejected the particle theory of material organization. With Epicurus, the pendulum swings strongly back to an interest in the sizes, shapes, and movements of the least subdivisions of matter.

Epicurus personally wrote a big work on nature in thirty-seven

[1] *LH*, 39–40 (see list of Abbreviations).

books. Only isolated parts of these have been recovered, and Epicurus' physical system is best followed in two other expositions: a letter written by him to Herodotus and preserved by Diogenes Laertius,[2] and a superb poem *On the Nature of Things* by Epicurus' Roman disciple Lucretius.[3] We shall depend on these two sources and on the important twentieth-century interpretations of them[4] for information concerning Epicurus' ideas of life and matter.

A caution: Epicureanism taken in its entirety is a total philosophical system—a comprehensive analysis of cosmogony and cosmology, of celestial and terrestrial physics, of man's history and present condition, of his ethics and esthetics, of his science and religion. One of the most signal contributions of Epicureanism is the complementarity it achieves between the disciplines of physics and ethics. We shall not endeavor here, however, to treat the total system, or to add to the scholarly interpretation of Epicurus. We shall focus, rather, on selected ideas that show his position in the development of the life-matter problem, including a note on the relations within this system of physics, physiology, and ethics, since they are more explicitly linked by Epicurus than any author we have studied thus far.

Lucretius was personally temperamentally worlds apart from his master. We think of Epicurus as affirmative and benign,[5] of Lucretius as pessimistic and aloof. Even the humor of Lucretius tends to be gloomy. Yet the two had the same central purpose: to free men through knowledge of nature[6] from the fear of death[7] and from obeisance to primitive religion.[8] Moreover, we have reasons to feel sure that in its principles, if not its illustrations, Lucretius' poem is

[2] *LH.*

[3] *DNR.*

[4] The basis of much modern scholarship is Herman K. Usener, *Epicurea* (Leipsig, Bonn, 1887). C. Bailey interprets Epicurus exhaustively in *The Greek Atomists and Epicurus: A Study* (Oxford, 1928). More useful in certain ways is N. W. DeWitt's *Epicurus and his Philosophy* (Minneapolis, 1954). See, also, n. 8 below.

[5] *DL,* bk. 10, 8–9, p. 537, where "But these people [the derogators of Epicurus] are mad. For our philosopher has abundance of witnesses to attest to his unsurpassed goodwill to all men...."

[6] *DNR* 1. 147–49, etc. (cited by book and line).

[7] E.g., at *DNR* 3. 830 where we find the famous *nil igitur mors est ad nos,* and elsewhere.

[8] At *DNR* 1. 62–778, 80–135, and especially 932 where *religionum animum nodis exsolvere pergo;* also 2. 650–58, 3. 46–54, etc. An alternative interpretation of Lucretius' purpose is suggested by J. A. L. Munro who, surprisingly, would have Lucretius choose Epicureanism merely as that philosophy most suited to poetic treatment. Surely, William Ellery Leonard came closer to the truth in thinking that Epicurus "chose" Lucretius, not vice versa. See Munro's introduction to his edition of Lucretius (London, 1900), p. 6; and Leonard, ditto (Madison, 1942), p. 28.

faithful to the master. What is more, it makes wonderful reading. Lucretius' verse flows (though not in most English translations) with lyrical power and excitement. The task he sets himself—to teach science in verse—is truly formidable. The degree of success he achieves is a literary miracle.

Matter

The historical evolution of particle theory revolves about a number of fundamental problems which are "classic questions" of physics: are the particles contiguous, or separated by empty space? how many kinds exist, and what are the distinctive features of the several kinds? are the differences primarily qualitative and dynamic or structural and dimensional? is there a link between the peculiarities of particles and their behavior in substantial aggregates? what regularities are inferable as to the way particles aggregate (and how is this regularity related to the characteristics of the particles themselves)? does aggregation occur among the different kinds of particles freely and randomly, or only in a limited and determined way? are there hierarchical levels of aggregation or can aggregates of all sizes occur indiscriminately? most importantly, does or does not analysis lead ultimately to particles that are indivisible (atomic) and nontransmutable?

It will be interesting in the following précis of Epicurus' theory to observe how many of these questions he identifies and how many of them he attempts to answer.

Atoms. The atoms,[9] of Epicurus, are invisibly small,[10] hard,[11] and solid (that is, they have no empty space *inside* them).[12] They do have space *between* them, however,[13] though this space is not quite the void envisioned by Democritus. It is, rather, a "thing"; intangible, but existent.[14] There are more than Plato's few sorts of atoms, we learn, though less than Democritus' infinite number. Dozens of kinds are actually mentioned. An "incomprehensibly great" number are postulated.[15] Of each kind, the number is infinite.[16]

9 "Atoms" since indivisible, *LH* 40–41; 56; *DNR* 1. 551–55.
10 *DNR* 2. 110–15.
11 *LH* sec. 41, and *DNR* 2. 87 where they are spoken of as *durissima*.
12 *DNR* 1. 483–550 where *sunt igitur solida ac sine inane corpore prima*.
13 *LH* 39; *DNR* 3. 329–97.
14 Bailey, *Greek Atomists*, p. 294.
15 *LH* 42, 56, *DNR* 2. 478–521.
16 *LH* 42; *DNR* 2. 521–68.

The differences in atoms are differences in shape, weight, and size;[17] of these, shape is primary, differences in size and weight being mere expressions of differences in shape. Atoms differ also in "interval, path, attachments, weights, impacts, clashes, and motions."[18] Size and shape differences depend on the numbers and arrangements of certain "minimal parts" that atoms are made of.[19] Apparently, different atoms may have similar minimal parts (as Plato's particles had triangles and modern atoms have subatomic particles). But, unlike Plato's corpuscles and modern atoms, the atoms of Epicurus are immutable; they won't resolve, won't break up into *minimas partes,* nor can one sort be converted to another. Each atom's existence is permanent.[20]

Lucretius supplies much detail about the way the size and shape of its atoms determines the sensible character of an object.[21] Water and oil are both liquids, but water atoms are smaller and rounder.[22] Atoms of agreeable music are smoother than atoms of unpleasant noise.[23] Harsh-tasting salt atoms are rougher than the pleasant-tasting water atoms with which they mix to form brine.[24] Rock and iron atoms have surface projections that restrict their individual motions; the atoms keep moving, however, even when thus intertangled (they vibrate).[25]

Aggregations. A mother cow knows her calf. If the latter is led to the altar, its mother accepts no substitute but stands alone, lowing irreconcilably in the willows. Animals that look alike, thus, are actually individually different.[26] And so it is with *all* aggregates of atoms. Lightning and ordinary fire bear a certain resemblance, but the atoms of lightning are smaller.[27] Nor is either of them really pure in the sense that its atoms are all alike. Although Epicurus acknowledged both elements and compounds,[28] Lucretius believes that even

---

[17] *LH* 54–55; *DNR* 2. 333–41.
[18] *DNR* 2. 725–27.
[19] *LH* 41, 57–59; *DNR* 1. 599–608; Bailey, *Greek Atomists,* p. 286–87.
[20] *DNR* 1. 790, 2. 294–307; the twin facts of spontaneous decay and induced fission of modern "atoms" make their designation as such a technical misnomer.
[21] *DNR* 2. 308–32, 730–865.
[22] *DNR* 2. 394, 451–52.
[23] *DNR* 2. 410–13.
[24] *DNR* 2. 466–77.
[25] *DNR* 2. 100–104, 444–46.
[26] *DNR* 2. 342–50.
[27] *DNR* 2. 381–83.
[28] *LH* 40–41.

seemingly pure things are in reality complex aggregations. Among these must be included the classical "four elements."[29] In any case, those who thought the whole cosmos comprised just four elements— or two, or one—were mistaken.[30] Nothing visible is simple.[31] (We shall see later, however, that Lucretius, like his mentor Epicurus, does not quite give up the theory of four elements).

Motion. Atoms weigh something. Hence, their natural motion is downward.[32] All, irrespective of weight, tend to move downward with inconceivably great, equal speed.[33] But which way, for the Epicureans, is down? The question is easy to answer with respect to the individual person; down is where the feet are with respect to the head when one is standing.[34] But with respect to the cosmos, the problem is more difficult. Downward cannot mean centripetally, because, being infinite,[35] the cosmos has no locatable center.[36] Indeed, direction and location have no absolute meaning in infinite space. Epicurus warns Herodotus that ". . . in the infinite we must not speak of 'up' or 'down' as though with reference to an absolute highest or lowest."[37] At least one modern student of Epicurus considers his idea as, perhaps, "a first essay in Relativity."[38]

As for the atoms, they do not move far downwards without interruption. When enmeshed in a compound, they vibrate.[39] Their downward motion is accompanied by a slight sidewise dislocation or swerve (*declinare solerent*).[40] This swerve results in atomic collisions, and collision may bring about attachments among atoms[41] or such results as rebounds,[42] and further collisions. Where a strong enough

[29] *DNR;* thus, fire particles exist *nec singillatim—sed complexa,* 2. 153–54; also, elements but not atoms are transmutable; hence different elements may contain some of the same atoms, 1. 763–802; fire, air, water, and earth are neither simple nor stable, *DNR* 5. 235–317; see E. W. Leonard, p. 52.

[30] *DNR* 1. 712–21, 763–81.

[31] *DNR* 2. 581–85.

[32] *LH* 61; *DNR* 2. 83, 184–215.

[33] *LH* 61.

[34] Bailey, *Greek Atomists,* p. 312.

[35] *LH* 41.

[36] *LH* 60; *DNR* 1. 1052–82.

[37] *LH* 60.

[38] Bailey, *Greek Atomists,* p. 312; DeWitt, *Epicurus,* pp. 167–68.

[39] *LH* 43; see also DeWitt, *Epicurus,* p. 164.

[40] *DNR* 2. 221; the swerve is not discussed in *LH,* but scholarship ancient and modern agrees that the notion is part of the original corpus of Epicurean ideas.

[41] *DNR* 2. 100–104.

[42] *LH* 43; *DNR* 2. 222–24.

entanglement occurs, a solid (earthlike) substance materializes. But atomic motion—of a vibratory nature—is still possible in such a substance.[43] Such motion is in some cases sufficient to require a retentive covering if the association is to persist; the result is a waterlike substance, or if the atoms rebound vigorously enough, an aerial one.[44] These graded attachments look backward, partly, to Empedocles' elements (fire, air, water, earth). They also look forward—in a limited and partial way—to the three states of matter (gas, liquid, solid) of the much later kinetic molecular theory. Lucretius asks his reader to observe the dance of motes in a sunbeam. He interprets this dance—correctly—as caused by collisions with still smaller particles.[45]

Compounds. Between atoms and visible objects Lucretius envisions entities (small aggregates) that are intermediate in size. Some students liken these aggregates to the molecules of 2,000 years later (e.g., to the "compound atoms" of Dalton).[46] The text of Lucretius' poem is less than convincing on this point, nor is there any real evidence in the surviving fragments of the master. It is true that invisible aggregates are mentioned as jostling the dust in a sunbeam. And heat particles are described as *complexa.*[47] But in neither of these cases do we hear anything of their actual makeup. And where we do learn something of the composition of these invisible, or barely visible, corpuscles they are not much like the molecules of our day. Some of them, for example, are invisible but fully developed little animals that Lucretius credits with constituting one-third of all living species! These animalcules have souls, and stomachs![48]

Seeds. A more difficult problem is posed by what the Epicureans call seeds (in Epicurus, *spermata,* in Lucretius, *semina*). These are, in some parts of Lucretius, equated with individual atoms.[49] Elsewhere—for both master and disciple—they are complex aggregates seemingly specific for each object containing them.[50] In some way they seem to be viewed as actually generating objects. They are

43 *LH* 62; *DNR* 2. 88–142.
44 *LH* 43–45; Bailey, *Greek Atomists,* p. 340.
45 *DNR* 2. 116 ff.
46 See Bailey, *Greek Atomists,* 342, and his refs.
47 *DNR* 2. 154–55.
48 *DNR* 4. 116–22.
49 *DNR* passim.
50 *DNR* 4. 645–68; see also Bailey, *Greek Atomists,* p. 343.

corpora genitalia which somehow "beget things" (*gignant res varias*),[51] and in reproduction, they act as genelike hereditary determiners.[52] But their generative function is applied also to nonliving objects.[53] In what sense are inorganic things created or generated by seeds? What is meant, presumably, is that seeds are spontaneously formed small atomic complexes of such a nature that they aggregate or differentiate or both into visible systems with specific, emergent properties. Aristotle notes that Democritus defined generation or begetting as entirely a matter of a proper aggregation, or combination, of atoms. There was in antiquity and in the middle ages a tradition to the effect that minerals evolve in the earth in a manner analogous to germination.

We have already observed in Anaxagoras and in Plato the tendency of Greek thought to extrapolate from living to physical reality. The idea of inorganic seeds would have seemed less strange to the Greeks than it does to modern scientists whose tendency is to extrapolate in the opposite direction—from physical to vital. The idea that minerals develop from seeds will recur in medieval and Renaissance science (see also van Helmont).

Life and Matter

Epicurus' treatment of the life-matter question is best viewed against the background of earlier solutions, those especially of Democritus, Plato, Aristotle. The solutions of these three thinkers had differed importantly with respect to the nature and interrelations of *soma* (the body) and *psyche* (the life-soul). Democritus appears to have distinguished between them by considering the fiery soul atoms—but not the body atoms in general—to be animate or alive. With Plato, the life-soul was a quite real but immaterial existence "bonded to" the presumably inanimate and material particles of the body. With Aristotle, the life-soul was (in the sense suggested in chapter 8) the form of the body. Epicurus' formulation owes something to each of these ideas but differs significantly from them all. The living microcosm appears to him to consist entirely of individually inanimate atoms, some of which are collectively designated body

[51] *DNR* 1. 58, 2. 62–63, etc.
[52] *DNR* 3. 741–75.
[53] *DNR* 2. 718–29.

(*soma*) and the others collectively designated soul (*psyche*). The animal body comprises, in short, two interlocking patterns of atoms, which it might be appropriate to designate psychic and somatic respectively.[54] It is inconceivable to Epicurus that the soul be incorporeal. Incorporeality is a characteristic only of the void.[55]

Disruption of either pattern, psychic or somatic, disrupts the other, or tends to.[56] Normal alterations of the psychic pattern are the basis of sensation,[57] but in a curious way, to be explained, somatic atoms likewise participate in sensation. When the psychic pattern alters the somatic, the result may be, under proper circumstances, willed locomotion.[58]

Pain involves a disagreeable dislocation in the psychic atomic system.[59] In drunkenness, the atomic disorder is temporary; in disease, it is sustained;[60] in death it is irreversible and progressive. In another sense, the relation of soul to body is like that of a volatile fluid to its containing vessel (*vas*).[61] The figure is apt, since soul comprises the most mobile of atoms. An analogous idea occurs elsewhere in biblical and other ancient writings.[62]

In death—caused by gradual or sudden disruption of either the psychic or the somatic pattern or both—the soul disintegrates. Its atoms disperse like those of smoke. When the soul disperses, since the integrity of each pattern depends on the integrity of the other, the body disintegrates also.[63] There is no immortality; no reincarnation.[64] No souls in transit work feverishly to patch together new bodies,[65] nor do they crowd about jockeying for position when something or someone is born.[66]

Epicurus' "protoplasm" is thus a strictly material, diphasic sys-

[54] The terms psychic and somatic are ours; both Epicurus' and Lucretius' identification of the two systems, however, is perfectly clear; *LH* 63.

[55] *LH* 67.

[56] *DNR* 2. 976–79.

[57] *LH* 63–64.

[58] *DNR* 4. 880–910; or, by a curious mechanism, localized feeling; see below, n. 87.

[59] See the "Prolegomena" in Bailey, *DNR,* pp. 60–61.

[60] *DNR* 3. 429–555.

[61] *DNR* 3. 548–57.

[62] Ecclesiastes 12: 6; see DeWitt, *Epicurus,* 199.

[63] *DNR* 2. 944–61.

[64] *LH* 81; this is also the burden of most of *DNR* 3.

[65] *DNR* 3. 718–40.

[66] *DNR* 3. 776–83.

tem comprising psychic and somatic patterns of atoms, existing in a kinetic condition of causal interdependence.[67] During their sojourn within the organism the proper atoms must be properly spaced and must undergo the proper "vital motions" (*vitalis convenientis motus*).[68] Life is, in the strict sense, emergent.[69] Worms emerge from dung; the living body, from inanimate nutriment. "Therefore nature changes all foods into living bodies, and from them brings forth all the feelings of animals, very much in the same way as she expands dry sticks into flames and turns them all into fire. Now do you see then that it is of great moment in what order the first beginnings [the atoms] of things are placed and with what commingled they cause and receive motion?"[70]

Microorganization

The somatic atomic pattern. It was a common, almost universal, assumption of Greek biology that only those systems can display life-as-action (nutrition, motion, etc.) that are composed of proper matter properly configured or proportioned. We have already considered evidence, though not always clear and conclusive, that even the hylozoists (who viewed the life-soul as immanent) made this assumption; and there is no reason to doubt that Plato (for whom the life-soul was imposed) did so as well. For Epicurus, the configuration of the body's atoms is central. His theory of the somatic atomic pattern, however, is complex and rather vaguely formulated. Nothing so sophisticated as Aristotle's levels of organization (element-tissue-organ) is acknowledged. As Lucretius views this problem, bodies are made of such materials as "blood, veins, heat, moisture, flesh and sinew."[71] The classical elements are thus indiscriminately bracketed with such complex components as blood and sinew. We do hear, however, that the different parts of animals differ in atomic makeup.[72] Different kinds of animals, too, are

[67] *Haec igitur natura tenetur corpore ab omni ipsaque corporis est custos et causa salutis; nam communibus inter se radicibus haerent nec sine pernicie divelli posse videntur.* Therefore, this nature (the soul) is contained in the whole body and is in turn the body's guardian and the cause of its well-being; they (soul and body) cling together by being rooted each in the other, nor do they seem able to be torn apart without destruction [trans. ours].

[68] *DNR* 2. 886–901, 942–43, 3. 847–52.

[69] *DNR* 2. 865–85, 4. 822–57.

[70] *DNR* 2. 878, trans. W. H. D. Rouse.

[71] *DNR* 2. 669–71.

[72] *DNR* 1. 859; 2. 671–72.

atomically different.[73] Is Lucretius suggesting the existence of species-specific "cow-" as opposed to "horse-atoms" (or small atomic aggregates)? Something of the sort seems suggested, but one could wish for a more precise formulation.[74]

Lucretius does have a definite notion of a cycle of atoms—from soil to foliage to cattle to man and thence back to soil either directly or by way of some animal that attacks the human cadaver.[75] He likewise sees the body as engaged in continual atomic erosion and replacement,[76] selecting the needed atoms from its food and rejecting the rest.[77]

The psychic atomic pattern. The psychic atomic pattern is more precisely described than the somatic. Epicurus makes it "a structure of tenuous [and mobile] corpuscles spread through the whole body and most like wind mixed somewhat with heat, in some ways like the former, in others like the latter."[78] According to Epicurus, three, and according to Lucretius four, sorts of atoms are involved in the composition of the soul. The three are fire (or heat), air, and wind atoms (or aggregates); the fourth kind Lucretius leaves nameless.[79] The three (or four) kinds form a tenuous net through the body. This net *is* the soul.[80]

In chapter 2, it was suggested that three formulations of life-matter relations have been more often adhered to than others, viz. the relations of immanence, imposition, and emergence. Epicurus was the first out-and-out emergentist. The seeds of emergentism were sown by Democritus and Empedocles whose theories were troubling, respectively, to Plato and Aristotle. But the view of life as merely the motions of lifeless matter was first clearly set forth by Epicurus. The debate over the correctness of this view has been central to theoretical biology ever since. Awareness, like life, is emergent, according to Epicurus. Individual atoms are insensate, but sensibility arises when the right kinds are assembled in the right configuration and undergo, within that configuration, the right sorts of motions. "If," as Bailey

---

[73] *DNR* 1. 820; 2. 661–68; 698–99, 711–13; 4. 645–48.

[74] *DNR* 4. 645–48.

[75] *DNR* 2. 875–79, and 3. 966 where we hear that there can be no hell since hell would accumulate matter needed as food by rising generations!

[76] *DNR* 4. 858–76.

[77] *DNR* 2. 711–17.

[78] *LH* 63.

[79] *DNR* 3. 117–35; *LH* 63; see also DeWitt, *Epicurus*, p. 198.

[80] *DNR* 3. 370–95.

observes, "the atoms of the right shape are in the right arrangement and perform the appropriate motions, the *concilium* of insensible atoms may itself obtain sensation: this is the origin of the soul of living creatures and the mind of man." Lucretius, in an outburst against hylozoism, ridicules the thought of atoms that are individually loquacious, learned, laughing, lamenting.[81]

As the picture is drawn by Lucretius, the nameless fourth, round-atomed ingredient furnishes the clue to soul's action. It dominates all vital functions. Its atoms are the very soul of the soul (*anima animai*).[82] Here, however, we encounter a fundamental refinement in the detailed depiction of soul.

With the Epicureans, as with most predecessors, it is soul that sets off living from nonliving. But in their system, as Lucretius develops it, there are two souls to be reckoned with, termed respectively *animus* and *anima*. *Animus* is equivalent to mind (*mens*), corresponding roughly to the rational soul of Aristotle and Plato. *Animus* is localized in the mid-breast. *Anima* is strung out through the body. The two are not "biochemically" distinguished; they are "held in conjunction together and compound one nature in common."[83] In movement, *anima* acts in obedience to *animus;* the impulse travels from *animus* to *anima* to the moving parts of the body.[84] When a limb is severed, *anima* lingers in it briefly, making it temporarily "animate," that is, living.[85]

Soul and Sensation

Effluence-psyche-soma. Among the classic questions of physiology, Epicurus was perhaps more interested in the nature of sensation and thought than in the nature of other major manifestations of life-as-action. His analysis of sensation is ingenious, if scientifically uneven. The notion of effluences from objects streaming in through the sense organs is unobjectionable enough,[86] and one can see how, once inside, these effluences, being of atomic constitution, would excite the psychic atomic pattern of the body. Their first act, as Lucretius spells it out, is to disturb those small round nameless atoms that,

[81] *DNR* 2. 976–79.
[82] *DNR* 3. 275.
[83] *DNR* 3. 136–44.
[84] *DNR* 4. 880–910.
[85] *DNR* 3. 624–69.
[86] *LH* 46, 49. Effluences as the real object of sensation is the subject of the bulk of *DNR* 4.

along with fire, wind, and air, constitute soul. The "nameless" sort of atoms next agitates (*ferit*) fire, fire agitates wind, wind agitates air, and air finally agitates the somatic atoms so that sense seems to reside in the tissues. The sequence *effluence-psyche-soma* seems strange to us who are accustomed to the sequence *effluence-soma-psyche.*[87]

In retrospect, the sequence is understandable in the sense that what *seems* to have sensation is often far from the head (or heart) where awareness is said to be centered: it is the wounded foot, the burned finger, the injured back that *seems* to feel the pain. When Greek thought came to Europe in the Middle Ages it came with the idea that the role of sensory nerves is to extend the visual, tactile, auditory and other faculties of the sensitive soul, from the brain to the part in which the sensation occurred. The ultimately effective argument against this idea of peripheral awareness was offered by Descartes in the first half of the seventeenth century.

Perception-intellection. As one modern scholar has it, body and soul are coterminous, coextensive, and cosensitive in the Epicurean system. Psychic action by either implies the presence of both. The activities of the two systems are, however, not identical. Sensory perception is a property of both, but intellection is reserved for the psychic system. Stated differently, *soma* becomes involved in the activity of *animus* but not of *anima*.

Cognition-recognition. Under certain circumstances (especially in dreaming and daydreaming) tenuous films can bypass the sense organs and impress the mind directly, the mind thus serving, in De-Witt's phrase (1954), as a super sense organ.[88] Whether impressions reach the mind thus directly or via the sense organs, repeated perceptions leave a cumulative impression on the psychic system; such a cumulative impression serves as a standard of comparison for subsequent perceptions. In other words it permits recognition. The implication is that a sensory perception is not meaningful or complete until it has been consciously tagged as conforming to a particular category. This concept of cognition—or recognition—is penetrating and, to a certain extent, valid.[89]

The simulacra. The credulity of the modern reader tends to balk

[87] *DNR* 3. 245–51; the treatment in *LH* (at 64) is less detailed, and is cryptic, but it is quite consistent with Lucretius' more explicit account.

[88] DeWitt, *Epicurus*, pp. 207 ff.

[89] *LH* 50–52; DeWitt, *Epicurus*, 204–9.

at those fast-flying films[90] envisioned by our authors, surface coats continually cast off by objects,[91] moving through space, and also penetrating glass,[92] and bouncing off mirrors.[93] These films—*eidola* or *simulacra* since they keep the form of object they are cast off by— are actually a heroic attempt to explain the everyday phenomena of sensation and many other phenomena.[94]

As for Lucretius, he admittedly rather abuses this invention. When horse film and human film merge in space and excite the soul to imagine a centaur, the whole thing seems scientifically preposterous.[95] The *simulacra* are fairly durable, it appears; the gossamers of the dead linger on to trouble the sleep of the living.[96] And atomically constituted images of the gods are made responsible for the visions of seers.[97]

Thus *simulacra,* we can see, have many everyday applications. To have them about the world is a conceptual convenience. We can appreciate the temptation of Lucretius, his need even, to believe in these atomic "idols."

Physics and Ethics

It has been our practice, in this study thus far, to isolate the scientific aspects of each thinker's intellectual system from the religious and ethical aspects. To do this is in many cases rather arbitrary; it is to project backward on to Greek thought a conceptual division of labor taken for granted today but only imperfectly realized then. The interrelations of physics and ethics were especially conspicuous, as we have already noted, in Plato's world view. So they are in the Epicurean-Lucretian theory of the atom which goes hand in hand with a theory of human nature and conduct.

In opposition to Aristotle, Lucretius exorcises formal[98] and final causes;[99] and he takes the gods completely out of everyday nature.[100] No plan, no *aidos,* no divine reason is needed. Natural organization

---

[90] *LH* 47; *DNR* 4. 26, 209–63.
[91] *LH* 46; *DNR* 4. 145.
[92] *DNR* 4. 147, 601–2.
[93] *DNR* 4. 98–108, 269–323.
[94] *LH* 46; *DNR* 4. 42–43, 98–101.
[95] *DNR* 4. 732–52; possibly also suggested at *LH* 48.
[96] *DNR* 4. 31–44, 757–61.
[97] *DNR* 5. 1169–71.
[98] *DNR* 5. 181–86.
[99] *DNR* 4. 823–57.
[100] *DNR* 2. 167–83, 645–47, 1090–1104; 5. 146–73.

as we know it through the senses arises inevitably out of the inter-action of atoms. In all of this, Lucretius is faithful to the precepts of the master. Celestial bodies are not controlled by blissful, im-mortal beings, nor are they soulful. Belief in such fallacies is disturb-ing to one's peace of mind, and should be avoided.[101]

The origins of natural order. In the case of Lucretius, few schol-ars have missed the point that there is something religious about his very irreligion. Indeed, the true Epicurean is not a disbeliever. His gods, ideal Epicureans, lead a detached existence—peaceful, aloof, "between worlds."[102] In Timaean terms, however, Lucretius is all for necessity (inevitability) in opposition to reason (guidance or plan-ning). Had Plato known Lucretius he would have thought him dangerous—and deranged[103] (an epithet Lucretius applied, in his turn, to those whose views he disagreed with).[104]

We begin to see in this opposition of necessity to reason a rift which would split human thinking vertically in history. The issue: whether *nomos,* whether order, is implicit in nature, or whether it enters *ab externo.* Whether law, in Whitehead's phrase, is *immanent* or *imposed.*[105] Lucretius, with almost bitter exultance, espouses nat-ural in opposition to supernatural order.[106] A powerfully determi-nistic materialism such as that of Epicurus and Lucretius poses ethical problems which we have already encountered in our brief study of the *Timaeus.* Along with the will of the gods[107] there is a tendency for human will to go out the window. But how, without human freedom, can we have any real human ethics?[108]

Freedom and the swerve of the atom. What one seeks at this juncture is a certain *in*determinacy somewhere in the physical system —an opening wedge, as it were, for the reintroduction of freedom. (In the twentieth century, too, indeterminacy will be seized upon as a pretext for the reinsertion of freedom.)[109] With Epicurus, in-

---

[101] *DNR* 5. 416–31; *LH* 76–77. Lucretius shared the belief, usual for his time, in spontaneous generation.

[102] *DNR* 2. 645–47.

[103] Plato *Laws* 10. 888.

[104] *DNR* 1. 692, 698, 704, 1068–71; 2. 985, etc.

[105] A. N. Whitehead, *Adventures of Ideas,* part 2, chap. 7, sec. 5.

[106] E.g., at 1021 ftnt., etc. In taking this position, he becomes the outspoken antagonist of the Stoic transcendentalists.

[107] *DNR* 2. 1090–1104.

[108] For a helpful discussion of the problem of freedom in Epicurus, see DeWitt, *Epicurus,* pp. 171–96.

[109] See, e.g., A. H. Compton, "Do We Live in a World of Chance?" *Yale Review,* 21 (1931):86–99.

determinacy exists in the sideward swerve of the atom.[110] The swerve's direction, perhaps even its occurrences, is not fully physically determined. This fact seems to Lucretius—and presumably to Epicurus—to liberate man from causation.

Exactly how the swerve does this is nowhere adequately detailed, and it is difficult to choose between scholarly interpretations of Lucretius' treatment of this subject. The interpretation closest to his text is that, by swerving, the mind atoms evade the tyranny of what would otherwise be an imposed and absolute determination.[111] Whether or not this is Lucretius' precise intent, he sees the swerve as somehow permitting volition. Willed action begins when a mental image of the body in motion excites the spirit atoms which, in turn, set the body moving. In some way, soul's impact is viewed as augmented by air entering through the pores of the body and spreading through it. The body is thus partly moved by air as a galleon is partly moved by wind.[112]

What to choose. If the swerve of the atom offers opportunity for choice, the all-important question remains, *what shall be chosen?* Here, too, we find ethics connected with science. Epicurean science commences, as we have seen, with the senses.[113] It begins with the impact of environment on the psychic system of atoms. Since too much atomic dislocation is painful,[114] such dislocation, and the pain that goes with it, must be assiduously avoided. The avoidance of pain is the very basis of the Epicurean ethic. The absence of pain is pleasure, and pleasure is the object of existence.

If the formula seems overly simple—even simple-minded—it acquires texture and substance when we ask what things are conducive of pleasure. Not hedonistic excesses, Lucretius advises us, not passion,[115] not the fulfillment of worldly ambition. All these are too often followed by satiety, violence, and frustration.[116]

The Good is that serenity and freedom from anxiety that are attainable through moderation, simple pleasure, contemplation, and above all a wise understanding of nature of which man himself is

[110] *DNR* 2. 250–93.
[111] *DNR* 2. 292.
[112] *DNR* 4. 876–906.
[113] *DNR* 4. 469–521.
[114] *DNR* 2. 963–68.
[115] *DNR* 4. 1074 ff.; Plato makes excessive pleasure *or* pain the cause of mental illness. *Tim.* 86B.
[116] *DNR* 3. 31–90.

a part.[117] In particular "the principal disturbance in the minds of men arises because they think that the celestial bodies are blessed and immortal . . . and because they are always expecting or imagining some everlasting misery, such as is depicted in legends, or even fear loss of feeling in death as though it would concern them themselves; . . . But peace of mind is being delivered of all this, and having a constant memory of the general and most essential principles."[118]

The fear of death is easily avoided. To allay it, religion vainly offers immortality with all it betokens of uncertainty and possible unpleasantness. Epicurus offers the opposite. He offers the assurance that after the final dispersal of the psychic pattern of atoms there will be no soul, no awareness, in short nothing to fear and no soul to do the fearing.[119]

Finally the soul, we remember, comprises four sorts of atoms. In different animals these are present in different proportions. Cool wind predominates in the deer whose heart is fearful and chilly; heat predominates in the courageous and passionate lion.[120] Men's souls also evince certain differences in atomic equilibrium, whence men's differing natural tempers. Yet these differences are not so pronounced but that a man may through discipline and study come very close to the blessed state of the gods.[121]

SUMMARY

Epicurus builds an integrated system of conceptual micromodels to account for the familiar phenomena of life-as-action. In so doing, he follows what we have identified as the classic strategy of both earlier and later biology; in many ways he is the outstanding practitioner of that strategy in the ancient period. First of all, his models explain the difference between living and nonliving things generally, by stipulating that only living things have an intercalated pattern of psychic atoms. They likewise account in detail for selected aspects of nutrition, generation, sensation, and volition. The slightest acquaintance with post-Renaissance physiology (from Descartes to the present) will make Epicurus seem closer than any other ancient

[117] *LH* 78 ff., 81; *DNR* 3. 1071–75. See also *DL, Pyth.,* 85.

[118] *LH* 81.

[119] Most of *DNR* 3, beginning at 418 and culminating at 830; the same idea, at *LH* 81–83.

[120] Who is capable of breaking his heart with passion.

[121] *DNR* 3. 299–322.

scientist to the emergentism and mechanistic materialism of the modern era. We shall see later that Epicurus is himself indirectly a founder of modern biology, since the revival of his system by Pierre Gassendi will contribute to the new micromechanical outlook that develops in the seventeenth century.

10

Physical and Physiological Powers

GALEN [A.D. 130 · 200]

Our studies of life and matter bring us next to that Pergamene "prince of physicians" who wrote scores of books and was attached to the courts of three Roman emperors. Galen commands our attention as the last major representative of ancient science, as a rich source of knowledge concerning that science (whose biomedical aspects he reports critically and in detail), and as a dominant influence in the new biology that will appear in Europe twelve centuries later. His position in late medieval and Renaissance medicine will compare with, and complement, that of Aristotle in science generally. At the same time, the points of disagreement between the two will be salutary in helping to maintain in medical dogma a degree of openness conducive to its further development.

The interpretation of Galen is complicated by the fact that he treats each cardinal biological problem many times and from several viewpoints. An urgent task for Galenic scholarship is to discover to what extent and in what areas his manifold interpretations permit a common synthesis that may be thought of as really his. That task lies mostly outside the range of the present study; but by selecting representative elements of Galen's thought we can gain at least a tentative idea of his place in the history of ideas about life and matter.

Galen's theories are partly original and partly derived—though with modification and, often, extensive elaboration—from eminent predecessors including the pre-Socratics, the Hippocratics, Plato and Aristotle, the Stoics, the Alexandrians, and later sectarians of varying persuasions. Galen is in general helpful in letting us know whose ideas he is espousing or opposing.

Galen is well known for his mistakes in descriptive morphology and for unwise extrapolations from animal to human anatomy. These flaws are highlighted, perhaps, because of their contrast with the stronger elements of his endeavor. His anatomy is impressive not so much for its accuracy as for its exhaustiveness and for the author's enlightened practice of interpreting structure in the light of function. Galen's creative and observant work as a surgeon gave him a far better grasp of the human body than is suggested by the easy statement that he anatomized animals only. He was an avid (though not an unprejudiced) observer—never missing an opportunity during surgery to notice the consequences of damage to a nerve or muscle, of pressure applied to this or that organ, or of injury or ablation.

He surpasses all predecessors in his extensive use of experimentation. He performs not just a few experiments such as the famous ones on the flow of the urine but, literally, hundreds. On the other hand, he seems not to realize fully the power of the experimental way of asking questions of nature. Favorite hypotheses that cannot be experimentally proven are flatly asserted, or defended in a formal-logical way, or by an appeal to common sense, or by citation of authority.

Admired predecessors, especially Aristotle and "Hippocrates," are praised as long as Galen happens to agree with them. "Enemies of the truth" are treated with irony or at times with ridicule and invective. Erasistratus and Asclepiades come in especially often for polemical treatment. Yet it must be admitted that Galen's arguments are typically substantive and not personal. He praises Erasistratus on points of accord, and finds occasion to disagree on important matters with Aristotle—especially on the role of the brain which he says Aristotle, though ashamed to say so openly,[1] renders totally useless.

Teleological Aspect of Galen's Physics

The early history of physics is marked by certain issues on which Greek thinkers take sides with varying degrees of explicitness and clarity. Are the fundamental differences in things qualitative, or merely dimensional? Is the world perfectly continuous, or is it interspersed with a certain amount of emptiness? Are physical events guided in some manner, or is everything left to circumstance and

---

[1] *U.P.* 8.3 (cited by book and chapter) (K3: 625). See list of Abbreviations. See, also, M. T. May's precise and welcome translation of this work (*Galen on the Usefulness of the Parts,* Ithaca, 1968) which was published after the present chapter was printed.

blind necessity? These are questions on which we shall expect Galen to have outspoken opinions.

Galen's physics is primarily Aristotelian. The perceptible differentiations of things are explicable in terms of Aristotle's (and Hippocrates') qualities (*poiotetes*)—hot, cold, dry, moist.[2] Change, in Galen's view as in Aristotle's, is motion (*kinesis*). A change in the quality of a thing (e.g. from cold to hot) is an "alterative motion." A change in place is "motion of translation."[3]

In developing these ideas, Galen takes sides on two central issues of ancient science. First, he vigorously rejects the view that would limit change to mere relocation, that is, to the motions of separate and unalterable particles in a vacuum. Galen's world is everywhere both continuous and alterable. Second, he repugns the idea that chance movements of atoms have produced the world as we see it. Galen tells us repeatedly that nature is an artist more ingenious than Phidias or Praxiteles or Polyclitus. Galen's physical world, like Aristotle's, is expressive of purpose and design.[4]

Galen does acknowledge a kind of microphenomenon rather like semipermeability; air can get through into the interior of fishes' gills whereas water is kept out; rare blood can get through from the right to the left side of the heart. We even hear that the decisive factor is the size of the pores through which the substance in question must pass. But this sort of "semipermeability" is not interpreted by Galen as based on atomism or consistent particle size; rather he probably thinks that the somehow rarer continuum of the air is more finely subdivisible and can get through smaller pores than the denser continuum of water; and so with rarer and denser blood.[5]

Purpose and design are especially apparent in the harmony of the animal body: in the nice proportion of the four qualities in each of its parts; in the structural conformance of adjacent parts (marvelously represented, for example, in the upper and lower teeth);[6] in the aptness of structure to function, and of function to usefulness in meeting the requirements of life itself.

[2] *N.F.* 1.2 (K2: 5).

[3] Ibid., 1.2 (K2: 2–3).

[4] Ibid., 1.12 (K2: 27–30); U.P. 17.1 (K4: 346–62).

[5] *U.P.* 6.9 (K3: 443). See also Siegel's translation of the passage about the gills (n. 86 below). Aristotle thought fishes took in no air, but only water, which affords them the same cooling effect that air affords in mammals. See also, on this and other physiological topics, R. E. Siegel's valuable book on *Galen's System of Physiology and Medicine* (Basel and New York, 1968), which was published after the present chapter was printed.

[6] *U.P.* 11.8 (K3: 868–69).

To ascribe these marvels to chance movements of atoms in a void—in the manner of Epicurus and Asclepiades—is to Galen an act of inconceivable irrationality. Refuting the idea of simply mechanical determinism, Galen insists that the primary cause of any object or part is its final cause or purpose, i.e., the function or role for which it has come into existence. To argue otherwise is as absurd as it would be for a man in the market to give as the reason for his being there not that "he came to buy a tool or a slave, or to meet a friend, or to sell something" but "that he had two feet which he could easily move and plant firmly on the ground and by which, resting first on one and then on the other, he arrived."[7] Galen acknowledges that such a man would be stating a cause but only an incidental (efficient) and not a primary (final) cause.

Composition of the Body

Our whole inquiry into the life-matter problem is meaningful only because of a constant assumption that runs through the history of physiological interpretation: life-as-action presupposes a proper and indispensable physical organization. Even for the dualistic Plato, who sees the life-soul as something distinct from the body, that soul expresses itself according to the physical system to which it is bonded.

As for Galen, he builds the body out of internally homogeneous components designated, after Aristotle, "homoeomerous" (alike in their parts). These include fiber, membrane, flesh, fat, bone, cartilage, ligament, nerve, marrow, "and all those others whose subdivisions are precisely identical in form." These body components (corresponding approximately to modern tissues) are "born of the elements that are closest to them," viz., the *four humors*. The humors have their origin in the things that we eat and drink, and these are engendered, in turn, of fire, air, water, and earth.

No sane person can doubt that grasses, plants, and fruits are begotten of the four elements nor that these vegetable matters become the aliments of animals. Nor will anyone in his senses dispute the fact that animal humors arise out of these vegetable aliments. It will be equally obvious that the humors first "come together" to form the "simplest sort of instrument" (tissue), namely that built to carry out a single kind of action. And that, out of these "simple instruments" (tissues), "greater instruments" (organs) are perfected.

---

[7] Ibid., 6.12 (K3: 464–65), trans. ours.

Finally, "from the joining of the latter the integrity (*systasis*) of the body as a whole is established." Thus presented, Galen's system appears as an orderly synthesis of Hippocratic and Aristotelian concepts. We shall note later that the picture is more complicated than this outline (taken from Galen's book *On the Elements*) suggests.[8]

In addition to the foregoing hierarchy of material components, Galen equips the body with two (or, according to some interpreters, three) "spirits" (*pneumata*) borrowed from various predecessors and modified for incorporation into his own system. We may temporarily defer our consideration of these.

Galen's Physiology

We have already noted that Greek physiology creates a plan, a master strategy, which will be used with modifications by the physiologists of all subsequent periods. It identifies certain classic problems and investigates these by moving back and forth between a world of sensible macrophenomena and a world of insensible but inferable microphenomena or cryptomena. Galen's cryptomenal world is a world of *powers*—the physical powers or dynamic qualities of the elements and, as we shall see, the physiological powers or faculties of the parts of living bodies. An interesting aspect of powers is that at least some of them are separable from the parts that possess them. We shall hear later that it is a power, not the matter, of air that sustains the vital heat; and a power of testicular semen, not semen as substance, that makes hair grow on the chin; that the heart transfers to the arteries not only blood but a faculty or power of pulsation.

In the standard pattern of both earlier and later physiology, Galen moves between this world of invisible powers and that of the visible aspects of nutrition, generation, motion, volition, and sensation considered in both their normal and abnormal aspects. He is in his own special style, then, a master practitioner of what we have identified as the main strategy of physiological interpretation.

Galen's powers or faculties have interesting ontological implications that suggest themselves when we ask what a faculty *is*. Much later (in the eighteenth century A.D.) we shall hear physiologists evoke "sensibility," "irritability," and "contractility" separately or in various combinations as fundamental to all vital manifestations.

[8] *El.* 1.8–9 (K1: 479–81); see also *De Plac.* 8.4 (K5: 671).

This sort of thinking will culminate with Bichat (ca. 1800). During the later nineteenth century, irritability will take the stage as the most fundamental "property of protoplasm" (see the discussion in chapters 27 and 46).

Galen's taxonomy of functions leads to a certain artificiality of interpretation. Every part has (*a*) attractive, (*b*) retentive (*c*) alterative, and (*d*) repulsive or eliminative faculties. But Galen deceives himself into clumping together certain actually unrelated phenomena. Attractive faculties, for example, include the pull of "proper juice" to each tissue, of food into the stomach, of blood to the heart, of black bile to the spleen, and of urine to the kidneys. Retentive faculties cause the stay of juice in the tissue, food in the stomach, and semen (and later the child) in the womb. Repulsion eliminates urine, discharges the gall bladder, empties the stomach, aborts an abnormal fetus, or delivers the normal child at term.

We should perhaps not tax Galen too gravely for his attempt to create a taxonomy of bodily functions. Some of his categories are fairly "natural"; for example, many of his examples of expulsion are properly recognized by him as varieties of fibrous contraction. And his accounts of the associated macrophenomena are often marvelously graphic and accurate. It is where he moves too facilely from macro- to unrelated microphenomena, from organ function to tissue function, that his groupings become artificial.

Triadic plan. Galen gives his physiology a triadic organization, basing it on the physiological primacy of heart, liver, and brain. Each of these three organs is compared to that central part (*rhizosis*) of a tree which obtains material from the roots and sends it into the branches. The heart obtains air (or air partially converted into *pneuma*) from the lungs and sends *pneuma* (mixed with arterial blood) to the rest of the body. The liver obtains chyle from the digestive system, converts it to blood and distributes it through the veins. The brain obtains exhalations from arterial blood and converts these into psychic *pneuma* for distribution to the nerves.[9]

The soul. An issue exists in later (Platonic and post-Platonic) Greek thought concerning the unity vs. multiplicity of the life-soul (*psyche*). Are there three *sorts* of soul (or three *distinct parts* of it, as Plato suggests)? Or are Aristotle and Posidonius right in thinking

[9] *De Plac.* 1.7 (K5: 199); 3.8 (K5: 356); and 6.3 (K5: 519–20). See also *U.P.* 1.16 (K3: 44–46).

of soul as a single entity with separate powers? Galen aligns himself on this question with Plato. "Our purpose from the start," he says, "has been to prove that one part of the soul is situated in the head, another part in the heart, and the third in the liver." Damage the brain and you impair sensibility and motion; the heart, and you interfere with the pulse. Again, anger alters the heart; but not the liver.

The part of the soul that resides in the brain is rational, sensory, and motor; the part in the heart is spirited; the part in the liver region, nutritive. The power possessed by the liver is possessed also by plants.[10] The heart ". . . is the source of the *pneuma*-carrying and boiling blood in animals, as it is of their arteries; and this fact indicates that the spirited part of the soul resides in it. . . ." As for the liver, ". . . it makes no difference whether the liver be said to be the source of the veins or blood, or of the desiderative soul, but it is more appropriate somehow for the physician to present his teaching in terms of physical organs, and the philosopher in terms of powers of the soul."[11]

Mind-body. A cardinal question in the history of Greek physiology was whether, and if so to what degree and in what way, the life-soul and its faculties or parts are conditioned by the composition of the body. This problem assumes a special interest where *soul* is used in the sense of ego-awareness, consciousness, mind, giving rise to what has come to be called the mind-body problem.

Galen insists that powers of the soul so conceived follow from body temperament and hence differ innately from person to person. Children show themselves, from birth, to be well or ill disposed. Wine, taken in moderation, renders the soul less fierce yet more courageous. It does this by altering the body temperament; the soul responds. Too much wine can affect the soul adversely, as can various poisons. All act through the intermediation of temperament which can "drive the soul forth (in death), can drive it mad, deprive it of memory and intelligence, render it sad, timid, as it appears in melancholia. . . ."[12]

Life and the triad. The overriding significance of the physiological triad is suggested in several passages in which Galen equates it with life itself. "Animals are ruled by three faculties different in

[10] *De Plac.* 6.2 (K5: 516).
[11] *De Plac.* 6.8 (K5: 573–77).
[12] *Quod animi mores corporis temperamenta sequantur* 1.3 (K4: 772–79).

kind, each distributed to the whole body as if from a fountain; Plato calls them souls and ascribes a peculiar substance to each." Having elaborated briefly on the three, Galen adds that "to preserve these three faculties is to preserve life itself. . . ."[13]

Faculty and use. One of the most telling differentia of living systems is the pattern of correlations—of "fitnesses"—that they display, the appropriateness of structure to function and of each individual function to the operation of the whole. It is arguable that this pervasive characteristic—the pattern of suitability—of living things will only begin to be scientifically analyzed in the nineteenth century when Darwin will give a new meaning to the phenomena of adaptation. But the problem is present, in more than inchoate form, in Galen's physiological theory. In Galen's physiology, every part of the body is distinguished by three characteristic features; first, a *power* or *faculty* for a particular action; second, the *action* in question; and third, the *effect* of that action. The three are linked causally (faculty causes action causes effect). In the veins, for example, a blood-forming faculty leads to blood-forming activity with blood as the effect.[14] Galenic physiology is partly an elaboration of this threefold way of interpreting the functions of the body.

Each body part has, in addition to its proper function, a specific usefulness to the economy of the whole animal. If we would truly grasp the nature of a part, we must think of its uses (*chreiai*) as well as its function.

What an organ can do depends ultimately on the essence, or blend of qualities, of its constituent tissues. *How well* it does it depends on the way the variously blended tissues are put together; it depends on the organ's "position, size, contexture, and conformation." The excellence of an organ, its beauty in the truest sense, depends, then, upon the fitness of its visible form (as determined by texture and conformation) to its function (as determined by its intrinsic essence or blend).[15]

The marvelous aptness of form to function evokes religious feelings in Galen. For him, true piety lies not in sacrifices or in the burning of incense and perfume but in understanding and teaching the

[13] *De methodo medendi* 9.10 (K10: 635–36).
[14] *N.F.* 1.4 (K2: 9–10); 2.3. (K2: 80).
[15] *U.P.* 1.9 (K3: 24).

wisdom, foresight, and bounty of man's founder and maker. Do we marvel at the beauty and order of the cosmos? No less marvelous is the craft revealed in the human microcosm.[16]

Krasis. We noted earlier that the central theme of our entire study is the persistent supposition that vital activities presuppose the proper physical conditions—the presence of the proper elements in proper amounts and proper juxtaposition. In confronting this problem Galen uses old materials but gives them a partly new and unprecedentedly extended application.

The functions of the parts vary according to the blending (*krasis*) of the four qualities in each. Health reflects a proper proportion (*eukrasia*); ill health, a disproportion (*dyskrasia*).[17] The formulation is familiar enough, but Galen complains that earlier writers have never adequately defined eucrasia. For a nonliving thing to be well-tempered means simply that it consists of equal parts of the combining opposites (dry-moist, warm-cold). For a plant or animal to be well-tempered requires that it possess that balance of opposites which best fits a member of its species to fulfill its proper functions. The best-tempered fig tree is the one whose temperament permits it to produce the best and most figs.[18] Well temperedness is a harmony of opposites in accordance with function.

Not only did his predecessors fail to define eucrasia, Galen protests; they also listed only half of the possible dyscrasias. Normal individuals can deviate from the mean condition by an excess of one or both of the pairs of opposites; they may be relatively warm, dry, moist, cold, warm *and* dry, warm and moist, cold and moist, or cold and dry. This gives a total of nine temperaments—the mean, and the eight sorts of deviations. The deviations in more extreme cases amount to dyscrasias of which there are, then, eight and not four.[19]

Having insisted upon the distinction between absolute and relative eucrasia, Galen makes the skin of the hand, as a tactile organ, well-tempered in the absolute sense. It is exactly equidistant between the extremes of heat (boiling water) and cold (numbing ice or snow) and between extremes of dry and moist.[20] The skin, further, is the

[16] Ibid., 3.10 (K3: 236).
[17] *N.F.* 2.8 (K2: 120–22); *Temp.* 1.4 (K1: 534).
[18] *Temp.* 1.6 (K1: 547).
[19] Ibid., 1.8 (K1: 559).
[20] Ibid., 1.9 (K1: 559–71).

best-tempered part of man and man the best-tempered of all animals.[21] The skin's absolute well-temperedness permits it to act as a
standard or measure by which to assess less well-tempered objects.
The other tissues are less evenly tempered than the skin and Galen
supplies considerable detail about the varied blends of the several
simple parts. The individual life span is a progression from the
warmth and moisture of blood and semen to the cold and dryness of
old age and ultimate death. But it is not true to say that children are
simply warmer than adults. Rather the warmth of children is "more
moist, abundant, and pleasant, that of adults sparse, dry, and less
pleasant."[22]

Nutrition. "Surely," Galen asserts, "this, if anything, is characteristic of living creatures, viz., that they can form other things similar
to themselves."[23] This they do in two ways, through generation and
through nutrition. Nutrition, in Galen's system, is a qualitative approximation of the nature of food to the nature of the body: a step
by step assimilation of that which feeds to that which is fed. "For
just as fire makes wood like itself, taking from the wood both its beginning and its nourishment, in the same way both plants and animals
are observed to assimilate their food to themselves."[24] The sequence
of events is superficially much as we should list them today: mastication and salivary action,[25] swallowing,[26] gastric alteration,[27]
absorption from the stomach,[28] pyloric expulsion,[29] absorption from
the intestine,[30] blood formation,[31] vascular distribution,[32] selective
absorption[33] from blood by the various parts of the body and finally,
inside the body parts, ultimate or actual nutrition involving three
steps: presentation, adhesion, assimilation.[34] The final transforma-

21 Ibid., 2.1 (K1: 572).
22 Ibid., 2.2 (K1: 598).
23 *De Sem.* 1.11 (K4: 553).
24 *De Plac.* 6.6 (K5: 554).
25 *N.F.* 3.7 (K2: 163).
26 Ibid., 3.8 (K2: 168–74).
27 Ibid., 3.4,7 (K2: 155, 162–66); *U.P.* 4.7 (K3: 275–82).
28 *N.F.* 3.7 (K2: 160–61); and especially *U.P.* 4.2 (K3: 268–70).
29 Ibid., 3.4 (K2: 153–54).
30 Ibid., 1.13, 16, 17 (K2: 39, 62–63, 69–75); *U.P.* 4.17 (K3: 323).
31 *U.P.* 4.2,3 (K3: 268–70).
32 Ibid., 16.10–14 (K3: 313–45).
33 ". . . each part is nourished by a nutritive fluid most similar to itself in substance, and . . . its nature equips it to prepare the fluid for itself by altering and
transforming it. . . ." *De Plac.* 6.8 (K5: 571).
34 *N.F.* 1.11 (K2: 24).

tion involves a resolution of the aliments into their elements as an important step in the act of assimilation.[35] Galen is at pains to point out that nature "shapes, increases and maintains (the parts) through and through, not on the outside only" unlike "Praxiteles and Phidias and all the other statuaries [who] used merely to decorate their material on the outside. . . ."[36]

In elucidating the nutritive processes, Galen at several points compares food with fuel. Different foods warm the body either immediately (as pitchy wood quickly feeds fire) or only after suitably prolonged preparation (as is necessary where wet wood is to be burned). But food has likewise a plastic function in that it replaces tissues displaced through wear and tear. It was suggested in chapter 5 that biology has a thematic character, and that one of its most persistently recurrent themes is that life is intimately involved in simultaneous processes of destruction and reconstruction, emptying (*anachoresis*) and filling (*plerosis*), waste and repair. Galen tells us that "part of the substance of every (living) thing is continuously flowing away; whence every such thing needs to be nourished and that which is added from the blood must resemble what was displaced."[37]

Galen's accounts of all these activities, given in various works, are empirically and theoretically impressive. In the stomach occur separation of good from bad and a qualitative alteration and first approximation of the food to the character of the body substance. Part of the end product is used to nourish the stomach itself. The rest is partly absorbed and partly passed along to the intestine.[38]

There is one respect in which Galen's idea of a nutritive chain of events differs radically from our own. He sees each step as valuable not primarily because it permits the next step but because it benefits the organ in which it occurs. Gastric *pepsis* feeds the stomach; the unused, residual chyle is then converted into blood. Blood formation feeds the liver; unused leftover blood is available for general distribution. But this is not to say that the stomach does not contribute to—indeed, exist for the sake of—the liver. Galen is insistent that each of the organs both functions intrinsically and contributes to the

[35] *De Sem.* 1.11 (K4: 553–55).
[36] *N.F.* 2.3 (K2: 82); *U.P.* 17.1 (K4: 352).
[37] See also T. S. Hall, "Life as Opposed Transformation," *Journal of the History of Medicine and Allied Sciences,* 20 (1965): 262–75.
[38] *N.F.* 3.7 (K2: 161–65); *U.P.* 4.5, 6 (K3: 272–75).

whole. This distinction between function and use occasions a whole separate book on the latter topic (*The Use of the Parts*) and there are frequent accounts (e.g. in the *De Semine*) of the remarkable reciprocity of the parts.

The humors. We have already suggested briefly the historic role of the humoral doctrine. It is in Galen's formulation that the doctrine reaches and influences Renaissance medical and physiological thought. Galen's humors follow those of the Hippocratics in that each humor is characterized by two of the four fundamental *poiotetes*. Yellow bile, like fire, is dry and hot; blood, like air, is moist and hot; phlegm, like water, is cold and wet; and black bile, like earth, is cold and dry.[39] The humors are formed out of absorbed nutriment primarily in the liver. The ratio of yellow bile to blood to phlegm depends, in general, on heat. Yellow bile appears in greatest quantities "at the warm periods of life, in warm countries, at warm seasons of the year, and when we are in a warm condition; similarly in people of warm temperaments, and in connection with warm temperaments, and in connection with warm occupations, modes of life, or diseases."[40] Again, temperate heat conduces to blood formation, and cold to the production of phlegm. These activities take place through the agency of the innate heat of the liver.

As absorbed nutriment moves from the portal vein into the liver it is resolved into three components comparable to *wine* (blood), the *foam* or *flower* of the wine (yellow bile), and the *leas* (black bile).[41] Yellow bile is attracted to the gall bladder where it is altered and assimilated, and some is sent to the intestine in order to drive off any excess phlegm that may be there and—Galen seems to say— to accelerate the onward movement of the contents of the alimentary tract. Black bile is assimilated by the spleen but some goes on to the stomach and causes that organ to contract upon and retain the food long enough to effect a thorough concoction.[42] The liver attracts the thus purified blood and partly assimilates it to its own nature by a process not dissimilar to clotting.[43] Blood leaving the liver

---

[39] Ibid., 2.9 (K2: 129–30); *De Plac.* 5.4. (K5: 676).
[40] *N.F.* 2.8 (K2: 122).
[41] *U.P.* 4.3 (K3: 270) and 12 and 13 (K3: 296–311).
[42] Ibid., 5.4 (K3: 358).
[43] Ibid., 4.12 (K3: 298).

contains enough water to give it needed fluidity; this will later be partly separated from the blood in the kidneys as urine.[44]

These activities illustrate what we have already noted to be a basic organizing idea that recurs often in Galen's physiology. Each organ has an attractive power, a transforming or assimilative power, and an expulsive power to rid itself of unwanted residues. The stomach attracts food from the mouth; the liver, chyle from the intestine; the gall bladder, bile from the liver; the tissues, nutriment from the blood. Each subjects what it attracts to an assimilative transformation, expelling what is unused and potentially toxic.

From the foregoing and other considerations it follows that the term blood can be used in two ways. In one sense, blood is just one of the four humors. In another, it is a mixture of them all. In total blood, the sanguineous humor predominates over the other three, a fact which accounts for the blood's color, but proper smaller amounts of the others are present as well.[45]

Of the blood entering the heart from the vena cava, part is drawn across the interventricular septum from the right ventricle to the left whence, as a more golden sort of blood whose ingredients are rarer (*leptomeres*), it is distributed through the arterial system.[46] This arterial blood contains *pneuma*—to whose origin and significance we shall return.

Galen writes of humors many times and in a variety of contexts. The results, though not always easily fitted to a common scheme, are not ultimately inconsistent. Humors appear now as balanced constituents of the tissues including even the solids; now as constituents of whole blood or of the maternal blood that turns into or at least nourishes the solids in the embyro;[47] now as noxious superfluities that the physician must extract with proper chymogogues;[48] now as useful secretions that travel to the intestine and help it perform its several tasks in an efficient manner.

A point which offers difficulty in Galen's thought—and which

---

[44] *U.P.* 4.5–6 (K3: 272–73). See also O. Temkin, "A Galenic Model for Quantitative Physiological Reasoning," *Bulletin of the History of Medicine,* 35 (1961): 470–75.

[45] *De Plac.* 8.4 (K5: 671).

[46] *De Plac.* 6.8 (K5: 572).

[47] *Hippocratis de natura hominis* 18 (K15: 59).

[48] *Temp.* 2.6 (K1: 630).

one hopes that a good modern Galenist may clarify—is whether he sees humors merely as nutritive precursors of the tissues or as, also, persistent components of them. We hear that, as Hippocrates rightly realized, each part attracts a juice as similar as possible to itself—"juice" meaning blood, thicker or thinner, warmer or cooler, more or less phlegmatic or atrabilious.[49] Yet in discussing the physical differences between tissues, Galen often omits any reference to humors; it is the blend of qualities (hot, cold, dry, moist) and not of humors that determines the function of the part in question. In Galen's important work *On Temperaments* (blends), the humors are scarcely mentioned, and, where they are, they appear as excrements or pathological end products. The role of the humors is apparently secondary, or instrumental to the end of producing the proper qualitative blend.[50] Yet the tissues can become unblended, so to speak; the blend can be resolved, on occasion, into pure humoral end products that constitute a pathological, often painful condition.[51]

Generation and Development

Reproductive physiology after (and for that matter before) Galen is based on the effort to answer two major questions. Through what mechanism is the developing individual transformed from a simple to a complex condition? And, what mechanism assures that the resulting complexity will be of the proper sort (i.e. like that of the parents)? Each new theorist answers these questions in terms set partly by his own fund of embryological knowledge and partly by his ideas of life and matter and their relations. We shall expect Galen, for example, to answer them in terms of a substantial empirical knowledge of embryos, but also at least partly in terms of his own interest in physical and physiological "faculties" or "powers."

Three such powers or sets thereof are required, he informs us; one set for the *de novo* production, or genesis, of the body parts; another for their subsequent growth; and a third for their maintenance after they stop growing.

[49] *De Sem.* 1.11 (K4: 554).
[50] Though at times (e.g., *De Plac.* 8.5 [K5: 686–87]) he seems to say that the humors compose man in the sense of continuing constituents.
[51] *De Plac.* 8.4 (K5: 677).

Genesis is a compound operation involving alteration (*alloiosis*) and molding (*diaplasis*). Galen has in mind here processes not totally unlike those to which twentieth-century biology will refer by the terms *chemodifferentiation* and *morphogenesis*. Indeed, in his approach to embryology, Galen generates or reinforces some subsequently powerful and durable traditions. Among them, one of the most prescient is the conception of an "underlying substance" on which the diaplastic faculty can do its work, a substance which "one would be justified in calling the material of the animal, just as wood is the material of the ship and wax of an image."[52]

The idea of an initially more or less homogeneous substrate which is capable of differentiation has been vaguely foreshadowed in Plato's *panspermia,* and Aristotle has assigned a comparable role to the menstrual blood of the mother. The idea looks ahead to the crystalline colliquament of William Harvey (1651); to the homogeneous formative substances that will be postulated by Wolff, Treviranus, and Mirbel; to protoplasm in the sense originally proposed by Purkinje. These substances will differ radically in conception but all will meet the demand for an undifferentiated substrate capable of developing into a differentiated individual. They may be grouped roughly into two classes depending on whether their differentiation is thought of as spontaneous or imposed. Galen makes the latter assumption, though there are differences in the details of his several accounts of generative activity. It is relevant to our present interests to review briefly each of two major theories that he puts forth on the subject of generation.

In treating this subject in his work *On the Natural Faculties,* Galen (like Aristotle) makes the starting substance the mother's menstrual blood (emphasizing elsewhere that this would be total, not merely humoral, blood).[53] It differentiates not spontaneously but only if acted upon by "alterative faculties" introduced with the semen of the male. There is not one alterative faculty, merely, but rather a whole set—one for each homoeomerous part or tissue. An alterative faculty creates a tissue by effecting differential local blendings of the four fundamental qualities, since it is with respect to these that tissues differ. It is to be noted that menstrual blood is "a single uniform

[52] *N.F.* 1.5 (K2: 10–11).
[53] *De Plac.* 8.4 (K5: 672).

matter" subjected to the artificer,[54] for "it is not for the wax (the blood) to discover how much of it is required: that is the business of Phidias," that is, of the alterative and plastic faculties of semen.[55]

In other treatments of generation and early development, Galen gives a more detailed and somewhat variant picture of initiative events. He insists upon an empirical approach and expresses contempt for those who "rely on conjecture."[56] The basic questions for him seem to be: what is the visible succession of events? and what are the controls?

He asserts that semen with its content of *pneuma* is coagulated by contact with the uterus which contracts protectively around it.[57] To this initiative event there are two sequels. First, surface coagulation encloses that portion of the semen that is destined to form a fetus. Galen twice quotes Hippocrates' story of the singing slave girl who is induced on the sixth day of pregnancy to abort an encapsulated mass looking like a raw egg with the shell stripped off.[58] Second, the membrane formed from coagulated semen, the chorion, comes into communication with the terminal orifices of uterine blood vessels of the mother, and at each point of contact a tiny vessel is formed within the seminal mass.[59] Drawing blood, *pneuma*, and heat from the mother, these small vessels grow inward and unite to form the larger vessels of the umbilical cord: these now penetrate an inner capsule, the amnion, whose contents will become a fetus. The umbilical vein arborizes in the manner of a tree trunk putting forth branches. The process of arborization is like a hollowing out of the semen.[60] On the subdivisions of one major branch, the liver develops by a coagulation of uterine blood; along the subdivisions of the other are formed the spleen, omentum, and abdominal divisions of the alimentary canal.[61] Beyond the liver, the vena cava branches further to permit the formation of the other viscera.

The heart is formed later, the exact time being unknown, from blood coming from the mother either arterially or through the umbilical vein to the liver and vena cava. Galen stresses frequently that

[54] *N.F.* 2.3 (K2: 83).
[55] Ibid., 2.3 (K2: 84).
[56] *Form Fet.* 1 (K4: 652).
[57] *De Sem.* 1.4 (K4: 524–25).
[58] Ibid., 1.4 (K4: 525–26); *Form Fet.* (K4: 654).
[59] See *U.P.* 15.4 (K4: 224–25).
[60] Galen also describes the allantois; see *De Sem.* 1.8 (K4: 540).
[61] *Form Fet.* 3 (K4: 660–61).

the heart is the warmest of the viscera, whence its origin is probably due to warm arterial blood. That the liver should be formed first is reasonable since the fetus lives at first a plantlike (merely nutritive) existence and has no need of a heart or pulsing arteries. The third major organ in Galen's triad, the brain, is formed from semen still later than the heart—a fetus having no need for sensation, movement, memory, imagination, or reason.[62] From the liver, then, the veins grow out like branches of a tree; from the heart the arteries do likewise; so do the nerves from the brain.[63] In general, everything fleshy is made from blood, everything solid or membranous from semen.[64]

Two related problems that dominate the interpretive biology of every era are the questions (*a*) how the organization of the individual is established and (*b*) how, once established, it is sustained. These are questions of control, and they are questions to which two sorts of answers are ultimately available: either the controls are imposed on the living system by some external agency of a psychic or teleological nature or else they are an emergent or immanent property of the system itself.

In the matter of controls, Galen weighs a series of alternative explanatory concepts—the unsatisfactory Epicurean idea of improvident atoms (emergentism), the idea of seed as divinely created and abandoned by its creator to develop on its own, the idea of the parts as animate beings that know the will of the governing *psyche,* of semen as matter used by the formative soul, or as the vehicle through which the formative soul uses catamenial blood. Galen finds none of these nor any other solution entirely worthy of credence. He can state only that the formative cause is highly skillful and wise and that the fully formed animal is governed during its life by those mysterious motive principles that arise from the brain, the heart, and the liver.[65] Galen's embryology like his cosmology in general expresses the belief in natural design.

Having built the fetus out of male semen and maternal—though not catamenial—blood, Galen adds another fluid component. From the ovaries, the spermatic vessels (oviducts) bring a female semen into the uterus. This substance serves to nourish the male semen and

[62] Ibid., 3 (K4: 672).
[63] *De Sem.* 1.9–10 (K4: 545–52).
[64] Ibid., 1.11 (K4: 551).
[65] *Form Fet.* 6 (K4: 699–701).

enter with it into the formation of membranes that anchor the fetus to the uterine wall.[66]

Galen pays detailed attention to the castration syndrome in men and animals, including birds, using it partly to refute the insistence of Aristotle and others that semen is formed in the spermatic vessels rather than the testes. Not only do the testes produce semen, in Galen's view; they also are a fountain of strength and augment the activity of the heart in supplying the whole body copiously with heat. The heart is a source merely of life; the testes, of the good life.[67]

As to the manner in which the testes might cause distant effects— the growth of hair on the chin, the deepening of the voice, and the like—Galen proposes that they distribute not the semen itself but a quality or power of the semen, just as the brain distributes to the nerves a power of sensation or motion, and the heart to the arteries a power of pulsation. "This power produces strength and virility in males," we are told, and "femininity in females. . . ."[68] Galen compares the distribution of a power to the sun's radiation of light.

In immoderate intercourse the testes place inordinate demand on their blood vessels for the juice out of which they make semen. As these vessels are exhausted they communicate the demand to other vessels until the whole body is depleted not only of the spermatic ingredients but of *pneuma;* the consequence is debilitation.

As the body wears away and must be replaced, the fleshy parts initially formed from blood are replaced from blood. The solid or membranous parts initially formed from semen are replaced from semen (presumably from a seminal juice that separates from blood). Fleshy parts regenerate; of those produced from semen, only—and rarely—the veins are able to do so.[69]

As to sex, male embryos are found in what we should call the right horn of the uterus (Galen presumably was extrapolating here from a goat or like animal), the females are found in the colder left horn. Because of the lack of warmth the female develops less perfectly, with the organs inside for reception of the sperm and for care of the developing foetus. If the pregnant female were fully warm the foetus would cause her to dry up and disintegrate. Nature thus wisely pro-

---

[66] *De Sem.* 1.7 (K4: 536–38).
[67] Ibid., 1.15 (K4: 573–75).
[68] Ibid., 1.16 (K4: 584–85).
[69] Ibid., 1.15 (K4: 588–89); also 1.11 (K4: 551–52) and 13 (K4: 557).

vides for every contingency.[70] Both right ovary and right testicle are productive of males because the blood they receive is warmer and less excrementitious. It should be noted thus that in Galen's system the male produces both male- and female-begetting seed. There is a tendency for the right hand tube of the female to attract seed from the right testicle of the male and vice versa.

The Pneumata *and Respiration*

Historically, the *pneumata* may be regarded as a useful fiction that seemed (to their inventors) to explain a variety of vital phenomena. The term inventors, in this connection, is perhaps misleading since, as noted earlier, scientific "pneumatism" evolves out of prescientific intuitions associating air with life. A modification of Galen's ideas on this subject will reach Renaissance Europe via Arabia, and will seem useful to physiologists and physicians until about the middle of the eighteenth century.

Galen's own pneumatology may be approached by contrasting it with that of Erasistratus in opposition to whom, partly, Galen offers his own views on this subject. (It is approached in this fashion by L. G. Wilson whose recent study of the subject is illuminating).[71] According to Erasistratus, vital spirit (*pneuma zōticon*) is derived from inspired air in the lungs and finds its way into the arborizations of the pulmonary veins. Erasistratus compares the heart to a brazier's bellows that can suck in air through one opening and squeeze it out through another. (Harvey will later use the bellows analogy but will see the heart as pumping a liquid; hence, Harvey's will be "water bellows"). When it dilates, according to Erasistratus, the heart draws vital spirit into itself from the pulmonary vein; when it contracts, it drives the same vital spirit through the arteries to the body.[72] Arteries carry *pneuma,* then, but, except under abnormal circumstances, they do not carry blood.

No single unequivocal summary can be given of Galen's pneumatology because his own accounts of it vary. One clear enough point—

---

[70] *U.P.* 14.6–7 (K4: 158).

[71] L. G. Wilson, "Erasistratus, Galen, and the *Pneuma*," *Bulletin of the History of Medicine,* 33 (1959):293–314. For the question whether there are two *pneumata* or three, see O. Temkin, "On Galen's Pneumatology," *Gesnerus* (Aarau), 8 (1950): 180–89. For a general summary of pneumatic doctrine, see G. Verbeke, *L'Évolution de la doctrine du pneuma du stoicisme à S. Augustin* (Paris, 1945).

[72] *De Plac.* 6.6 (K5: 548–50).

and one in which he differs from Erasistratus and Chrysippus—is that blood is present in both the ventricles of the heart and in the arteries as well as the veins.[73] We say "present in" rather than "flows through" because Galen views the arteries as filling not from one source so as to produce a simple current but from many sources: he insists that the venous and arterial systems are connected throughout the body by invisible anastomoses through which blood moves from veins to arteries whenever the arteries dilate.[74]

A similar situation exists in the heart where we noted that a certain rarer and "more golden" portion of the blood moves from right ventricle to left through these invisible pores which Galen notoriously imagined to be present in the interventricular partition.[75] Galen is, then, as much interested in the *filling* of the arteries as he is in the *flow* that occurs through them, though he regards both arteries and veins as distributors of innate heat and, to some extent, of vital *pneuma*. Where, then, does vital *pneuma* come from?

Sources of vital pneuma. Galen gives different (but not irreconcilable) answers to this question in his different discussions of the subject. He makes much of the fact that inspired air undergoes a first elaboration into vital *pneuma* in the lungs.[76] This *pneuma* enters the arborizations of the pulmonary veins and through them is attracted to the left ventricle for distribution by way of the arterial system.[77] Galen tells us, further, that the heart and arteries—especially those of the reticular plexus of the brain—continue the task begun in the lungs of elaborating air into vital *pneuma*.[78] We hear that the dilating heart attracts air (containing *pneuma*) from the lungs, uses it to sustain and moderate the innate heat of which the heart is the principal focus and, on contracting, sends it back to the lungs loaded with sooty waste vapors.[79] When blood and vital *pneuma* move out of the heart and into the arterial system they do so less because of the

[73] *An in Art.* (K4: 703); *U. Puls.* 5 (K5: 165); *U.P.* 6.17 (K3: 493).

[74] *U. Puls.* 4–5 (K5: 163–65); *U.P.* 6.10 (K3: 455) and 17 (K3: 492).

[75] *N.F.* 3.15 (K2: 207–8); *U.P.* 6.17 (K3: 496).

[76] *U.P.* 7.8 (K3: 541).

[77] For recent discussions of Galen on the movement of blood, see D. Fleming "Galen on the Motion of the Blood in the Heart and Lungs," *Isis,* 46 (1955):14–21; A. R. Hall, "Studies in the History of the Cardiovascular System," *Bulletin of the History of Medicine,* 34 (1960): and n. 86 below.

[78] *U.P.* 7.8 (K3: 541–42).

[79] *U.P.* 6.17 (K3: 412); *U.R.* 3 (K4: 491–92).

contractile and expulsive force of the heart than because of the attractive faculty of the dilating arteries.[80]

But the arteries have an additional source of vital *pneuma*. Whenever they dilate they aspirate not only blood from the veins but also air entering through imperceptibly small "mouths" that exist wherever an artery terminates on the body surface.[81] We hear that this occurs for the sake of three functions, namely (*a*) the cooling and (*b*) the ventilation of the innate heat and (*c*) the generation of psychic *pneuma*.[82] The arteries presumably alter this air into vital *pneuma* by the same elaborative process that occurs in the lungs and in the heart (Galen compares this change with that which the liver induces in food when turning it into blood).[83] For completeness we should add a third source of vital spirit, because Galen mentions without elaboration that vital spirits are nourished not only from respiration but also from blood.[84]

When the arteries contract they squeeze a certain amount of sooty vapor out through pores in the skin and of vital *pneuma* into the venous system. One might say that through their anastomoses with the veins, arteries trade a portion of *pneuma* for a portion of venous blood. In the lung a different situation prevails. When the lung collapses in response to thoracic contraction, vital *pneuma* is forced from the pulmonary vein with blood primarily into the left ventricle of the heart whence both proceed to the aorta.[85] Meanwhile, blood in the arborizations of the pulmonary artery does not return to the heart. It keeps moving unidirectionally through the lung from artery to vein. Galen had a definite idea that at least some blood passes from arteries to veins in the lung.[86]

Why respiration? As to the use of respiration, Galen says that it

[80] The chief effect of systole is to drive air loaded with sooty vapors back to the lung.

[81] *U. Puls.* 4–5 (K5: 163–66).

[82] *De Plac.* 8.8 (K5: 709).

[83] *U.P.* 7.8 (K3: 539–40).

[84] At *De methodo medendi* 12.5 (K10: 839) Galen says that "it is not unreasonable to suppose that the nourishment [of the vital spirit] stems mostly, to be sure, from [the air of] respiration, but also from blood."

[85] *U.P.* 6.10 (K3: 455–57).

[86] See R. E. Siegel, "The Influence of Galen's Doctrine of Pulmonary Bloodflow on the Development of Modern Concepts of Circulation," *Sudhoff's Archiv für Geschichte der Medizin,* 46 (1962):311–22.

"pertains not to a single action but to life itself."[87] Is respiration the genesis of the life-soul (Asclepiades), or in some sense a reinforcer thereof (Praxagoras), or a refrigerant agent (Philistion, Diocles), or both a nutrient and a refrigerant agent (Hippocrates [!]), or a way of getting *pneuma* into the arteries (Erasistratus)? These ideas meet, to some extent, in making respiration indispensable to innate heat, and this, in Galen's view, is its primary use.

He argues elaborately that some quality of the air, rather than its substance, is necessary to the innate heat.[88] Comparing heat production in the body to the production of flame in a lamp, Galen sees air as necessary for supplying just the right degree of (*a*) ventilation (fanning) and (*b*) cooling. The blood, or something in it, is comparable to the oil of the lamp. Exhalation serves to remove the physiological equivalents of soot. An American historian of our own times, Donald Fleming, has made it clear that Galen sees sooty waste as going from heart to lung by way of the pulmonary vein and not the pulmonary artery as many earlier interpreters supposed.[89]

We breathe more deeply during exercise because of the increase of warmth caused by vigorous movements, for "the greater flame needs more air and the smaller needs less." In growing children, innate heat is greater than in adults, and they consequently need more air. While growing, children produce more fuliginous vapors than adults produce, and these vapors must be removed through adequate exhalation. Finally, Galen considers it possible that air converted into vital *pneuma* nourishes the psychic *pneuma* which, during growth, is produced in ever-increasing amounts.

In addition to the vital *pneuma*, Galen acknowledges (along with Erasistratus) another, or psychic, *pneuma*, formed in the brain and distributed by the nerves. Some have equated psychic *pneuma* with the soul, but Galen thinks that the psychic *pneuma* can leave the body momentarily without irrevocable loss of life, and hence is the instrument (*organon*) of soul and probably not to be thought of as soul itself.[90]

Galen argues often and strongly for the brain and against the heart as the site of the motive-perceptive rational soul. Application of pres-

[87] *U.R.* 1 (K4: 470).
[88] Ibid., 3 (K4: 484).
[89] Ibid., 3, 4 (K4: 487); for Fleming, see note 77.
[90] *De Plac.* 7.3 (K5: 606).

sure to the brain may cause paralysis and anesthesia, but the heart can be vigorously pressed without any such effect. Psychic *pneuma* is distributed, then, from the brain and (contrary to Chrysippus) not from the heart.[91] Careful observation shows that nerves arise from the brain. Those like Praxagoras who have said that nerves arise from the heart either were blind or listened to other people who were. Praxagoras was wrong too in his supposition that nerves are terminal continuations of arteries. The brain not the heart is the seat of the soul.[92] Excoriated animals can cry out and even run whereas a bull when the walls of the brain ventricle are exposed or compressed loses voice, breath, and sense and motion in general.[93] The brain is the "source of all organs, called nerves, that transmit sensation and voluntary motion to all parts of the animal, and the heart is the source of the arteries." Disease or damage to heart and brain have independent consequences.[94] The heart is not without its own psychic aspect, however; it is the site of such emotions as anger, distress, anxiety, joy, and cheer.

The psychic *pneuma* has three possible origins of which Galen assigns importance now to one and now to another. We are told on one occasion that most *pneuma* is formed out of air inhaled directly from nasal passages to the brain, as suggested by Hippocrates' disciples.[95] Elsewhere Galen allows for a lesser amount from this source, and seems inclined to other ideas.[96] First, we noted earlier Galen's idea that "air, drawn in from outside by the bronchioles undergoes elaboration first in the lung tissue, then afterwards in the heart and arteries especially those of the reticular plexus (of the brain), and finally and most perfectly in the brain itself where it becomes psychic *pneuma, sensu stricto.*"[97] Galen seems to distinguish this idea, however, from still another (offered only as a possibility) according to which the psychic spirit needs for its nourishment an "exhalation" product of blood.[98] The exhaled vapor presumably passes from the brain's arteries to the ventricles of the brain itself. Exactly what Galen means by exhalation (*anathymiasis*) is not an easy question.

[91] Ibid., 1.6 (K5: 184).
[92] Ibid., 2.1, 6 (K5: 211).
[93] Ibid., 2.4 (K5: 238–39).
[94] Ibid., 3.6 (K5: 333).
[95] *U.R.* 5 (K4: 504–6).
[96] Ibid., See also *De causis respirationis* (K4: 466).
[97] *U.P.* 7.8 (K3: 541); *De Plac.* 3.8 (K5: 356) and 7.3 (K5: 608).
[98] *U.R.* 5 (K4: 502, 506); *U.P.* 6.17 (K3: 496).

It is the process that gives rise to dry or wet vapors. *Anathymiasis* is the word Heraclitus is quoted as using in his description of the formation of the cosmic soul.

Muscle action. There is much to admire in Galen's treatment of gross animal movement which he interprets in several treatises including one devoted exclusively to this subject (*On the Movement of Muscles*). He sees gross movements as resulting from nerve-incited contractions of antagonistic muscles.[99] The structures involved are nerve, ligament, muscle, tendon, and bone (or other, softer parts). Galen wishes to clear up the confusion that exists concerning the difference between nerve, ligament, and tendon. Nerves are sensitive; ligaments, insensitive. Within the body of a muscle the subdivisions of nerve and ligament combine, and emerge from the farther end as tendon.[100] Galen says—and we should agree—that the central event in muscular movement is the shortening of the muscle along its major axis. Muscle is to bone as mover is to moved, he says, and muscle action, like a lever, can be used to produce a mechanical advantage. Muscles are typically arranged in antagonistic pairs.

A fully contracted muscle must elongate before it can contract again but does so only passively, primarily in response to the contraction of its antagonist. If you sever a flexor, the limb is immediately extended by active contraction of the tensor. If you cut the tensor, flexion results from the unopposed contraction of the flexor. Hence, contraction (of either antagonist) is an active function, and elongation is merely passive.[101]

Whence, then, the source of contractile activity? Its source is double, in Galen's opinion. First, every muscle tends to contract almost maximally as a result of its intrinsic constitution; this tendency is shown by the prompt shortening of the two halves of a severed muscle, and by other experiments. Second, a psychic stimulus—conveyed through the animal spirit—can reinforce the muscle's intrinsic contractile propensity. "All muscles . . . require to receive a nerve from brain or from the spinal cord. And this nerve is small to behold but by no means slight in power. . . ." An alternate theory, Galen admits, would be that the psychic impulse *relaxes* one antago-

99 *De motu musculorum* 1.9–10 (K4: 411).
100 *De Plac.* 1.9 (K5: 204).
101 *De motu musculorum* 1.5 (K4: 391).

nist, permitting the other to contract; but this seems to be ruled out by the fact that neurosection produces a permanent relaxation of the innervated muscle.[102] One must admire the series of properly interpreted experiments on which Galen bases his theory of muscle action.

They permit, incidentally, at least a partial explanation of volition. If a movable bone is temporarily motionless, this will be due to the equal pulls of antagonistic muscles canceling one another mutually. But by reinforcing one pull or the other, the soul upsets the balance of forces and causes flexion or extension as needed. This analysis poses urgently, for Galenic as for later physiology, the question of the nature of nerve action.

Nerve action. Galen's theory of nervous function is more adequate than that of any earlier thinker. This point is clear in view of earlier developments that achieved a kind of culmination with him.[103]

Diocles (ca. 300 B.C.) and Praxagoras (ca. 300 B.C.) had given the psychic *pneuma* its source in the heart whose arteries distribute it to the body generally whereas the veins carry blood. Praxagoras supposed that as the arteries decrease in size their central openings are eventually obliterated; the arteries thus become solid *neura* through which *pneuma* can presumably move on to activate the muscles.[104] As to afferent conduction of sense perceptions to the central organs we have no sure testimony concerning Praxagoras' thought.

It remained for Praxagoras' pupil Herophilus to take the giant stride of identifying the motor nerves and their role in efferent conduction, as well as the sensory nerves which he probably regarded as conductors of a sensory *pneuma*.[105] The important point to note here is that Herophilus and his successor at Alexandria, Erasistratus, now returned the brain to the position of central organ of sensation and volition. Herophilus showed through dissection that nerves arise from the brain. Erasistratus's doctrine is already familiar to us. The

---

[102] Ibid., 1.8–9 (K4: 401–21).

[103] See Friedrich Solmsen, "Greek Philosophy and the Discovery of Nerves," *Museum Helveticum* (Basel), 18 (1961):150–97.

[104] *De Plac.* 1.6, 7 (K5: 184–89).

[105] Galen, *De Tremore*, 5 (K7: 605) (cited by Solmsen).

arteries distribute vital *pneuma* to the body generally, he said, including the brain. Here psychic *pneuma* is formed and travels outward over the nerves. Galen's ideas are notable especially because of the extent to which actual experimental evidence influenced his opinions.

Galen says that nerves arise from the brain or its extension, the spinal marrow.[106] Cranial nerves supply the head region as well as the intestines and certain other viscera. Other parts receive spinal nerves. Nervous action has its source in the brain since if a nerve or the spinal cord is severed everything peripheral to the cut loses its power of sensation, or motion. Nature distributes nerves with three ends in view—to give sensibility to sense organs, movement to muscles, and to all organs the ability to recognize any harm that they may experience. The brain is divided into anterior and posterior sectors, the former primarily sensory, the latter (cerebellum) exclusively motor. [107]

There are two sorts of nerves, "soft" (impressionable, or, as we should say, sensory) and "hard" (motor). To sever or ligate a nerve destroys its normal sensory or motor capacity. In the case of organs possessing double innervation (tongue, eye), damage to the motor nerve produces paralysis without impairing the sensory function, and vice versa.[108] We hear, however, that the distinction between hard (motor) and soft (sensory) is relative since, first, there is a third, intermediate type; second, everything harder than this intermediate type is hard and everything softer is soft; third, the harder a nerve, the better its disposition for motor action, and vice versa; and fourth, some sensory function may be exhibited by a hard nerve, and vice versa. The different sensory nerves resemble in quality the object by which they are aroused. Galen was not sure whether in motor action (*a*) *pneuma* flows on command from brain to muscle, or (*b*) cerebral pneuma "pushes" the resident pneuma of the nerve, or (*c*) there is only a "flow of potency" through the resident pneuma. He seems to discount the existence of sensory transmission. Rather, the nerve "is part of the brain" and the sense organ, receiving into itself the sensory power, is capable of conscious perception.[109]

[106] *De Plac.* 1.9 (K5: 205–6).
[107] *U.P.* 5.9 (K3: 377); 8.6 (K3: 637).
[108] *U.P.* 8.5 (K3: 633); 16.2 (K4: 270–72).
[109] *U.P.* 9.14 (K3: 740–41); *De Plac.* 7.4, 7 (K5: 612, 642).

Summary

Galen's endeavor was the finest fruit of the biology of antiquity, though it was less *generally* biological than the endeavor of Aristotle. This assessment seems fair despite the flaws in the Galenic legacy: his presentation was too elaborate and repetitive, his information often faulty, his basic assumptions and consequent judgments of others sometimes wrong. But beneath these weaknesses there remains a hard and massive body of information and interpretation that is for the most part as good as anything that antiquity has to offer. As for method, although Galen's experiments were often inconclusive, they were also numerous and relevant and were not to be matched in ingenuity until the seventeenth century.

There is much for the "present-minded" historian of the twentieth century to criticize in Galen's outlook—his teleology, especially, and his occasional infidelity to self-imposed standards of objectivity and experimental validation of his views. But it is instructive to compare his performance, as we shall have an opportunity to do later, with the performance fourteen hundred years later of, say, a Francis Bacon or a René Descartes whose protestations of scientific purity are much more emphatic than Galen's.

Perhaps Galen's most important contribution was his exhaustive working out of what we have identified in previous chapters as a standard, persistent program of physiological inquiry (a program in which physiologists are still engaged in the second half of the twentieth century). We have tried to show in this inadequate presentation of Galen that he deals with the "classic questions"—questions concerning nutrition, generation, perception, volition, and motion—and with the subquestions into which the general questions must be resolved if they are to be answered. He has an explicit—if derivative and synthetic rather than innovative—theory of matter (he is a biologist who uses physics, rather than a physicist who embraces biology). He quite definitely believes that vital processes depend on a special physical organization (involving qualities, elements, humors, tissues, organs and, as a special class, *pneumata*). Finally, he deals in explicit detail with the ways in which organization accounts for the visible manifestations of life-as-action.[110]

[110] The author is particularly indebted to Phillip De Lacy, Margaret T. May, Richard J. Durling, and R. E. Siegel for helpful suggestions in the interpretation of Galen's physiological ideas.

Part 2

TOWARD A NEW BIOLOGY

11

Emergence of Renaissance Science

Scientific inquiry is but one of the ways in which men of past centuries have acquired what seemed, to them at least, reliable and useful knowledge. Among other methods were augury and divination, revelation and mystical illumination, dialectic and disputation, and an uncritical assimilation of formal dogma or popular opinion. In our own day, a physicist or physician is scarcely expected to have side interests in alchemy or astrology, but his sixteenth- and seventeenth-century precursors enjoyed considerable latitude in such matters. It is sometimes surprising to discover how recently serious scientific thought has embodied what seem, from our perspective, nonscientific or pseudoscientific ingredients. Thus, even as scientifically oriented a figure as Robert Boyle (1627–1691) recommends for curing a nosebleed the placing of moss in the sufferer's palm—moss, preferably, grown on an Irishman's skull. Slightly earlier, the great William Harvey had treated tumors—or other "excrescences"—by the laying on of dead hands, while Harvey's senior contemporary, van Helmont, effected a variety of cures with magnets and other amuletic devices.

Since the above practices were still in vogue around or even after the middle of the seventeenth century, we shall not be surprised to find the considerably earlier physiologist, Jean Fernel (ca. 1550), prescribing such remedies as swallow ash and dog dung or the still earlier Paracelsus (ca. 1535) mixing his pioneering chemotherapeutic theories with substantial amounts of alchemy and astrology.

Yet in all of these thinkers and in most of their scientific contemporaries, elements of cold common sense coexisted with what would ultimately be considered crass superstition. In the sixteenth century, the situation was more than one of mutual tolerance between science on the one hand and a separate pseudoscience on the other. Often the two were used together in the interpretation by one and the same individual of one and the same phenomenon. This is especially apparent in the mixture of natural and supernatural interpretations in both biology and cosmology. Tycho's new star was a nova, but it was also, for Tycho, a sign. Kepler's elliptical orbits squared well with observable fact, but they were also governed by a World Soul resident in the central sun. We shall find that Paracelsus' pathological "beings" (*entia*) include both a natural sort (*ens naturale*) and a sort that is supernatural (*ens dei*). And that according to van Helmont natural objects develop from "seeds" which are "divinely imprinted with a sense of their own future nature." In biological books, many of them encyclopedic compilations, actual and mythical animals and plants are presented, often with little or no distinction between the two kinds. The mythical ones are derived from multiple sources, Greek, Latin, Alexandrian, Neoplatonic, Arabian, and cabalistic as well as from a vast reservoir of uncodified or semicodified popular belief.

A crucial feature of the scientific Renaissance, and this is the point we would chiefly stress, was to be the resolution of this mixture of science and the pseudosciences into separate components. There was a certain irony in this enterprise in that earlier medieval thought had been at deliberate pains, in some quarters, to create the syntheses which had ultimately to be dismembered. For the Schoolmen, and for none more conspicuously than Aquinas, a prime scientific goal had been to reconcile Greek psychobiology with ecclesiastical doctrine on soul. Friar Bacon had preached empiricism—but as a way of vindicating (and not of objectively testing) truth as revealed. Before it could be free to develop on its own, science had to be separated from its contaminants through a process of conceptual distillation.

There were in general four bodies of doctrine from which the new science had to disengage itself. They were: first, Christian dogma as it related to scientific questions; second, the revived legacy of Graeco-Roman biology and cosmology; third, the sister pseudosciences of alchemy and astrology; and fourth, the vast and confused imbroglio of medieval superstition as represented, for example, in magic, demonology, and witchcraft. The four cannot be thought of as separate and distinct. They formed a variously integrated doctrinal amalgam whose contents varied from decade to decade and from individual thinker to individual thinker.

We must reserve for treatment elsewhere, entirely outside the present studies, the extensive subject of medieval theories about the nature of life and of matter. We would make here, instead, a single point about these theories. Their most pronounced feature—and here science, religion, and magic come together—was that they based their image of the organism on their own characteristic conceptions of the latent or insensible causes of patent or sensible life-as-action. This point was important both theoretically and practically.

Theoretically it suggests a commonalty that unifies medieval thought and links it to the thought of earlier and later eras. Science, logic, revelation, superstition, and, one might justifiably add, esthetic intuition have all always assumed that essential reality was something hidden; all have sought to grasp some central secret or secrets, in the case of biology the secret or secrets of life.

Practically, this view is significant in that, in all periods, the control of visible nature has been attempted through the manipulation of its assumed invisible causes. If the cause of an epidemic is something having to do with penitence and redemption, the practical response will be one thing; it will be something quite different if the cause is, say, a *contagium vivum,* or, again, the handiwork of witches. What must be remembered is that in medieval and early Renaissance thought, the supposed latent causes of patent vital phenomena, normal and abnormal, were—in derivation—variously empirical, religious, formal-logical, doctrinal, magical, or usually some combination of these—the mixture varying from one thinker to another.

The Soul in Greek Biology and Christian Doctrine

In Christian dogma on life and matter, one influence deserves particular notice, namely, the peculiar elaboration of the official doctrine on soul. Many aspects of the church's conception of soul, including the aspects of penitence and redemption, were pre-Christian in derivation. But there was an evolving emphasis on sacred as opposed to other aspects, and a detailed elaboration of them that made the Christian soul resemble less and less the physiological life-soul of the surviving—and revived—ancient science. There was an extensive literature, initially patristic and later Scholastic, arguing the compatability vs. noncompatability of the two mainstreams of thought about the soul. That story is too complex to be developed here. But we may note at least the point that certain strains were imposed on both notions of soul, but especially upon the Greek idea that it is the motive cause of life-as-action. The irreconcilability of the Greek and Christian theories of soul was to culminate in the physiological philosophy of Descartes (ca. 1635) who endeavored, with ultimate success, to eliminate the physiological life-soul entirely. In the three immediately succeeding chapters, we shall witness in Paracelsus and Fernel some of the preliminaries, and in van Helmont some of the concomitants, of the denouement precipitated by Descartes.

Biology and Alchemy

Alchemy, like Christian doctrine and ancient science, had its theory—or complex web of theories—about the covert equivalents or causes of overt phenomena, vital and other. The alchemists sought, like other scientists, to simplify and order the phenomena of nature by referring their varieties and vicissitudes to one or a few material first principles. These were typically, for the alchemists, mercury as the *prima materia* with sulphur as a necessary concomitant. A kind of quintessence of mercury—or in a variant theory—a proper union of the quintessence of sulphur with that of mercury—constituted the putative "Philosopher's Stone," a substance capable of turning base metals into noble ones, silver and gold. Such transmutation, often termed the *Magnum Opus,* was the central goal of much—but not all —alchemical endeavor.

Transmutation was accomplished in some variants of alchemical dogma by inducing small amounts, or seeds, of noble metals to behave as if alive, that is, by inducing them to undergo self-multiplication under the genetic influence of the Stone. In some schemes, the Stone itself was considered capable of self-replication. In others, it facilitated the replication of the seed. The analogy between this process and organic growth was suggested in the fact that the Great Work was often carried out in an egg-shaped vessel, symbolizing generation. The steps in the Work were subject to innumerable variations dictated partly by practical or technical considerations and partly by the complex body of sacerdotal, magical, astrological and occultish dogma with which the techniques became associated.

Another alchemical idea with biomedical implications was the belief that objects—both living things and those that we today consider living—comprise both a body and a quintessential spirit, the spirit being somehow responsible for body's uniqueness. In some things (e.g. wine), the spirit was supposed to be separable through distillation. In living things, there was in particular a special *aqua vitae,* or *elixir of life,* which had on sick or moribund living bodies a corrective effect comparable to the ennobling effect of the Stone on ignoble metals. The idea of a duality involving a resident *pneuma* had antecedents in Greek Stoicism which, some historians believe, underwent amalgamation in Moslem science with similar elements of Far Eastern alchemical (and cosmological) thought. The idea acquired many variants in the sixteenth and seventeenth centuries, where we shall encounter the belief that the body parts differ by virtue not only of the configuration of elements they contain but by virtue of a resident spirit, humor, or archeus (see Paracelsus, Francis Bacon, and even Boerhaave).

Summary

We must view medieval and early Renaissance theorists in the light of their own procedural assumptions if we are to appreciate their contributions to the problem that concerns us. Their ideas were largely products of mixed procedures which both molded and delimited their ability to deal with this problem. Their incomplete self-detachment from superstition, authoritarianism, and religion both directed and restricted their analytical capacities, as will become apparent in our brief encounters in the chapters to follow with three of the more colorful and influential later thinkers of the period.

Most importantly, the life-matter problem still presented itself to medieval and early Renaissance science in a way very different from that in which it was to present itself increasingly to the science of a later day. The tendency of late seventeenth-century and later science will be to view the world of life in a physical way, whereas early Renaissance science still follows the Greek mode of viewing the physical world as pregnant with life. For a Bruno, the cosmos lives; for Francis Bacon, matter perceives. Indeed, our story as it concerns the sixteenth century is partly the story of a painfully slow and confused transition from the medieval idea (that life is intrinsic to matter) to modern idea (that matter is intrinsic to life). This transformation goes hand in hand, we shall find, with the slow liberation of science from superstition and religion, and from the tenacious influence of Plato and Aristotle, of the Stoics, and of Galen.

· 12 ·

A Medieval Amalgam

PARACELSUS
(THEOPHRASTUS BOMBAST VON HOHENHEIM)
[1490 · 1541]

Modern science expects its theory builders to follow certain conventions. It admires, in its theories, a certain economy of formulation. It wants the different pieces of a theory to fit together, expects religion and magic to be omitted, and is exacting in the matter of evidence.

Paracelsus observed none of these conventions. What he says about matter and life, subjects which interested him passionately, must be understood in terms of his own intellectual style, a style often given to complexity, disarticulation, contention, self-contradiction, and dogmatic assertion. As an alchemist he felt justified in keeping some matters to himself. At times, we find him openly contemptuous of other thinkers. At other times, his contempt takes the form of a kind of jovial hoodwinking aimed at a particular sector of his public as though the author were in league with the rest of us to make fun of certain special fools.

As for nature, and naturalism, Paracelsus was their outspoken advocate, but his advocacy was impure. He mixed it liberally with biblical doctrine, natural magic, cabalistic allusion, alchemy, astrology, and folklore. As for intellectual economy, Paracelsus seems in perspective to have been less interested, despite his dogmatism, in any one right way of explaining things, than in throwing off an opulence of explanations as if to show the many ways in which a subject may be treated. Thus the modern reader, to appreciate and enjoy Paracelsus, must adopt a special intellectual posture. He must prepare to be inundated by an exuberant mixture of objectivity with fantasy, humility with arrogance, naturalism with superstition.

Paracelsus' style has to be understood, too, in relation to his rebellious temperament. We find him frequently fleeing, if not for his life then for his personal welfare and intellectual freedom, from community to community. Arriving in a new locality, he typically secured the patronage of sympathetic and powerful benefactors, as if for the very purpose of disingratiating himself later. The evidence on Paracelsus' personality is contradictory and often unreliable, but it is perhaps not insignificant that one of his seven *Defensiones* was written "To Excuse His Wondrous Ways and Wrathful Manners." If it means something to speak of creative maladjustment, Paracelsus manifests it on a heroic plane.

Once we understand the special context in which Paracelsus is to be received, he proves marvelously productive of ideas (including ideas of life and matter) some at least of which seem to form a semi-integrated pattern. What we obtain from them is not a clear or ordered understanding of nature. Paracelsus' intent was not to convey anything of the sort. To do so would not have fitted his theory of knowledge—if it may be called that. He saw knowing as, rather, a felt or intuited unity between essential aspects of objects and corresponding aspects of a man's "astral" self. This seemed possible to Paracelsus because he thought of man as a microcosmic extract of the constituents of the cosmos. Moreover, knowing seemed to him to entail an identification of the subject with *the science implicit in the object*. Not only subjects but objects, it seems, have knowledge—which, in the case of objects, informs them how to be what they are.[1]

Paracelsus' thought, here, is not completely different from the Greek view of *logos* as immanent in nature and of knowledge as a harmony or concordance between the individual soul and the order of the cosmos. For Paracelsus, as for truthseekers since the beginning, the world was dual: sensible and insensible. "For nature [reality] is a light that outshines the light of the sun . . . beyond the purview and power of the eye. And in this self-same light the unseeable may be seen. . . ."[2] Yet this double world of sensibilia or phenomena and insensibilia or cryptomena differs in certain ways from that of Greek science. Note, for example, Paracelsus' extensive involvement with

[1] For an interpretation of Paracelsus' theory of knowledge, see W. Pagel, *Paracelsus: An Introduction to Philosophical Medicine in the Era of the Renaissance* (Basel and New York, 1958), p. 53. This work is the most enlightening and reliable source of information on the endeavor of Paracelsus.

[2] "Die Bücher von den unsichtbaren Krankheiten," in K. Sudhoff and W. Matthiessen, *Paracelsus Sämtliche Werke* (Munich and Berlin, 1922–33), 9:253.

the alchemical doctrine of correspondence. Sensible, temporal things are analogs, or signs, of corresponding insensible, eternal things. And this correspondence has the effect of making Paracelsus a vocal, but not really faithful, empiricist. His recommended way of grasping knowledge is very different from the dogmatic and dialectical procedures that he despises. What is required is a thoroughly direct and practical experience with things themselves. You must read nature's book with your feet (you must go and see). You must ". . . cling to the practice of finding out whence the disease arises. . . . This practice will teach you."[3]

Where does this complexly grounded scheme fit the developing picture of interpretive physiology? It will appear, as the story unfolds, that, in his intentionally bewildering manner, Paracelsus treats questions with which we are already familiar (we have called them "classic questions"), and that his mutiple answers to these questions display a complex and imperfect parallelism with answers given by many predecessors.

THE COMPOSITION OF THE BODY

Paracelsus' position in time, and the unique character of his indoctrination, enabled him to choose among many past modes of interpretation. Not a bookish person, he was nevertheless impressively familiar with the views of his predecessors, Greek, Neoplatonic, Scholastic, occultist, alchemical, cabalistic, astrological. How, then, will he build his microcosm? Will he utilize the elements of Empedoceles? the humors of the Hippocratics? the *pneumata* of the Stoics or Galen? the sulphur-mercury duo of alchemy (to which, not uninfluenced by earlier thinkers, he adds a third principle, salt)? a zodiacal or planetary conception? The answer is that he both assails and utilizes virtually every existing analytical system. He composes man with a rich eclecticism that selects (and often radically transforms) ideas from a host of earlier thinkers. We should perhaps note especially that, even in his most concrete and corporeal vein, Paracelsus' viewpoint is far from a materialist one. He makes the sensible body a temporal and superficial expression of a more fundamental body which is insensible and eternal.[4]

[3] "Volumen Medicinae Paramirum," trans. K. F. Leidecker, *Supplements to the Bulletin of the History of Medicine* (Baltimore, 1949), p. 21.
[4] Pagel, *Paracelsus,* p. 54.

Origins

In the beginning, according to Paracelsus, everything existed, un-formed, in a homogeneous matrix (*Mysterium Magnum, Ilyaster*).[5] From this, the world arose by differentiation, or *separatio,* a process anticipated in the pre-Socratic idea of *ekkrisis.*[6]

As to the steps involved, these are variously explained in various treatises. The account is partly biblical, partly Greek, partly al-chemical. According to one account, everything began when "God was pleased to create water an element. . . ."[7] From water, earth and heaven were created. The contents of the story—and to some extent even the language in which it is told—remind us of the biblical ac-count of creation. Elsewhere, we get something more Greek. Ilyaster gives rise by *separation* to air, fire, earth, and water, in that order. Again, in a difficult passage, we hear that the elements arise out of a single body founded on the *tria prima* (mercury, sulphur, and salt) in which all things exist without as yet being formed.[8]

If there is a common thread in these accounts it is Paracelsus' subscription to the notion of some sort of initially undifferentiated starting substance. Such substances reappear repeatedly in both biology and cosmology from Greek to modern times. In Paracelsus' treatment, out of earth God created plants, animals, and man.[9] Man he created from a single matter containing an extract of all creatures, elements, stars, powers, essences. Hence, man is a microcosm and a quintessence.[10]

Present Condition

Man's present composition, like his origin, is described several times by Paracelsus and in several different ways. In general, the

[5] "Paramiri liber quartus de matrice," Sudhoff and Matthiessen, *Paracelsus,* 9:190.

[6] "Philosophia ad Athenienses," Sudhoff and Matthiessen, *Paracelsus,* 13:389–402.

[7] "Das Buch De Mineralibus," Sudhoff and Matthiessen, *Paracelsus,* 3:33.

[8] "Wie aber nun got beschaffen hat die welt, ist also. er hats in ein corpus gemacht, anfenglich, so weit die vier element gênt. dises corpus hat er gesetzt in drie stück, in mercurium, sulphur, und sal, also das do seind drei ding, machen ein corpus; dise drei ding haben in inen ist gelegen die miner, der tag, manch und warm und kalt, stein und obs und anders, aber noch nicht gformiert." "Philosophia de Generationibus et Fructibus Quattuor Elementorum," Sudhoff and Matthiessen, *Paracelsus,* 13:9.

[9] "Labyrinthus Medicorum Errantium," Sudhoff and Matthiessen, *Paracelsus,* 11:178.

[10] "Erklärung der Ganzen Astronomei," Sudhoff and Matthiessen, *Paracelsus,* 12:452–55.

body comprises: a firmament; elements; complexions or temperaments; and humors—to mention only the more or less material
ingredients.[11]

Firmament and Ens astrale. The corporeal like the celestial firmament contains several planetary bodies (see table 5). Just as the heavenly bodies constellate the firmament "so is man constellated might-
ily. . . ."[12] That the microcosm corresponds in this way to the
macrocosm does not suggest to Paracelsus, however, that influxes,
conjunctions, and the like control a man's propensities, ability, or
luck. These are, rather, controlled by his seed. Nor do they determine
his native longevity. His own corporeal firmament decides that. If
long life is predestined, the cycles are run through slowly. If life is to
be short, the same cycles are accomplished more quickly.[13] Yet,
though the heavenly bodies "do not endow us, nor draw us on, nor
shape our characteristics, you ought to note well in what they disease
and dispatch the body." This is explained, approximately as follows.

Life may be likened (he admits that the comparison is clumsy)
to a fire—the body representing the log. The linkage of life with
innate heat and of that heat with familiar fire is a tradition that we
have already encountered in our studies, most explicitly in Galen.
Moreover (again Paracelsus associates himself with old and persistent ideas) there must be something to keep life from consuming
the body. This something is an Astral Being (*Ens astrale*) which
corresponds to air approximately but not exactly. Everything earthly
and heavenly is sustained by the primal matrix or *Mysterium
Magnum* to which the stars add exudations which influence men to
good or ill. The sum of these exudations constitutes the *Ens*. From
the *Ens* we obtain the requisite cold, heat, dryness, moisture—but
also a variety of poisons. Various stars tend during their ascendancies
to oversalt or overacidify, to mercurize, to sulphurize the *Mysterium*.
Where these exudations penetrate to the earth, they cause disease in
susceptible persons. One poison harms one organ; another, another;
and different poisons cause different diseases.[14]

Elements in the body. Paracelsus approaches the problem of elements in varying contexts with varying outcomes. Where he speaks

[11] Leidecker, *Paramirum*, p. 45.
[12] Ibid., pp. 35–36.
[13] Ibid., p. 38.
[14] Ibid., pp. 13–22.

TABLE 5

CORRESPONDENCE BETWEEN
PLANETS AND ORGANS

Jupiter	Liver
Moon	Brain
Sun	Heart
Mar	Gall
Saturn	Spleen
Mercury	Lungs
Venus	Kidneys

of the classic four, they differ from the familiar ones in several fundamental ways. He probably regarded them not as ultimates but rather
as composites (of mercury, sulphur, and salt). Nor are they
individually uniform; there are many sorts of each of the four. Nor
are they material components of bodies in the classic sense; they are,
rather, generative or directive or functional principles of which one
only is ultimately expressed in a particular body. Where he speaks
of three rather than four, he, in general, ascribes combustion to
sulphur, transmutability to mercury, stability to salt.[15]

There are many other ways in which Paracelsian elements differ
from the standard ones. Thus, fire is not ordinarily visible in the body
(but becomes so when, for example, the eye is struck and we see
sparks). As for water, it is "lodged in the entire body, . . ." flowing
through all its channels, permeating all its parts. "Air is in the body,"
interestingly, "owing to the constant movement of the organs which
produces wind in the body. Just as the four winds of the world take
their rise, so it is to be understood in regard to these. . . . In like
manner, there are four elements in man. . . ."[16]

Much more might be said concerning the role of the several elements and of the principles. Man is not only created out of earth,
but derives his nourishment therefrom. Air keeps him from suffocating. Air also acts as a balsam or sort of cement to prevent dissolution by vital processes which consume man as does a fire. Elsewhere,
however, salt, or salts, constitute the balsam. Mercury may be thought
of as having a special role, since a specially fiery and perfect form
of mercury is the Philosopher's Stone or elixir through which life
can be alchemically synthesized. Again, the mother is the earth, the
womb the tree, the child the fruit, and as the tree supplies the four

[15] Pagel, *Paracelsus,* pp. 92, 99.
[16] Leidecker, *Paramirum,* pp. 42–43.

elements to the fruit, so the mother to the child. "Thus the microcosm, then, is a lesser cosmos and has in its body all the minerals of the world." (*Also ist nun die microcosma minor mundus und hat in irem leib alle mineralia mundi.*)[17]

What are we to make of this disconnected sequence of interpretive thrusts? This raises a basic question of Paracelsian scholarship, namely whether by a strong effort to extract the probable meaning of these disjointed fragments, one can find they add up to some sort of whole, something that Paracelsus "really believes." The question has not been settled, and, for the time being, we must accept Paracelsus' effusions as a series of loosely integrated stabs toward some kind of reductive understanding.

Humors and complexions. The classical four humors are, like many traditional analytical schemes, denounced by Paracelsus but utilized in modified form in the various schemes he pretends to substitute for them. The temperaments manifest themselves as certain complexions or "tastes"; bitter, acid, salt, and sweet. The correspondences are as follows:

TABLE 6
TYPOLOGICAL CONSIDERATIONS

TEMPERAMENT	TASTE	CHARACTERISTICS
choleric	bitterness	hot and dry
sanguine	saltiness	hot and moist
melancholic	acidity	cold and dry
phlegmatic	sweetness	cold and moist

Mention of humors here is conspicuously absent. Association with the familiar elements is also explicitly denied. All complexions may "reside in the body, while one will come to the fore." Elsewhere we hear that not all necessarily coexist; a body may contain one, two, three, or four.[18]

Paracelsus has much more to say about humors, much of it in opposition to the Hippocratic idea of them as constitutents. Nor are they the causes of diseases. The "ancients" described hundreds, thousands, not merely four. The body has indeed four humors, we

[17] "Paramiri liber quartus de matrice," Sudhoff and Matthiessen, *Paracelsus,* 9:210–11.

[18] Leidecker, *Paramirum,* pp. 43–44; see also "Die Beiden Bücher De renovatione et restauratione und Vom langen Leben," Sudhoff and Matthiessen, *Paracelsus,* 3:206–7.

hear, but not the traditional ones. Paracelsus substitutes four others, *das blut in adern, die feuchte im fleisch, der viscus im geeder, die schmelze in der feist* (blood in the veins, humidity in the flesh, viscus in the nerves, and fluid in the fat).[19]

Quite aside from the four humors, he mentions two other and, for our purpose, more relevant moist substances, viz. the *radical humor,* and the *liquor vitae.*

Radical humor and liquor vitae. A recurrent theme in the evolution of ideas about life and matter is the association of life with a special single substance. Some principle or matter is regarded as— more than any other—*the* living substance. Many different materials are so designated as the life-matter problem evolves, and these function differently as supposed substrates of vital activity.

In the pre-Paracelsian period, for example, there was much interest in a supposed radical moisture (*humeur radicale, humidum primigenium,* etc.). A substance, so designated, appears in Avicenna's *Canon.* Arnald of Villanova writes a pair of treatises about it, makes it the vehicle of innate heat, and cites Rhazes, Galen, and Theophrastus on the subject. Closer than anything Galen said on this subject was a similar idea in Hippo of Samos (ca. 450 B.C.) who, according to Anonymus Londinensis, "believes that there is in us a natural moisture whereby we perceive and by which we live" such that when the moisture "is in its normal condition, the animal is healthy, but when it dries up the animal loses consciousness and dies."[20]

For Paracelsus, the radical humor is a nonmaterial entity having the relation to the body that substance has to matter. It is associated, but apparently not identified, with vital spirit and with a certain *liquor vitae* (*v. infra*). This trio (radical humor, vital spirit, *liquor vitae*) coexists in the limbs as the tone exists in a bell—heard but not seen. There is a reciprocal relation between radical humor and the body in that the former both generates and is generated by the latter. When the body is renovated this occurs through a replacement of its corporeal component, and in this way, the bell's tone is improved.[21] The body falls away and is restored, but the moisture

[19] "De peste libri tres," Sudhoff and Matthiessen, *Paracelsus,* 9:604–5.

[20] W. H. S. Jones, *The Medical Writings of Anonymus Londinensis* (Cambridge, 1947), p. 53. The history of the idea of a radical humor is the subject of a separate study planned by the author.

[21] "De renovatione" (see above, n. 18), p. 201.

persists. Elsewhere in medieval medicine we hear debated the issue whether the radical moisture is gradually consumed from birth to death (see Fernel).

It is not possible to be sure how far Paracelsus would go, if pressed, toward equating the radical moisture with his *liquor vitae*. We noted earlier how, in the Greek scientific corpus, terms designating "life," "soul," "breath," "spirit," "air" had multiple and changing meanings which partially overlapped in a complex semantic network. Paracelsus presents the same complexity, all within his own use of terms. He is less interested in hard and fast distinctions and consistent usage than he is in shattering the rigor of ideas which he disbelieves.

In any case he bids the reader to turn his "attentions to the humor which is equivalent to the *liquor vitae* because the body lives by it. Concerning this humor it should be noted that there is a moisture which permeates the whole body. It is the life of the organs. This humor is an *Ens* in its own right. . . ."[22] He goes on to credit it with determining the quality of the "ore," i.e., with deciding how sound a body's constitution is to be. And we shall later note that the *liquor vitae* provides a way of explaining inheritance.

Organization. The *liquor vitae* bears on the important question of the precise way in which elements, humors, "planets" are put together, how their order is (*a*) established and (*b*) governed. We have already singled out this as one of the persistent classic questions with which biologists of every era concern themselves. It is perhaps, also, the most difficult question, the most resistant to material-mechanist answers. No other biological problem has been as elicitive of vitalistic solutions at this dual problem of the developmental origins and subsequent preservation of the individual as an organized being.

One of the most pregnant of Paracelsus' several oblique answers to this problem of organization is his postulation of certain special "beings" (*Entia*). One of these, an organizing *Ens seminis,* is viewed as present in the animal germ and in plant seeds as well.[23] The *Ens seminis* is abetted in its organizing role by the *Ens astrale,* noted above, and, in ways less than clear by a certain "Alchemist" whose

[22]Leidecker, *Paramirum,* p. 44.
[23] Ibid., pp. 13–16.

acquaintance we shall make when we consider nutrition. For the time being it may be noted concerning the Alchemist that he "divides the evil [*Venenum*] in the food from the good [*Essentia*], changes the good into a tincture, conditions the body so it will live, attunes the subject to nature, conditions nature so she becomes flesh and blood." This convenient agent is reminiscent in its varied competences of a certain Elixir posited more than a century earlier by Arnald of Villanova.[24] If the alchemist is not on the job, the *Ens veneni* takes over and putrefactive digestion ensues, and this it is "which indicates to us the disease of a person."[25]

Still another determiner of organization—or another metaphor for the same determiner—is a certain *Archeus,* variously presented as an inner "blacksmith," "vulcan," or "alchemist" that forges and maintains the organizational specificity of the individual. Subordinate *Archei* preside in the several organs.

Archei are not limited to living organisms. Comparable "alchemists" operate in the elemental matrices. Earth's *Archeus* makes plants and animals—and minerals—grow. This terrestrial *Archeus* has, moreover, *subarchei* corresponding to those in the human body. To restore a given organ, the physician utilizes a drug whose *subarcheus* corresponds to that of the organ to be strengthened. The doctrine of correspondence is used repeatedly by Paracelsus and, as Pagel notes, is the dominant thread in his view of nature.[26]

Spirits. Paracelsus' medical experiences convinced him—and experience was, for him, the basis of conviction—that man's body has both a material and a spiritual aspect. Either of the two can become diseased independently and the diseased part then affects the other adversely. In diseases that have their origins in the mind, "treat the mind and the body will get well. For, the spirit is sore, and not the body."[27]

In a special treatise on the *spiritus vitae* we hear that this spirit is one and the same in every part of the body. But local corporeal differences cause it to confer apparently different powers on different organs. It should be free to circulate through open pores *hin und her daraus und darein* through every part of the body. If it is blocked,

[24] See Pagel, *Paracelsus,* p. 263.
[25] Leidecker, *Paramirum,* p. 29.
[26] Pagel, *Paracelsus,* p. 107.
[27] Leidecker, *Paramirum,* p. 52.

local putrefaction ensues. Its coagulation is associated with chills, its solution with fevers.[28]

The physiological role of spirits is diversely characterized. "Just as the air keeps creatures from suffocating, so the spirit subtends the body."[29] Spirit "situated in all the members . . . manifests itself in various ways according to the variety of its seats. . . ." It "originates from outside causes [the stars] not from the flesh." It acts now as preservative, now as virtue, now as elixir, now as ferment.[30]

Certain psychobiological properties of the spirits are more precisely defined than their purely biological properties. "I have a spirit, the other person has one also. Spirits know each other even as I and the next person know each other; they speak to one another, just as we do ordinarily, but freely, without resort to common speech."[31] This ability of spirits to bypass regular communications pathways is filled with interesting consequences. The will of one man can affect the body of another through spiritual intermediation. This can happen even at a distance if a wax image of the other man's body is used. Similarly, by abusing the image of a thief one can force him to return to the place where he last practiced his art. Physicians are advised not to "treat this as a joke. . . ."[32]

Quintessence. We noted in chapter 11 that medieval thought gave varied expression to a partly Stoic and partly Eastern dualism that imputed to every object both *soma* and a *pneuma*. Paracelsus' treatment of quintessences is undoubtedly a special case of this dualism. The "life" of every body (including ones that we should consider nonliving) is or resides in a special centrally important "fifth essence." In the case of minerals, this is, theoretically, separable. If it is separated from living things, they lose their vitality, and it loses its power as an elixir and preserver of health. It is variously characterized as a nature, a virtue of power, a remedy, a spirit "like the spirit of life."

Man is the Work of his parents and so on back to Adam who was God's Work. God made everything but man out of nothing. Man he made of the doughy substance (*Massa*), or dust, to be a quintes-

[28] "De Viribus Membrorum 6 Bücher, auch, De Spiritu Vitae benannt," Sudhoff and Matthiessen, *Paracelsus,* 3:13–18.
[29] Leidecker, *Paramirum,* p. 48.
[30] Pagel, *Paracelsus,* p. 118.
[31] Leidecker, *Paramirum,* p. 49.
[32] Ibid., pp. 52–53.

sence or *Minor Mundus,* a purified extract of the *Major Mundus.* As such, man was made immortal, but other mortal things affect him and confer mortality on him. He is the result of three marriages—that of his parents, that of the elements, and that of the stars.[33]

Definitions of Life

In a dozen passages, and in a dozen ways, Paracelsus wrestles with, or tosses off ideas about, the nature of life itself. "Life is a spiritual existence"—an existence common, incidentally, to organic and inorganic beings. "We must therefore know that God at the beginning and creation of all things created none [i.e. nothing] whatever without its own spirit"—its own life. Thus, the life of what we should call living things is, for Paracelsus, at best a special case of life in general,[34] and life in general is a property of the cosmos, of nature. In another more corporeal sense, life is "balsam," presumably a sort of physiological cement (in the Anaximenean tradition). Again (but signifying what?) life is "a fire, a contained air, a spirit of salt." Or, if we prefer, life is "an incorporeal volatile thing which like fire when wood is added burns furiously."[35]

From this loose postulation and piling up of imagery one obtains not a single concept of life but Paracelsus' usual outpouring of alternative concepts, some of them highly suggestive, many defying comprehension. We see here foreshadowings—primitive metaphors, as it were—of later theories of metabolism, respiration, and physiological combustion. If any single feature links these alternative, awkward interpretations it is that life itself *is,* or *resides in,* something occult, immaterial, quintessential.

Nutrition

Paracelsus is intensely concerned with a variety of nutritional questions, some of which throw light on the problem of the material vehicle of life. Grass feeds the cow; the cow, man. But when an animal uses grass, or the flesh of another animal, as food, part of that

[33] "Erklärung der ganzen Astronomei," Sudhoff and Matthiessen, *Paracelsus,* 12:452–55.

[34] The universality of individual life is treated often but especially at "Die 9 Bücher de Natura Rerum," Sudhoff and Matthiessen, *Paracelsus,* 3:329. See also J. R. Partington, *A History of Chemistry* (London, 1961), 2:141, and especially Alexandre Koyré, *Mystiques, Spirituels, Alchimistes der XVIe Siècle Allemand* (Paris, 1955), pp. 49–52.

[35] Sudhoff and Matthiessen, *Paracelsus,* 3:201.

food is poison. Nothing is poisonous to itself but all may be in relation to other things which they serve as food. Something in the
stomach segregates the poisonous from the nonpoisonous parts, eliminates the former, and converts the latter into directly useful
nutriment.

The agent of all this activity is, according to Paracelsus, "an alchemist." Thus "God has appointed an alchemist for us to convert the
imperfect which we have to utilize into something useful. . . ." The
alchemist "conditions nature so she becomes flesh and blood." We
hear—in a statement similar to one attributed to Anaxagoras—that
we do not have to eat hair to grow a beard. Further, "if bread be
conveyed into a man it becomes the flesh of a man, if into a fish the
flesh of a fish," and so on.

"This alchemist resides in the stomach, which is his instrument
wherein he boils and labors"—the labor that most interests Paracelsus being that of separation. "To the pig, excrements are proper
[food]. Although these are a poison (being eliminated for this reason
by nature's alchemist from man) they nevertheless serve the pig
as food. . . . The pig's alchemist extracts food even from the excrements, which man's alchemist has not been able to do. . . . For, there
is no shrewder alchemist that will analyze food more minutely than
the pig's alchemist." The peacock, which eats a snake, has its alchemist, too, to separate food from poison.[36]

The alchemist postulated by Paracelsus is admirably illustrative
of his personal style which tends to be loose, metaphorical, amusing,
occasionally vulgar, often penetrating. The passage reflects, too, the
varied external and internal influences working on Paracelsus—his
travels, his rural upbringing, his doctrinal eclecticism. The stomach's
alchemist corresponds to nature's alchemist, to that principle or
Archeus which in nature often works to prepare materials by acts
of segregation and purification. "Nature does not need all these; [he
has been speaking of the separators, preparers, liquefactors, et cetera,
in a metallurgical laboratory] but still needs *her own people*." The
Microcosm reflects the Macrocosm. The corporeal alchemist is, incidentally, not entirely unlike those of Galen's faculties that have to do
with the concoction and ultimate assimilation of food. Which is not
to say that Paracelsus obtained the idea from Galen, but rather that
the alchemist is Paracelsus' answer to a classic question—the ques-

[36] Leidecker, *Paramirum,* pp. 26–27.

tion of nutritive assimilation to which each generation of early thinkers offered solutions only partly new.

Inheritance

In an interesting speculation, Paracelsus embellishes his concept of the *liquor vitae* and develops it into a theory of inheritance (another classic problem). The *liquor vitae* exists, like an "inner shadow," in every part of the body. Under certain conditions, the individual is able to effect a willed enkindling of the *liquor* which converts it into seed (*Samen*). This occurs, principally, under the influence of a reciprocal *Phantasei* between sexual partners. To be successful, the conversion must realize God's will as expressed in the essence of the seed. The *Samen* must not be confused—as physicians have foolishly done—with the visible semen (*Sperma*). The latter is merely an egestive product of the *liquor vitae,* just as the *stercora* are in the intestines and the *menses* in the womb.

The true *Samen* lies quietly in wait exactly where it is formed, in the various parts of the body. Fetation depends upon an attractive power of the womb. Thither *Samen* is drawn from all parts of both participants in the generative process. Paternal and maternal seed for a given part engage in conflict. Each part, depending on the outcome, resembles the corresponding part of the father or mother. Sex depends upon whose seed, father's or mother's, is first to arrive in the womb. Exactly simultaneous arrival of both leads to a hermaphroditic fetation. With this mechanism Paracelsus manages to account for infertile matings, and a number of other phenomena —among them the nontransmission of mutilations.[37]

CONCLUSION

Our purpose, in the above outline, has not been to add to the scholarly interpretation of Paracelsus, but to see him in the light of our present concern with the history of a particular scientific problem. Like earlier—and, as we shall see, later—students of vital phenomena, he moves back and forth between the two realms of observation and explication. He develops ideas about the inner organization of the body and utilizes these to account for their outer

---

[37] "Das Buch von der Gebärung der empfindlichen Dinge in der Vernunft (Tractatus Secundus)," Sudhoff and Matthiessen, *Paracelsus,* 1:257–265; and the following "Buchlein (Philosophia) de Generatio Hominis," Ibid., pp. 287–306.

attributes and functions—nutrition, generation, psychic action. Like all comprehensive biological theorists he concerns himself with the origins of things and, especially, with their deviations from normality (their diseases).

Paracelsus practices, then, what we have called the persistent master strategy of biological interpretation. But his use of that strategy is highly idiosyncratic. It is redundant in that the differently denominated cryptomenal elements are used in a complexly overlapping way as explicatory devices. It is paradoxical in that Paracelsus typically denounces the analytical concepts of his predecessors without discarding them. Rather he radically transforms these concepts, gives them the stamp of his strong individuality, and uses them in his own special manner. The net effect is one with which we are already familiar. Paracelsus' solutions are not wholly new. They bear what we previously termed a metaphorical relation to earlier solutions, albeit the metaphors sometimes have, with Paracelsus, a somewhat extravagant quality.

How then is Paracelsus to be judged? By the staying power of his thought? By the reliability of his assumptions? By his own fidelity to these assumptions? By his provocativeness and heuristic impact?

To assess Paracelsus is as difficult and subtle as to comprehend him. To read him is like dreaming. Indeed, the spontaneous and bewildering explosion of his ideas is startlingly suggestive of the activity of the human unconscious. Within the limits of a single treatise one can often grasp an ordered mosaic of ideas. But in another treatise the same—or different and even contradictory—ideas are arranged in an entirely different mosaic. We move from book to book with the fascination with which we watch the changing pattern of kaleidoscope.

Paracelsus' influence was powerful partly because he flouted bookish authority with observations from nature. At Basel one cause of his downfall was his tossing of the works of Avicenna into a bonfire. Paracelsus was influential likewise in his incipient adaptation of alchemy—formerly mostly concerned with the production of gold, or of life itself—to practical pharmaceutical studies. Whatever else he accomplished, he precipitated a reconsideration of innumerable scientific problems, chemical, metallurgical, astronomical, physiological, cosmological.

The impact of Paracelsus, coming primarily after his death, was an impact of provocation. As Partington puts it, "He administered a rude shock to the conventional alchemists, and by his blustering profusion of abusive rhetoric he pushed aside their unintelligible jargon by one even less comprehensible but more modern. . . ." The shock was not to alchemists only but to Arabists and Galenists as well. Western science found itself challenged, partly by Paracelsus, to study new problems and to study old problems anew. This fact subjected the Paracelsian legacy itself to radical reformulation and, for the most part, repudiation. The coming science of the Renaissance was to demand something purer, something more Euclidean and Epicurean. Science, as Bacon soon would assert, spells power. But meanwhile magic seems to spell power too. In Paracelsus, we find the two ingredients together, and he makes their immiscibility painfully clear. In this way, he paves the way for the slow alchemy of history, which in time will separate science from magic.

Neotraditional Physiology

JEAN FERNEL [1497 · 1558]

It would be difficult to conceive minds more different than those of the intellectually impulsive Paracelsus and the deliberative and sober Jean Fernel. A revered teacher and respected practitioner, Fernel was ultimately physician-in-chief to Henry II of France. His role in the progress of medicine has not been satisfactorily settled. Here we shall venture a provisional assessment of one aspect of the influence he exerted. Fernel's chief twentieth-century champion, Charles Sherrington, acknowledging the archaic orientation of Fernel's pathology, says correctly that "assertions masquerading as facts are far less frequent in Fernel" [than in favored earlier pathologists, such as Benevieni]. Sherrington also says, of Fernel as a physiologist, that he "welcomed a new fact when it came his way. . . . But he did not, as did Harvey [three quarters of a century later], spend hours worrying out a fact in answer to some question it might settle."[1] Sherrington feels that the life endeavor of Fernel showed a difficultly achieved relinquishment of the medieval ideas that absorbed him as a young man.

Can we go further toward an evaluation of Fernel? A judgment that seems tenable, at least, is that his influence was partly one of provocation. Paracelsus was provocative because he was eclectic, iconoclastic, and tantalizingly obscure; Fernel, for different reasons. Fernel made himself conspicuous by a ringing restatement of traditional ideas at a time when traditional ideas, in nearly every aspect of art and learning, were being questioned. It is not insignificant that Fernel's essentially conservative *Natural Part of Medicine,* 1542, preceded by only about a year the appearance of the forwardly-

---

[1] Charles Sherrington, *The Endeavour of Jean Fernel with a List of the Editions of his Writings* (Cambridge, 1946), pp. 141–45.

directed *Revolutions* of Copernicus and Vesalius' founding work on the *Fabric of the Human Body,* both 1543.

The central doctrine proffered by Fernel, namely the Greek idea of a continuously acting causal physiological soul, was to be attacked less than a century later, especially by Descartes. Descartes, and other thinkers, were to confront the life-soul idea with a new emergentism and mechanicism that would change the whole course and aspect of biomedical thought.

The contents of Fernel's medical works (1542, 1554) are patently Greek in both approach and major assumptions, though among them we shall find a few concessions to Arabic and medieval doctrine including an idea about life and matter that comes close to the heart of our story (see the section below on the radical humor). Early sixteenth-century biotheorists were inevitably exposed to a manifold influence including (*a*) new discoveries in anatomy, geography and cosmology; (*b*) alchemical, astrological, and magical ideas that survived from the preceding centuries; (*c*) Scholastic metaphysicis; and (*d*) the modified variants of Greek science that were argued by Catholic scholars at the time. Fernel's intellectual practice, and here he differed from Paracelsus, was to separate these various components and to concentrate as a biologist (less so, we shall see, as a cosmologist) on Greek science, mostly without the embellishments of the Arabs. To this statement, if we are to do Fernel justice, we should add two qualifications.

First, his release from medievalism was neither immediate or total. In his much earlier *Monalosphaerium* (1526), he had "set forth such matters as the mansions of the moon, and what to do in each of them, the aspects of the planets, critical days, the division of the zodiac into astrological houses, revolutions, and nativities."[2] These topics reflect Fernel's early absorption in astrology. His *Natural Part of Medicine* (1542),[3] with which his serious medical publications

---

[2] Lynn Thorndike, *A History of Magic and Experimental Science* (New York, 1941), 5:557.

[3] Jean Fernel, *De naturali parte medicinae* (Paris, 1542). In 1554, there appeared in Paris Fernel's *Medicina,* a work in three parts treating respectively physiology, pathology, and therapeutics of which the part on physiology is a revised and simplified version of the corresponding sectors of the *De naturali.* In the following century, the physiological part appeared in a French translation by C. de Saint-Germain, *Les VII livres de la physiologie* (Paris, 1655). Our book and chapter references apply equally to the Latin original of the *Medicina* and the French translation.

began, was by contrast largely free from medieval contaminants. But a somewhat later work *On the Hidden Causes of Things* (1548) showed residual occultist inclinations. In dialogue form, this later book reconsidered many of the questions raised initially in the *Medicine,* reviewing them in the light of demonology, magic, astrology, alchemy, and Scholastic logic, and treating them in a soberly questioning but by no means incredulous or completely scientific way. In retrospect, Fernel's physiology as presented in 1542 (and again in 1554–55) seems an episode—a Greek phase, so to speak—in his slow and irregular movement away (but not ultimately far away) from the modes of thought of the Middle Ages.

Second, Fernel's reformulation of ancient ideas was far from a blind ascription to dogma. It was, rather, a critical reworking of Platonic and especially of Aristotelian and Galenic theory in terms of his own experience as a student and successful physician. If he did not move far from Greek thought, he was not afraid to criticize and revise it, and we can see from the perspective of today that, partly by his cautious and questioning habits and perhaps partly by his restatement of an irrevocably doomed theory of human nature, he helped pave the way for the real revolution to follow.

An interesting problem, thus far not successfully explored by medical historians, is summed up in the question: to what extent is the scientist's work affected by his philosophic image of himself— his epistemological grasp of his methods, and his ontological attitude toward the conceptual materials he uses? Fernel lived at a time when it was still fashionable for the scientific worker to raise such questions. He viewed the good scientist as one who, inspired by illustrious predecessors, first takes cognizance of sensible things, then pushes on to recognize their causal origins, and then moves farther still until he arrives at a point "where the mind stands as if finally fulfilled (*quo expleta mens tanquam in extremo consistat*)."[4]

Applying the foregoing canon to the study of man, Fernel recommended the combined use of both *resolutio* (which proceeds from wholes to parts, from composites to simples, from effects to causes, from the posterior to the prior) and *compositio* (which proceeds in the opposite direction). Through resolution or analysis we can see of what elements the visible parts are composed, and what the mixture

[4] For an appreciation of Fernel's responses to astrology, alchemy, magic and occultism, see Sherrington, *Jean Fernel,* pp. 33–59.

of elements and the temperament of qualities is, what faculties and forces lie hidden, and what heat or spirit sustains the whole. Through composition we proceed from the efficient causes of things to grasp what humors are generated by them, what the functions are of the simple parts, and what the natural administration of the whole. In this way we study "universal physiology," a science that pursues the path of demonstration to the natural study of man (*Sic universa contrahetur physiologia quae naturalem de homine contemplationem demonstrationis vi constituit*).[5]

Matter and Living Matter

Elements and temperaments. We have already noted that biologists have followed, from Greek times, a standard but flexible interpretive program involving (*a*) the construction of a partly microstructural and partly microdynamic model of the organism and (*b*) the use of that model in the analysis of visible organization and action. Fernel's philosophy of the organism was no exception to the classic procedures of his science. He considered the individual, we shall see, both as an integrated totality and as an assemblage of semi-independent parts.

He compounded the world and man out of the four primordia of Greek physics, making very little use of the alchemical two or the Paracelsian three.[6] He built the body itself on Aristotelian lines—of composite parts made up of simple or self-similar parts. He went to great pains to distinguish "true" simple parts (bone, cartilage, ligament, membrane, tendon, nerve, artery, vein, flesh, and skin) from others that were not, for him, parts in the strict sense (blood, milk, the humors, spirits, bone marrow, hair, nails, fat, etc.). Only the former qualify, he thought, because only they are stably affixed to the rest of the body and only they share its life.[7] This point is important because it bears on a perennial and henceforth increasingly explicit question, namely: what parts of an organism are and are not properly considered as living?

The simple parts are found by Fernel to be composed of the four elements differently proportioned in each. The uniform temper of a simple part is not due to a mutual total interpenetration of all ele-

---

[5] *Medicina,* bk. 2, Preface.
[6] Ibid., 2.4,5 (cited by book and chapter).
[7] Ibid., 2.2; 3.11.

ments by all others (as certain Neoplatonists pretended). Rather, the substantive elements of the parts are sufficiently finely divided and sufficiently intimately intermingled so that their dynamic qualities are uniformly mixed and blended to produce a self-similar whole.

Temperament is not the physical mixture itself but the proportion of qualities therein, somewhat as, in music, harmony is a proportion among tones. In both cases—in mixtures and in music—the ingredients, must "embrace" to produce a certain "moderate and concordant mediocrity" which is due in physical mixtures to a reconciliation and partial mutual inactivation but by no means a real annihilation of their contrary qualities.[8]

As to the number of temperaments, we remember that Galen had acknowledged eight intemperate deviations (hot, dry, cold, wet, hot-and-dry, hot-and-wet, cold-and-dry, cold-and-wet) from the "just" or temperate form. Fernel agreed with this dispensation, but he thought that in the case of the composite (two-quality) deviations, one of the two excess qualities could deviate more than the other and that either could deviate to any degree, making innumerable differences available for the composition of the self-similar parts.[9] He also noted, following Galen, that what should be considered a just temperament is not an absolute but a relative matter, depending upon the use to which the thing in question is destined.

A persistent goal of biologists has been to understand the underlying causes of sameness and difference as expressed in living things. To Fernel the explanation lay in temperament or crasis. Temperament was supposed by him to account for the commonalties and idiosyncrasies of all living systems: tissues (self-similar parts), organs (heterogeneous parts), and organisms—as well as species and genera. He also ascribed to temperament such differences as are associated with age. With respect to the organism, Fernel viewed the temperament of the whole individual as something more than the sum of the temperaments of the parts. Two possible integrative agents suggested themselves to Fernel. Could it be that motile fluids (humors and spirits whose qualities are partly affected by the stable organs) move about the body and mingle together to produce a kind of temperamental amalgam? Or, rather, is the well-diffused con-

8 Ibid., 2.8; 3.1,2.
9 Ibid., 3.3.

natural—and celestial—vital heat the chief integrative agent? Fernel adopted on this point, a synthetic position, he supposed that both factors—body fluids and heat—combine to establish the distinctive temperament of the whole.[10] An important study for historians, incidentally, would be to trace the evolution in physiological thought of ideas about the relations of wholes and parts.

To the eight temperamental deviations of Galen, Arabic medicine had added the idea that normal individuals fall into four temperamental types (sanguinary, phlegmatic, choleric, melancholic) based on the putative predominance of one of the four constitutive humors. Fernel weighed this idea, and finding little to support and much to oppose it in his own medical experience, decided against it. Among other dissuasions was his own rejection of the Hippocratic humors as persistent constituents of differentiated tissue. He looked upon the humors, he said, as in—but not of—the body.[11] A classic subdivision of the body parts had been into the containing and the contained. For Fernel only the former—the solids but not the liquids—were properly regarded as living.

Soul, Parts, Faculties, Functions

Fernel's causal-analytical system assumed a soul and in connection therewith a hierarchy of parts, faculties, and auxiliary faculties—the latter corresponding to (and causing) functions. Greek biology had left certain questions about soul unanswered. Are there several life-souls? just one soul with several parts? or one soul with several faculties? What are the relations of soul to faculty to part? and the relations of all three to function? Some thinkers had supposed such questions unanswerable, Fernel observed, but he was willing to hazard some answers. In the whole living world, he acknowledged three sorts of souls, natural, sensitive, intelligent; but he supposed that no two souls were ever present in a single individual. He saw the natural soul as confined to plants, the sensitive to beasts, and the intelligent to man. The sensitive soul, however, has two parts, one natural and one sensitive—the natural part of the animal's soul corresponding to the natural soul of plants. Similarly, the intelligent soul of man has three parts, natural, sensitive, intelligent. Its intel-

[10] Ibid., 3.7.
[11] Ibid., 5.1.

ligent part accounts for the functions that are uniquely human; its sensitive part for the functions men share with animals; its natural part, for the functions that men and animals share with plants.[12]

TABLE 7
KINDS OF SOULS AND THEIR PARTS

PLANTS	ANIMALS	MEN
A natural soul without parts	A sensitive soul with two parts, namely: natural sensitive	An intelligent soul with three parts, namely: natural sensitive intelligent

Fernel was not dogmatic about the foregoing scheme, but defended it with reasons, some rational and others empirical. For example, as to soul in general, he was persuaded of its existence by a comparison of the organism just before and just after death. Since the material body is still present after death, it follows that "the body is not the efficient cause of its own works, there is something in a living man by which he is more excellent and powerful than a dead man, by which he is properly prepared to undertake his proper work: there is in effect in man some ability and some efficiency of action. . . ." This cause is the soul, "the principle and cause of the functions of the living body."[13]

Although the existence of the soul is thus clear, its nature is shrouded in darkness and must be grasped through the sensible actions it occasions, just as all occult and hidden causes are known by their effects. To call the soul "the perfection" of the body is acceptable if that is taken to mean that the soul is the cause of all actions. Fernel seems, on the nature of soul and of its relation to the body, to stand closer to Aristotle than to Plato. We hear that "the soul is the species of the living body," that is, bodies "to which life is peculiar and in which action is the [soul's] companion and friend."[14]

Fernel's analysis of the soul and its faculties betrays a certain meshing of medieval with classical influences in his personal outlook. To say, as he does, that the soul can only be known in the way that occult causes are known is within the context of his slightly later

[12] Ibid., 5.1,2.
[13] Ibid., 5.1.
[14] Ibid., 5.2.

book-length work *On Hidden Causes*. In that work (1548) we are given a fuller treatment of topics merely intimated in the *Natural Part of Medicine* (1542). In the earlier work, faculties are related to soul in the way that attraction is related to the magnet. In the later work, magnetism becomes an "occult property" comparable to the purgative powers of certain drugs, or to the ability of quails and starlings to handle foods that are poisonous to man, or to the ostrich's supposed ability to deal digestively with iron (a favorite subject of speculation in medieval literature). Indeed, Fernel invoked a "tacit and recondite" property of the stomach in general, a property permitting digestive activities other than those made possible by its ordinary concoctive or assimilative faculties (to which Fernel ascribed in addition).[15] Historically, the "occult property" theme was to undergo important later developments. Elsewhere we shall touch on the role in interpretive biology, of what we shall call inexplicable explicative devices (see chapter 33), which have their roots, in part, in medieval occult causes.

Fernel was emphatic about the ontological difference between (*a*) the parts of the soul and (*b*) its faculties. He saw the parts as, in the Scholastic sense, substantial; the faculties, as accidental. Parts, that is, are inherent in the soul and inseparable from it without its destruction. Faculties are related to the soul (or its parts) in the way that color is related to an apple, or attraction (and the tendency to point north) to certain needles. "A faculty is a virtue and power that the soul produces as from within itself (*de sinu suo*), and which it employs to perform its several functions. . . ."[16] Again, the soul causes the faculties to cause the functions, there being a separate faculty for each separate function. The net function of a part depends upon the combination and relative strengths of the faculties resident in it. The soul, then, is the cause of active life, and its faculties are adjuvant causes; or, to express it differently, soul is the ultimate cause, faculties the proximate causes, of life-as-action.[17]

The three major parts of the human soul—natural, sensitive, intelligent—are causes not of single functions but of constellations thereof. This is possible because each part gives rise to faculties and subfaculties so that a sort of hierarchy results (see table 8).

---

15 Ibid., 6.1; see also Fernel's *De abditis causis rerum* (first publ. Paris, 1548) (many editions), 2.2.

16 *Medicina*, 5.2.

17 Ibid., 5.3.

TABLE 8
PARTIAL RECONSTRUCTION OF HIERARCHY
OF PARTS AND FACULTIES OF THE SOUL AS DESCRIBED BY FERNEL

PARTS	MAJOR FACULTIES	AUXILIARY FACULTIES

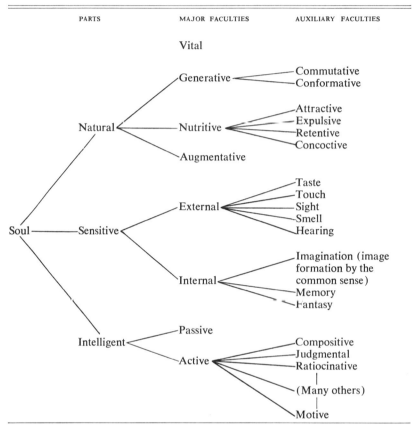

The intelligent part differs from the others, in being simple, individual, incorporeal, immaterial, immortal, and of an essence approaching the divine. It is given divinely from without. It permits man to grasp universal ideas and the essences of things as distinct from their matter. It is disjoined from body and matter and can theoretically function without them, though the objects of its cognizance come to it initially through the senses. As the qualities of external things are the subject matter of sensation, so the images of these things as conceived in the internal common sense, and separated and purged of all corporeal taint, are the subjects of intelligence.[18] This analysis derives mostly from Neoplatonic and neo-

[18] Ibid., 5.11.

Aristotelian elements in medieval and early Renaissance psychology.

The Vital Faculty

In addition to the classic major faculties—natural, sensitive, intelligent—Fernel posited a further hypothetical "vital" faculty related to the others in special and critical ways.[19] He supposed the vital faculty to be present in men and animals but absent from plants. Aristotle had equated the vital with the natural (or nutritive) faculty, according to Fernel, whence for Aristotle life was defined by nutrition. With this position, Fernel found himself in partial disagreement. When he compared plant with animal nutrition, the latter seemed more elaborate and seemed to demand a faculty of a more excellent nature. Even ignorant and idiotic mentalities must be convinced that living creatures include plants, animals, and men—as well as heavenly beings. The terrestrial three live respectively by natural faculties (shared by them all), sensitive faculties (shared by animals and men), and intelligent faculties (exclusive with man). Of these, the chief seat of the first is the liver and of the second and third, the brain. Why, then, do men and animals have hearts?

The heart seemed to Fernel to exist for the generation of a vital faculty that is distributed with the vital heat to the liver, the brain, and the body in general where it, or its subfaculties, permit and sustain the varied functions and offices of the parts in question. Considerable detail is given by Fernel about the mutually supportive and sustentive roles of the vital faculty and the three constellations of other faculties. Fernel insisted that the vital faculty is what its name implies—a faculty, and not a part, but that it belongs to the soul as a whole which it "implements and perfects, maintains and preserves, assembles and unites as being both its union and its bond." Somehow, Fernel saw the vital faculty as activating and integrating the others. In making this point, he vouchsafed a definition of life. "For the life of animate beings," he said, "is the conservation of all faculties and actions, accomplished by the [vital] faculty of the heart along with the heat flowing thence: to live is to use [all] faculties through the vital [faculty] and so to flourish; death is the extinction of the vital force (*vitalis roboris extinctio*)."[20]

[19] Ibid., 5.12–16.
[20] *De naturali,* see above n. 3, 5.15. The comparable passage in the *Medicina* is found at the end of 5.16. The French translation of this passage (*Physiologie,* see above n. 3) is corrupt.

Heat, Soul, Spirits, Humors, Body

The core of Fernel's physiological theory is found in the relational nexus he envisioned among soul (and its faculties), heat, spirits, humors, and the body solids. To understand him on their relations presents interpretive problems, but the kernel of his thought was as follows.

We may commence with the comparisons he drew between the natural or vital heat and heat production through combustion. The two differ, he supposed, in that the vital heat is divine. At death the form of the body persists; the elements, including elementary fire are still there; but the vital heat disappears. Hence, the heat "flows initially not from a mixture of elements, but from some other hidden source."[21] Fernel's position on this subject is significant historically because, like his position on the life-soul, it was destined to be challenged, and from the same direction. Descartes would soon both exalt the importance of heat (specifically the heat in the heart) and strip it of its ultimate mystery (by comparing it with the sort of heat formed in wet hay).

Although the source of the vital heat differs, in this system, from that of ordinary heat, the two are alike in what they need to sustain or feed them: both require an access of air to which Fernel—in the manner of Paracelsus and in anticipation of Boyle—assigned the role of a pabulum or nutriment. This question—whether air is actually consumed in heat production—was to be central, as we shall see, to respiration theory in the period of its active investigation in Oxford about a century later (chapters 19–24). Fernel agreed with Galen that air also acts to moderate or cool the heat and to absorb its smoky exhaust in addition to serving as a precursor of the spirits. He saw the vital heat, finally, as similar to ordinary flame in still another important respect, namely, in requiring in addition to air another aliment which is fatty or oily. Nothing that burns—or lives— can do so without containing such an oily substance.[22]

In addition to air and oil the vital heat, to be connected effectively with the body, "must be contained in some . . . fluxile body . . . a matter very subtle, prompt, and light in which it is placed and which is its friend and familiar. . . ." Fernel considered this substance as,

[21] *Medicina*, 4.1.
[22] Ibid., 6.16; 4.3.

like the vital heat itself, "aërial or, if you prefer, aetherial . . . ," i.e., it is a spirit. This insistence on spirits as intermediators between the vital heat and the body was an extension, Fernel acknowledged, of a similar idea that had had a wide currency in both Arabian and Scholastic biomedicine, namely that spirits have the function of connecting body with (not fire but) soul. This theory appears, for example, in the introduction to Avicenna's *Canon* and had been discussed repeatedly in the context of Christian dogma (see e.g., St. Thomas Aquinas, *Summa Theologica,* seventh article of the eighty-sixth question). There was nothing new in Fernel's conclusion, after a review of ancient and more recent ecclesiastical thought on the subject, that "each part of the soul is grounded in a certain spirit as a firm basis through which it both resides in, and administers all the functions of, the body." Fernel thus made the spirit both a basis for the faculties of the soul and a vehicle of the heat through which those faculties work.[23] He further posited, in the Stoic (and alchemical) tradition of *soma* and *pneuma,* an immanent spirit in each organ, derived from natural spirits which the organs in question receive, by way of the veins, from the liver.[24]

Natural spirit, in turn, is made in the liver out of a specially pure part of the nutriment brought to that organ. Passing to the right ventricle of the heart, and thence through pores to the left, it is converted into vital spirit, with air as an adjuvant substance. Vital spirit, raised by arteries to the brain, undergoes a third elaboration in the *rete mirabile,* the air coming direct from the nostrils. After that, it enters the ventricles and is finally changed again into a spirit that is at once animal and almost celestial. This spirit disposes the body to movement and feeling and makes for all the rational functions.[25] It will be readily apparent in what ways this picture is similar to and dissimilar from that drawn 1400 years earlier by Galen on which it is obviously based but from which it departs in its details. These departures are partly Fernel's but also reflect, to some extent, Arabian and medieval reconstruction of Galen's system.

Generation

The classic questions in the history of reproductive physiology are:

[23] Ibid., 4.2.
[24] Ibid., 4.11.
[25] Ibid., 4.11.

what does each parent contribute to the offspring? how do the two contributions interact? what is the nature of the resulting fetation? how does it develop into a full-fledged animal?

Fernel supposed that both parents contribute semen. He used his pneumatology and his faculty theory to develop a pangenetic concept of semen formation. Semen he viewed as matter endowed with requisite faculties. The material component seemed to him to be "a certain useful and pure portion of the [final] aliment assigned to the solid parts." This substance is separated from each such part by the attractive force of the male or female testes to which it descends. Out of it, the testes and spermatic vessels, through their endowment of vital heat, engender semen. This semen, however, is not considered fecund or fertile unless—and here spirits (and faculties) enter the picture—"there flow also, from the whole body, an abundance of . . . spirits which are diligently mixed with the matter." These spirits, bringing soul faculties with them, come via veins from the liver (natural spirits), via arteries from the heart (vital spirits), and via nerves from the brain and spinal marrow (animal spirits). Also, at the moment of orgasm, local spirits from the body as a whole are perhaps transferred to the semen, producing a lively sensation.[26] The principal fact that Fernel wished to emphasize was the origin of our faculties not from the mixture of elements in our bodies but from the semen with which our individual existences began. The semen is a mixture of matter with spirits, the spirits being responsible for the presence of the faculties of the soul.

A classic concern of physiological biology has always been the complex of problems surrounding embryonic differentiation. Fernel's theory followed Galen's in its general lines, envisioning commutative faculties for transforming semen into similar parts and conformative faculties for arranging the latter into heterogeneous parts. Generation was thus for Fernel (as for so many predecessors) similar to nutrition with the important difference that in nutritive commutation the end product is like something that was already there (nutrition is a like-making or assimilation) whereas generative commutation and conformation give rise to a progression (except of course in the previous generation) of things that were not there before.[27] Fernel made it a cardinal point that the essence or rationale of the

26 Ibid., 7.5.
27 Ibid., 5.3.

organism does not stem from the elements "condensed and congregated there . . ." but from spirit-borne faculties derived from the parents. Moreover, even these would remain "suppressed and mute" in the semen except for the presence there of another spirit permitting the effective action of a special and celestial heat comparable to that of the stars. On these subjects, Fernel acknowledged his indebtedness to Aristotle.[28]

Fernel reviewed the evidence for the notion of male as contributing form and female as contributing matter. He rejected this Aristotelian view in favor of the view that effective semen is formed by a mixture and mutual moderation of antecedent male and female semina, a mixture in which the faculties or powers of male and female form a perfect union.[29] The catamenia he viewed as a source of sustenance but not as the beginning building matter of the fetus.[30] A particular power of the womb awakens the sleeping potency of the mixed semen, he thought, and causes it to start its work.[31]

The development of the fetus is described on epigenetic lines and in considerable detail, specific days being assigned to many events. In general, a new virtue arising in the seminal *mélange* causes its cold and earthy ingredients to be concentrated externally around a hotter, subtler internal portion. Partly under the influence of contact with the womb the membranes are formed in the order: allantois, chorion, amnion. Apertures in the outer membrane permit uptake of nutrient from maternal blood. Next a spirit is generated with very broad formative faculties—a "moderator of heat and of all faculties and the first author of procreation. . . ." Through it, the generative faculties build the organism in steps, the first morphogenetic event being the materialization of three transparent, fluid-filled bubbles representing the future brain, heart, and liver.[32] The subsequent stops are too detailed to be treated even synoptically here. They show evidence, however, of at least some empirical acquaintance with human embryos and have attracted less attention than they should, from historians, as constituting a full-fledged and more than merely derivative theory of differentiation. On this subject, Fernel says that "from the twentieth to fortieth day, the [fibrous] solid parts being

28 Ibid., 7.5.
29 Ibid., 7.6.
30 Ibid., 7.7.
31 Ibid., 7.8.
32 Ibid., 7.9.

formed, they commence to be nourished, and to be supported with flesh, this with the aid of the liver, which makes and administers nutriment for the parts in general. Nature then gradually fills the available empty spaces that separate the spermatic fibers—after the example of the painter who first shadows forth the image in crude strokes and later renders and perfects it with various sorts of pigments."[33]

Humors

Thus far, we have touched the nexus of relations, as depicted by Fernel, among the body, spirits, soul faculties, and heat. We have seen how he built the body along lines employing (*a*) the Aristotelian *stemma* (elements, simple parts, and composite or heterogeneous parts), (*b*) a synthesis and modification of Galenic and Stoic *pneumata,* and (*c*) an adapted version of Galenic faculty theory. We have said nothing, so far, about humors.

In treating these, Fernel had four traditions to draw upon, and he freely utilized all. First, he reused and only slightly revised the analogy Galen had drawn between vinous fermentation and the formation of the Hippocratic four humors out of chyle in the liver.[34] In Fernel's formulation, the biles appear either (*a*) as aliments, or (*b*) as superfluities (drawn to gall-bladder and spleen), or (*c*) as harmful preternatural matters needing to be voided. Second, Fernel agreed with earlier authors who specified four secondary humors viewed as intermediate stages in the conversion of finally nutrient blood into tissue (the fleshy packing substance that fills the interfibrillar intervals of the similar parts). Third, Fernel continued the old idea that certain humors under certain circumstances are harmful and must be removed from the body. These included the preternatural biles, the excrementitous substances urine and sweat, and the nasal mucous discharge still derived, according to Fernel, from the brain.[35] Fourth, Fernel posited a set of three humors, two of which we have already mentioned and one that is new. The three were: first, the alimentary humor (actually, we noted, a transformational sequence of four) that turns into the fleshy matter of tissue; second, the oily humor (the one that nourishes the natural or vital heat);

---

[33] Ibid., 7.10.
[34] Ibid., 6.3.
[35] Ibid., 4.6.

third, an aqueous humor that permeates and amalgamates (the perennial idea of a balsam or cohesive). The oily one is the first-born or "radical humor" with which we are now ready to round out our account of Fernel's ideas about life and life's relation to matter.[36] In medieval thought, the putative "radical humor" ("primigenial moisture," etc.) had been assigned a variety of roles as simultaneous sustainer and moderator of vital heat, as a defender of the body against decomposition, as an agent determining the idiosyncrasies of body components, and so on.

Fernel approached the radical humor idea by first revising slightly his earlier definition of spirit. He had already depicted spirit par excellence as a combination of (a) a celestial, aerial substance and (b) the vital heat which that substance joins effectively to the body. But it seemed to Fernel that the resulting composite spirit, a substance very similar to flame, required a further vehicle and this, he supposed, must be the radical humor. He saw this humor as fatty and as entirely identical to the oily humor already mentioned above that serves as fuel. This sustainer of the heat and of the solids is airy, is invisible, and is inserted into the self-similar parts of the body (the tissues) from which it may be separated "by industry and art."[37]

We hear that the ". . . [radical] humor is the basis and first substance of the spirits and of natural heat . . . , without the aid and assistance of which . . . [the latter two] cannot subsist or last very long, . . ." The radical humor exists in each particle of each similar part. It prevents them, in a way to be mentioned, either from drying out or from becoming too cold. In our studies of life and matter, we have encountered many explanations, beginning in pre-Socratic times, of the living system's capacity to resist the entropic dissipation which sets in immediately after death. Fernel pointed out that though bones and other tissues resist consumption by the vital heat during life, they easily burn, because of their residual oily content, after death (because the radical humor is no longer present).[38]

Fernel developed his version of the radical humor in considerable further detail. A part of radical-humor lore had been the supposition that we are given a certain quantity of this humor at birth which is gradually consumed without being replaced; in this respect it would

36 Ibid., 4.3.
37 Ibid., 4.4.
38 Ibid., 4.4,6.

be unlike the rest of the body components that are generally subject to displacement and replacement. The presumed irreplaceability of the radical humor was debated during the Middle Ages, and Arnald of Villanova (1276[?]–1312) devoted a chapter to discussing the pros and cons of the question. To Fernel, a stable continuity of the radical humor seemed important as the vehicle of that sine qua non of life, the vital or natural heat. The radical humor had for him, as for earlier thinkers, the dual role of sustaining this heat and yet by its humidity tempering it and preventing it from desiccating the body. As old age approaches and the humor declines, both its sustentive and antidesiccative properties are lost. The body grows progressively both colder and more dry. Men of the Middle Ages and the Renaissance were preoccupied with the question of life's duration and of the reason for its limitation and its culmination in death. Death wore some special—some specially dreadful—aspects then, partly because of the epidemic forms in which it often appeared. Alchemy, astrology, and Christian theology had their solutions to the problem of death. The radical humor was a physiological answer. When the innate endowment of an indispensable yet irreplaceable humor gives out, it was thought, then death inevitably ensues.[39]

RECAPITULATION AND CONCLUSION

Fernel believed that the four elements constitute various mixtures, and that the resulting blend of their qualities determines the characeristics of the resulting similar parts whose composition is (*a*) partly solid (in the sense of being built on a fibrous mesh), (*b*) partly fleshy (and flesh is a consolidated humor), and (*c*) partly spirituous. Under humors Fernel included: the Greek four that are derivatives of chyle; the secondary nutrient humors that are precursors of tissue formation; certain harmful humors, all mentioned as such by earlier authors; and three that, while not new with Fernel, were of special interest to him, viz. the oily, the aqueous, and the radical or primigenial.

He saw the simple parts as assembled (as they had been by Aristotle and Galen) into complex parts or organs. He regarded the successive steps in the ontogenic construction of the body to be governed and activated by generative faculties. He supposed that other faculties control the function of the parts once created. Fernel's faculties—

[39] Ibid., 4.8.

a modified adaptation of Galen's—were in the language of Scholasticism, "accidental" properties of the three "substantial" parts of the soul, viz. the natural, the sensitive, the intelligent or rational. To do their work, in Fernel's system, the faculties employed vital heat, and this he viewed as connected to the body by spirits. Finally, he believed that heat, spirits, and the soul faculties have a common vehicle, namely a radical or primigenial humor, the indispensable material fundament of life itself.

Fernel appears in retrospect less an innovator than a recoverer. He gave mostly classical and some medieval ideas an eclectic and only partly new interpretation. His special interest in one "vital humor" was not original, but his retention of it as a primarily medieval and early Renaissance component of an otherwise essentially Greek physiological system is significant. This "radical humor" was to be reinvoked in various modifications by other authors including, importantly, William Harvey (see chapter 17), and was thus to take its place in the series of models constructed in the effort to associate life with a special single substance. The radical humor was fated also to come under attack (for example, by van Helmont, who wrote a treatise against it). It disappeared, in its classic formulation, during the second half of the seventeenth century, but the idea it contained was often revived and often reformulated during the next three hundred years.[40]

In reworking inherited doctrine, Fernel proceeded with cautious forthrightness and apparent good sense. The explicitness and directness with which he espoused his own version of older ideas was well calculated to elicit opposition in the era—just dawning—of skepticism and incipient anti-authoritarianism. This was to prove especially true of Fernel's central physiological axiom, namely the Greek idea that life-as-soul is the cause of life-as-action. More than any other development, the assault on the life-soul was to mark the birth of a new biological era.

[40] See, on this subject, T. S. Hall, "Life, Death, and the Radical Moisture: A Study of Thematic Pattern in Medieval Medical Thought," *Clio Medica,* 6 (1971): 3–23.

· 14 ·

The Iatrochemical Approach

JOHANNES BAPTISTA VAN HELMONT

[1577 · 1644]

Van Helmont was an influential Belgian chemist and physician, born in the sixteenth century, active in the seventeenth, intellectually a member of both. Certain sectors of his chemical and biochemical work make him an indirect forerunner of modern chemistry, and he was the principal founder of the iatrochemical movement. In these two respects his work is pointed toward the future, but his medical ideas are mixed with mysticism and magic and in this respect still bound to the past.

Van Helmont begins by rejecting the four causes of Aristotle,[1] the four elements of ancient physics,[2] the four humors of Galenic medicine,[3] and the three principles of Paracelsus.[4] In theorizing about the world and life, his self-set task is one of almost total reconstruction.

As to causes, only two (of the Aristotelian four) seem to him to be needed, namely the material and the efficient. His repudiation of the other two, the formal and the final, seems attractive from the perspective of later science. Actually, his promise to exclude them is not effectively realized. Before long, we shall see, formal and final causes, or something very much like them, reenter van Helmont's physiology by the back door.

---

[1] Johannes Baptista van Helmont, "Causa et initia naturalium," *Ortus Medicinae* (Amsterdam, posth., 1648). This, van Helmont's magnum opus, edited by his son, is a collection of 123 treatises dealing with special subjects. Hereafter only the title of the treatise will be given, followed by the marginal reference number. It exists in an English translation, J. Chandler, *Oriatrike* or, *Physick Refined* (London, 1662).

[2] "Elementa," 2; "Terra," 3.

[3] "Ignotus hospes morbus," 17; "Elementa," 9, 10.

[4] "Complexionum atque mistionum elementarium figmentum," 11; "Causa et initia naturalium," 13.

Matter

As to matter, "The first beginnings of bodies . . . are two: an element . . . and a ferment."[5] The element is water;[6] the ferment, something which "predisposes water to become something else" (this turns out to be practically everything). The element water is "that out of which" (*ex quo*), the ferment "that by means of which" (*per quod*) the world, nonliving and living, is made.[7]

Specifically, the effect of ferments is to convert the simple, indifferent element water into what van Helmont terms "seeds." Seeds, he thinks, are the indispensable precursors of specifically organized matter—including living matter. The idea has alchemical and Greek antecedents. In our own limited sampling of earlier opinion we have encountered similar ideas in Paracelsus and Lucretius.[8]

The transformation of water into seeds is plainly a process of crucial importance. In this process, a sample of primordial water gets imprinted with an image of its future self (*imago rei*), with a "predisposing notion of what has to happen" (*notitia dispositiva rerum agendarum*).[9] If it seems remarkable that a ferment should be able to make such an extraordinary impression on indifferent matter, van Helmont has a ready explanation. The ferment is acting as an agent of "Divine Light," that is, of God.[10]

The next step in the creative process is the further transformation of seed into ordinary objects. Van Helmont explains this by suggesting that seed (matter divinely impressed with an idea of what it should do) possesses an organizing principle, or "*Archeus*," which transforms it into definite, concrete objects.[11] Such objects may, of course, be alive. The role of *Archei* in living things is important, then, and van Helmont's view of their mode of action is central to

[5] "Causa et initia naturalium," 23.
[6] "Elementa," 11, 17. Strictly, van Helmont postulates two elements, the other being air. Most familiar objects, however, he believes to be transformations of water. Van Helmont's relative lack of concern with air is notable.
[7] "Causa et initia naturalium," 23.
[8] See J. R. Partington, *A History of Chemistry* (London, 1961), 2:141.
[9] "Imago fermenti impraegnant massam semine," 12.
[10] "Causa et initia naturalium," 25.
[11] "Archeus faber," 3; see also Walter Pagel, "The Religious and Philosophical Aspects of van Helmont's Science and Medicine," *Supplements to the Bulletin of the History of Medicine,* no. 2 (1944), p. 20.

his ideas about matter and life. We learn, in this connection, that the *Archeus* in seed is a sort of "spiritual Gas."[12]

Gas

The term gas is van Helmont's own introduction. He is, historically, the first to establish that burning charcoal yields a substance different from air and identical to the substance yielded by alcoholic fermentations. This substance (today, carbon dioxide) van Helmont names *Gas sylvestre*.[13]

Gas becomes, for its inventor, an entity of prime importance and very general application. He sees gases as present in all objects. The gas in anything is in fact the material expression of that thing's physical individuality—it is what makes it different from everything else. This idea, namely that the specificity of things is due to something other than their tangible contents, that it is due to the presence of an additional material (or immaterial) substance, is not new with van Helmont. Hippo assigned a similar role to water, we remember, as did the Stoics to *pneuma,* the alchemists to quintessential spirits, and certain medieval and Renaissance theorists to the "radical humor"—though van Helmont rejects the classical concept of the radical humor in a treatise devoted to the subject. Van Helmont makes his gases, not surprisingly, special differentiations of the universal element, water.[14]

This much information about gases makes possible a better definition of seed. "Seed, then, is a substance in which there already exists an *Archeus* which is a spiritual Gas containing in itself a ferment, an image of itself, a predisposing notion as to what must be done."[15] "Generation," by which seed becomes a concrete object, "means nothing but the flow of the seed towards perfection, the emergence of hidden qualities, the consummation of the schedule down to its end."[16]

---

[12] Pagel, "Religious and Philosophical Aspects," p. 11, 18; see also "Imago fermenti impraegnant massam semine," 12.

[13] "Complexionum atque mistionum elementarium figmentum," 14.

[14] Ibid., 14. See also "Humidum Radicale."

[15] "Imago fermenti impraegnant massam semine," 12.

[16] "Ignotus hospes morbus," 38; trans. Pagel, "Religious and Philosophical Aspects," p. 30.

Life

Van Helmont says that life is "a light and a formal first principle through which a thing acts as commanded to act." We receive some further items of information about life. We hear that it is given by the Creator as fire by flint. It is limited (*clauditur*) by a unity and identity. It is distinguished into genera and species. Life's light is not that fiery combustive sort that consumes the radical moisture. It is as lively in a (cold) fish as in a (hot) lion, in a (cold) poppy as in a (hot) pepper. It is not a balsam, not the mummy, not vital spirit. It is neither creature, substance, nor accident, but is, in fact, the *Form* of the thing identified as living. Van Helmont further acknowledges not one life, merely, but five. There are: a mute and subdued life of minerals; a slightly more explicit life in seeds; a life of plants, symptomized by growth and reproduction; a life of beasts, entailing motion, sense, and choice, along with a certain "discourse of imagination"; a life, finally, hidden in man's mind but to some extent exposed in mind's "vicaress," the sensitive soul (for the relation of mind and soul see below).[17]

Living Bodies

The generation of living bodies is, for van Helmont, only a special case of generation at large. We obtain here a preliminary clue to van Helmont's viewpoint on "living matter." He is, in the first place, more interested in life's variety than in its uniformity. To understand living matter, for him, is to understand the origins of its entirely admirable diversity. The problem, as he develops it, is to grasp how this diversity happens to exist.

In the human body, for example, food is subjected to six "ascending digestions" each governed by a special ferment: (1) *gastric* and (2) *duodenal* digestion; (3) a *preliminary sanguification* (in the mesenteries and liver); (4) a further and more *advanced sanguification* (in the heart); (5) conversion to *vital spirits* (also in the heart); and (6) a *final transformation* (occurring in the "kitchens"[18] [*culinis*] of the individual organs).[19]

---

[17] "Vita."
[18] "Catarrhi deliramenta," 31; elsewhere, *coquinis*.
[19] "Sextuplex digestio alimenti humani"; see also "Blas humanum," 23.

Archei. Of all questions in the history of physiology, perhaps the most evocative and elusive is the question how form is established, and once established, maintained. According to van Helmont, each of the six ascending transformations is guided by an appropriate *Archeus.* Of these, the body has many. A centrally presiding *Archeus influus* governs a company of secondary *Archei insiti.* The central *Archeus* is enthroned, curiously, at the pyloric (lower) orifice of the stomach.[20] The others are stationed in various parts of the body. As to their mode of action, the emphasis is upon elaboration or differentiation, that is, the establishment of organic individuality in different parts of the body. The *Archei* hammer out an organized body. The *Archeus* is a "blacksmith."[21]

Blas. The process is helped by certain kinetic impulses termed by van Helmont, "Blas."[22] These form the subject matter of two treatises dealing respectively with their celestial (*Blas meterion*) and physiological (*Blas humanum*) varieties. They are means by which motion is initiated or communicated. They thus partake partly of the nature of forces, but they are not the forces that physicists deal with and measure. *Blas* emanating from the stars have much to do with the weather and a variety of other terrestrial and human oc currences.[23] In the body, *Blas* are responsible for both natural and voluntary motions and are thus intimately mixed up with all physiological transformations. Van Helmont's *Blas* seem, in effect, to be highly serviceable fictions which, provided one does not inquire too suspiciously into their real counterparts, are helpful to their inventor in solving certain practical and theoretical physiological problems. Science has always depended on such "useful fictions"; but later scientists were to find relatively little use for van Helmont's *Blas*.

Ferments. Of the six "ascending digestions" none is more important than that occurring in the heart in whose left sinus (ventricle) "resides a special, maximally vital, luminous ferment."[24] It is this

[20] The location of the soul is considered in the treatises "Demens idea," "Sedes animae," "Jus duumviratus," etc.

[21] The relevant treatise is entitled "Archeus faber" (compare Paracelsus' "alchemist").

[22] Sing., *Blas,* plur., *Blas;* an undeclinable neuter noun.

[23] Michael Foster properly points out that van Helmont does not clearly distinguish *Blas* from *Archeus.* At times, van Helmont gives *Blas* the role of causing motion, *archeus* that of organizing it. Elsewhere, *Blas* seem to be—like *archei*—directive. Michael Foster, *Lectures on the History of Physiology* (Cambridge, 1924), p. 130.

[24] "Blas humanum," 22.

ferment which converts preliminary blood (*cruor*) into finished blood (*sanguis*) and the latter into *spiritus vitae*.[25] This ferment—if van Helmont can be understood on the subject—is itself a sort of vital spirit capable of increasing itself in the blood.[26]

Van Helmont goes further than most predecessors in specifying the character of the vital spirits. He views them, not surprisingly, as a gas,[27] and what he has to say about these gaseous vital spirits squares properly with his general ideas about matter and the source of its differentiations. In producing the vital spirits, the heart's ferment "illuminates" the blood with the "light of life."[28] Like other ferments, it gets its ability to do this from the Father of Lights (God). Blood, thus vitalized, is ready for its ultimate transformation into tissue substance.

Van Helmont answers in advance our possible protest that this characterization of life is supernatural rather than scientific. "God renders life in the abstract intellectually incomprehensible,"—unless we are to know it "ecstatically," through a mystical union of the mind with its object.[29] We might agree with van Helmont that mere science has not thus far produced a clear definition of life; but we should wish to add that the *light* with which he equates *life,* is itself merely a stab in the direction of a definition.[30] It is one of those momentarily useful metaphors which constitute much of the substance of every science as it unfolds. If van Helmont's metaphor mixes religion with science, it differs from most of those which we shall consider later but not from most of those considered thus far.

Soul and Mind

If van Helmont's scheme is to be meaningful, we shall wish to grasp more directly the nature of life's "light," and in order to do so must take account of an entity new to our analysis, namely *soul*. Life is not by nature flaming, blazing, or fiery. "It is," rather, "a formal light, of the character of the sensitive soul."[31] The introduction of

---

[25] "Sextuplex digestio alimenti humani," 62.

[26] Ibid., 63; see also Foster, *Lectures,* p. 138.

[27] 'Scripsi tandem de vita longe, spiritum vitae nostrae arterialem de natura Gas esse,' in "Complexionum atque mistionum elementarium figmentum," 40.

[28] The topic is explored by van Helmont in "Spiritus vitae," 21–24.

[29] "Spiritus vitae," 19; "Blas humanum," 23.

[30] Moreover, Pagel, "Religious and Philosophical Aspects," p. 34, points out that similar ideas appear in cabalistic and Neoplatonic teaching.

[31] "Spiritus vitae," 21.

soul brings us to the final step in our outline of van Helmont's solution of the life-matter problem. Three entities demand attention under the general idea of soul, and these are not without a certain similarity to the tripartition of the soul in Greek biology.

1. *Ens vitale*. In plants and nonhuman animals, there is, strictly, no soul but instead a somewhat analogous vital entity (*ens vitale*) which is, perhaps, a precursor or primitive analog of soul.[32]

2. *Sensitive soul*. In man, a *sensitive soul* presides, from its central position at the pyloric end of the stomach, over the totality of bodily functions.[33] The sensitive soul has, as its immediate organ of action, the *archeus* which occupies the same site. It also uses the *archeus* which resides in the neighboring spleen,[34] and the resultant dual governance constitutes a physiological "sheriffdom" or duumvirate.

The sensitive soul has a luminous character. It is "a descending light, mortal and clear, nor is it to be explained in any other way. . . ."[35] It spreads its influence through the body in the manner in which the sun, another luminous body, spreads its light through the cosmos.[36] It exerts its effects on the body substance, using *Archei* as instruments.

3. *Mind*. Of greater dignity than the sensitive soul, and separate from but linked to it, is man's immortal mind. "The immortal mind is an incorporeal and lucid substance directly reflecting the image of its God."[37] Before the fall, man was furnished with a mind but not a soul; the mind, at that time, worked through the *Archei*. When man fell from grace, the sensitive soul was added to the mind like a husk or envelope. Mind is "chained" to soul[38] in the sense that when soul is destroyed, mind returns to its Maker. When the sensitive soul was supplied it, the mind acquired a certain independence. It was no longer required to use the *Archei* directly, whereas the sensitive soul was bound willy-nilly to use them.[39]

Van Helmont places understanding, along with soul, in the pyloric end of the stomach. He would not propose in this formulation (of which we have sketched only a few of the principal ideas) a total

---

[32] "Formarum ortus," 66, 82; Foster, *Lectures,* p. 141.
[33] "Jus duumviratus," 22; "Sedes animae."
[34] "Sedes animae," 27.
[35] Ibid., 20.
[36] Ibid., 1.
[37] Ibid., 23.
[38] Ibid., 17–19.
[39] Ibid., 29.

return to Aristotle's dispossession of the brain. The cerebrum is important in his system, but is secondary. Via its connections with the stomach on the one hand and the rest of the body on the other, it serves as an intermediate "office" of motion and sensory action. As a center of sensation, it has the important faculties of memory, will, and imagination; and this dispensation (since mind *qua* understanding is far away in the stomach) is a considerable concession to the importance of the brain.

Van Helmont's Experience with Wolfsbane

Van Helmont's discovery that the soul and mind are located in the stomach came to him during an episode of drug-induced illumination. Convinced that poisons generally conceal "virgin powers," he tasted, without swallowing, a crude preparation of wolfsbane (*Napellum*). His first reaction was as of a tight girdle around his skull, but after a short interval devoted to normal business, he noticed that something in his midregion rather than in his head seemed to be performing the tasks of understanding and imagination (the brain occupied itself, meanwhile, with sense and motion)! The experience was exceedingly lucid and pleasant and left van Helmont with a number of scientific convictions about the mind, the soul, and the body. Thus, understanding occurs in the *duumvirate* or "sheriffdom" of stomach and spleen (this can be demonstrated not only under the influence of drugs but also in ecstasies and during silent prayer). Madness proceeds from the same part of the body. Volition resides in the heart (whence murders, adulteries, etc.). Memory resides in the brain. Understanding is the essence of the mind; will and memory belong to the mortal sensitive soul. In short, van Helmont drew many of his central conclusions from his experience with wolfsbane.[40]

Van Helmont in Relation to his Times

Van Helmont's formulation of the life-substance, and his general physiological and pathological outlook, have to be understood in terms partly of his own powerful individuality and partly of the crisis of his century, the hundred years ending with his death. He is often casually considered a successor to Paracelsus, and the point is undeniable despite van Helmont's recurrent and assiduous repudiation of his famous predecessor. Daremberg says that van Helmont "was,

[40] "Demens Idea."

like Paracelsus, a mystic but more learned; an enemy of tradition, but more erudite; an empiric, but a better clinician and observer; a violent polemicist, but more of a gentleman; as obscure and pretentious an author, but with fewer divagations."[41] The two asked many of the same questions. The differences in the answers they gave are the less surprising in view of the drama that had unfolded in the period intervening between them.

Between the death of Paracelsus (1541) and that of van Helmont (1644) there appeared such founding works of the new science as Copernicus' "Revolutions," the *Fabrica* of Vesalius, Tycho Brahe's "New Star," Gilbert "On the Magnet," and Harvey's *Exercitatio de Motu.* Also Fallopio's "Observations," Rondelet on fishes, Aldrovandi's *Ornithologia,* Belon on both fishes and birds, and Cesalpino's "Questions." The period opened, approximately, with Gesner's *Historia Animalium* and ended with Aselli on the lacteals. Fabrizzio, Piccolomini, Eustacchio, Severino, Realdo Colombo, Bauhin, the Bartholins, the elder Riolan—all were of this period. And the same hundred years witnessed the publication of all but certain posthumous works of Francis Bacon, Galileo, Kepler, Ronsard, Spinoza, Montaigne, and Descartes.

During the earlier part of this period the great Italian basilicas and palazzi were built or rebuilt in the Renaissance style. (It is worth noting, however, that the naturalistic revolution in the arts—as represented by da Vinci, Botticelli, Donatello, and Raphael—preceded that of the sciences by a century or more.) The same general era saw the penetration of the Atlantic barriers of North America and other widespread explorations. It saw the founding of a scientific society, the *Accademia dei Lincei.* Finally, it saw Servetus burned by the Calvinists (in 1553) and Bruno by the Catholics (in 1600). Indeed, van Helmont himself and his contemporary Descartes had their difficulties with religious authority as did, notoriously, Galileo.

Not all of the creative productions of the era were equally enlightened. Indeed, the second half of the sixteenth century saw a vast outpouring especially of astrological but also of other occult writings, and the Inquisition was as concerned with its fight against these as with its fight against the rising tide of natural science.[42] At Sala-

[41] C. Daremberg, *Histoire des sciences médicales* (Paris, 1870), 1:471–72.

[42] See Lynn Thorndike, *A History of Magic and Experimental Science,* 6:chaps. 33–35; van Helmont's own difficulties with the authorities surrounded the publication of a treatise on the curative power of the magnet.

manca, the chair in astrology would still be occupied in the eighteenth century.[43]

Van Helmont himself was somewhat caught up in the mixture of enlightenment with benightedness that marked this period. His notions were of such cosmological and general biological range that only a religious mind could have conceived them, according to Pagel, and thus he is "a classical example of the active role which religious motives played in the birth of modern science during the seventeenth century."

A precaution is needed at this point. Our account of van Helmont's physical and physiological doctrine may give the impression of a consistent and rational, if mystical, mentality. Indeed, if we compare him to Paracelsus, he seems a model of sense and rational compunction. But, when we see the man in the totality of his works including his letters, we are differently impressed. What are we to make of the wounds he thinks the moon is capable of inflicting? of the healing powers of stones, iron rings, toads, and amulets? of his belief in the basilisk and in the *magnum oportet*?[44] For the purposes of our story we have abstracted from this farrago of sense and nonsense the nucleus of sense, and examined this for the role it plays in the evolution of biological thought. To know the *man,* we should have had to take account of the entirety of his thought which is very imperfectly disengaged from its alchemical and magical antecedents.

Van Helmont and Descartes. The principal problem confronting van Helmont—and, incidentally, Descartes—is how the devout person is to deal with the dislocations caused by stubbornly encroaching materialist and mechanist ideas. The drama of creation tends in the newer thought to be snatched from the skies and relocated in the created object. To the spiritual problems thus posed Descartes and van Helmont gave precisely opposite answers. Neither's answers were destined to fit the frame of later serious science; but of the two, Descartes' were vastly more indicative of the way the wind was blowing.

Descartes' solution was: to embrace a thoroughgoing mechanism —but to save the soul by granting it fully separate status. In his famous posthumous *Traité de l'homme,* we shall hear him describe

[43] Ibid., 6:166.

[44] Among the best recent interpretations of van Helmont are those of Pagel and of Thorndike, *History of Magic,* 7:8; see also J. Mepham, "Johann van Helmont," *Early Seventeenth Century Scientists,* ed. R. Harré (Oxford, 1965), pp. 129–57.

a wonderful machine, a robot so marvelously wrought that it could walk, breathe, digest, perceive objects in optical perspective, do everything, in fact, that a human being can do—except reason. Descartes leaves it to the reader to decide how far to go in identifying this robot with man himself. As for Descartes, he makes man, too, a machine but with the reservation that man's separate soul can interact with the machine in limited ways through the medium of the pineal gland.[45]

Van Helmont gives an opposite answer. Instead of isolating the soul, he does everything possible to imbed it directly in those seeds which differentiate into concrete objects, nonliving as well as living. In Pagel's phrase, van Helmont "spiritualizes" matter itself. With van Helmont, soul, light, life—all are intensely immanent in the physical substrate.

SUMMARY

In sum, van Helmont's solution depicts the living body as the differentiated product of generative precursors, or seeds, imprinted in advance with a sense of what they must become, and guided in the act of becoming by gases or *archei* that act as organizing instruments of the sensitive soul. The soul likewise serves, as van Helmont assures us, to preserve the generated body from the opposite eventuality of *de*generation.[46] Working immanently through *archei,* or gases, and by way of these through the vital spirits, the sensitive soul thus acts to create life and preserve it.[47]

This formulation, we may note in conclusion, differs importantly from that to be given later—for example, in the nineteenth century when a certain school of opinion will demand, as substrate for life, a common material, an "identical general medium" (Pagel). For van Helmont, life burns with an infinite variety of flames. To discover the source of this variety, we must seek beyond mere matter itself; we must go back to God. Each seed receives from on high and specified in advance the image of its future identity. Variety, in short, is originally metaphysical rather than merely material.

We may finally agree with Pagel who suggests the irrelevance of

[45] René Descartes, *Traité de l'homme,* various editions; first publ. posthumously, in Latin trans. (Leyden, 1662); see chap. 30.

[46] '*Gas ergo vitale, quia lumen est atque balsamus, praeservam a corruptione . . .*' in "Complexionum atque mistionum elementalium Figmentum," 42.

[47] "Spiritus vitae," 21.

assessing van Helmont from the point of view of a later and differently motivated (deliberately nonreligious) biochemical science. From a slightly different point of view, we may say that certain elements of van Helmont's endeavor are parallel to and reinforce the forward movement of that science whereas other elements are oblique to and do little to further it. Thus his work on gases, his quantitative studies of plant growth, his recognition of the acidity of the stomach and the alkalinity of the small intestine are endeavors that help chemistry along the path that starts in pseudoscience and ultimately leads to science. But his endowment of matter with a transcendental and to a degree psychistic causality sidesteps the essential chemical problem and leaves to later exponents of that science its effective application to the interpretation of life.

· 15 ·

Mechanism, Mechanics, and Mechanistic Biology: Introduction

A new century, the seventeenth, brought with it a revolution in biological thinking. The new biology was part of a wider transformation embracing, notoriously, every aspect of science. And this wider revolution resulted, in turn, from a vast outpouring of talent set free by new ways of obtaining, ordering, and interpreting information about nature. These procedures—empirical, inductive, experimental, deterministic, secular—though not individually new now first came together to form a powerful methodological synthesis.

To be sure, the theorists who most eloquently recommended the new strategy were not always its best practitioners. Thus Bacon preached induction, but his biology is elaborately deductive. Harvey's work viewed in its entirety seems the issue of a curious mismarriage between experimental objectivity and committed Aristotelianism. Descartes unwisely applied to physiology an abstract and semiempirical rationalism which, well-suited to mathematics, had no place in scientific biology.

So inherently powerful were the new strategies, however, that even imperfect applications of them were sufficient to transform traditional ideas of nature. Not that the problems were new. Old issues persisted and the same stubborn questions were brought back for reconsideration. How do living systems differ physically from non-living systems? How do they maintain their organization? Is this organization essentially the same wherever life is found? How is micro- related to macroorganization? What is the role of psychic factors in biological phenomena? But the answers to these questions were affected by new information and newly accepted axioms.

These new answers were not in any sense final; on the contrary, they were subjects of sharply divergent opinion. We find Descartes espousing, William Harvey scornfully rejecting, a materialistic interpretation of nature. Bacon insists upon, while Descartes dismisses, the idea of a common substrate for vital phenomena. Bacon makes "perception" the most intrinsic of all properties of matter, while Descartes purges matter of psychic attributes.

These and other issues were to be fought over for at least another century. Ultimately, however, if there was to be any real biological science, a degree of harmony had to emerge from such extreme conceptual discord. The ordering process, working more swiftly in physics than in biology, lay largely in the gradual acceptance of certain new modes of natural interpretation. Of these interpretative modes, three—the mathematical, the material, and the mechanical—were of paramount importance, and among the three it was perhaps chiefly the mechanical that affected biology (though we shall see that each of the three was intimately—though not totally—involved in the other two).

Mechanism, Machines, Mechanics

The term *machine* had, for Renaissance and post-Renaissance science, a dual signification. Machines were, first, *facts of life:* that is, such practical realities as waterwheels, looms, pumps, and presses and the windmills in view everywhere from the British Midlands to the Peloponnesus.[1] But machines were also, for this era, *hypothetical constructs* used for explaining everything from atoms, to animals, to the cosmos. These abstract constructs had, in turn, two sources. They arose, first, as extrapolations from familiar, concrete machinery and, second, as extensions or applications of abstract mathematical mechanics.

Practical Mechanics

The five-hundred-year period, A.D. 1000–1500, saw an uneven development of technology and of the machines that technology depends on. Improved water mills in the eleventh century, windmills in the twelfth, the spinning wheel in the thirteenth, mechanical clocks in the fourteenth, movable type in the fifteenth, matchlock and wheellock firearms in the sixteenth—these were only a few of the transforming technical developments.

---

[1] Windmills are, more often, features of low-lying terrains; in the mountains water does not need to be raised; its source is the uplands whence its flow downwards can be used as a source of power.

Leonardo. As for actual machinery in use in the sixteenth century, information is available from illustrated treatises published at the time. Evidence of a quite different kind is found in the earlier notebooks of Leonardo da Vinci with whom the century opens (d. 1519).[2] Leonardo offers a vivid array, or disarray, of new and old physical theory to which he adds innumerable suggestions not only for applying this theory but also for testing its soundness. His experiments were, however, mostly paper experiments. Leonardo's inventions—his airships, his pumps and valves, his automobile (spring-driven), his hydraulic screws and turbines, his lathes, drills, presses, and scores of other devices—would not be actually constructed in most cases until long after his lifetime, if ever.

Agricola. Three-quarters of a century later, Agricola, in contrast with Leonardo, limited his interest primarily to things tested through actual usage. Agricola's machines were, exclusively, ones useful in mining, mainly pumps (for drying the mine shafts); hoists and derricks; and ventilating and conveying equipment. These devices were driven by wind- or waterwheels, handcranks, or treadles, the latter powered by men or animals. They represent evolving applications of principles which, far from being new, all dated from Graeco Roman or Egyptian antiquity.[3]

Ramelli. The interests of Agostino Ramelli (1588) were broader. He concentrated principally on ways of raising water and harnessing its power. But his illustrations, more refined than those of Agricola, also show systems for milling and sawing; for lifting big stones and dragging artillery pieces; for digging and draining foundations; for crossing moats; for battering fortifications; and for hurling such heavy projectiles as giant darts, rocks, and fireballs. Equally useful militarily are his amphibious pontoons or landing barges and a formidable lever to be used for prying open a heavy portcullis. Among Ramelli's more esoteric devices is a "beautiful and artificial machine for all persons delighting in study especially the ill-disposed or those

[2] For technical and applied mechanical developments, see Abbott P. Usher, *A History of Mechanical Inventions* [with an excellent bibliography to 1929] (Boston, 1959), A. Wolf, *A History of Science, Technology, and Philosophy in the Sixteenth and Seventeenth Centuries* (New York, 1950, 1959); and Charles J. Singer, E. J. Holmyard, A. R. Hall, T. Williams, eds. *History of Technology* (Oxford, 1956 [vol. 2] and 1957 [vol. 3]).

[3] Georg Agricola, *De re metallica,* trans. Herbert and Lou Hoover (London, 1912).

that are subject to gout."[4] The machine in question is a wheel, about half a man's height in diameter, that can be spun round to present as many as a dozen books to view without the reader's having to rise or to shift the heavy tomes into reading position (many books were quartos or folios in those days).

De Caus. The above works of Agricola and Ramelli were mostly straightforward descriptions little concerned with abstract physical reasoning. In 1615, however, there was published at Frankfurt Salomon de Caus' *Raisons des forces mouvantes.*[5] This begins with a brief, clear treatment of abstract mechanics, illustrated with engravings. The abstract materials are followed by a presentation of thirty-five ways of raising water and of using it—to operate a clock, to measure temperature, to build a self-moving machine, to saw wood, to cause mechanical birds to move and sing and, especially, to operate a variety of elaborate fountains. A second part deals with these fountains specifically, and a third part with the manufacture of musical organs.

To summarize, Leonardo created an unrealized, unpublished paper-revolution in practical mechanics; his revolution was primarily conceptual. It was for the most part not concretely realized by himself or others of his period. Agricola and Ramelli, and other authors of published technical works described a realized, or realizable, applied mechanics—but one which was nonrevolutionary.

Seventeenth century. As the sixteenth century gives way to the seventeenth we witness if not a revolution a steady evolution in practical mechanics. By far the most striking developments were in the area of scientific instrumentation. The microscope, useful telescopes, air thermometers, mercury and water-barometers, airpumps, the slide rule, and the pendulum clock were among the developments of the hundred years 1575–1675. We shall note presently that many of these stemmed from theoretical discoveries of a basic order. The majority, however, were not machines in the sense of doing work or producing a mechanical advantage.

[4] Agostino Ramelli, *Le diverse et artificiose machine del capitano Agostino Ramelli dal Ponta della Tresia* (Paris, 1588), p. 316.

[5] Salomon de Caus, *Les raisons des forces mouvantes avec diverses machines tant utiles que plaisantes aus quelles sont adjoints plusieurs desseings de grotes et fontaines* (Frankfurt, 1615). De Caus derives much from earlier revivals of the *Pneumatics* of Hero of Alexandria; see e.g. Giovanni Battista Aleotti d'Argenta, *Gli artifitiosi et curiosi moti spiritali di Herrone* (Ferrara, 1589).

Of commercial application were innovations in spinning, knitting, and weaving. Windmills and water mills and pumping devices were gradually improved during this period. The year 1663 saw the patenting of the Marquis of Worcester's apparatus in which steam was used for raising water from mineshafts. More esoteric, but significant for our story in having attracted Descartes' attention, were animated representations of men and animals.

"Automata" (or pseudo-automata)

A curious way in which the organism-mechanism analogy expressed itself was in artificial figures of men and animals made with movable parts. Aristotle, discussing the cause of motion in living bodies, compares them with mechanical devices in which a small initial input can produce a major motile response. But such figures, by Aristotle's time, were already a subject with a history. In Greek myth, Vulcan appears as an early fabricator of animated figures, and Homer speaks of the intelligent silver and gold dogs at the palace of Acinous. Actual (as opposed to mythical) automata appeared early in the Mediterranean cultures, and in Africa, America, Oceania, Southeast Asia, China, and India. In their primitive or very early manifestation they served as toys, tomb figures, magical and amuletic devices, religious images, marionettes and other theatrical or ritual artifices. The term automaton needs to be qualified, however, since motion was imparted by men or animals, water, wind, air under pressure, weights, magnets, and springs. In other words none were self-moving in quite the sense an internal combustion engine or a living organism moves itself, that is, by the conversion of stored chemical energy into work.

A note on pre-Renaissance automata. We can trace the history of automata, or pseudo-automata, in early times from Greece to Alexandria (where they achieved a remarkable development), and by way of Byzantium and Islam to Europe. Certain Greek fatidic statues would nod when prophesying—or to indicate which way they wished to be carried on the shoulders of the priests. In Alexandria, Ctesibius, Philo, and especially Hero (in his extant *Pneumatics,* ca. 250 B.C.) describe a variety of ingenious moving figures, sound-producing devices, and theatrical automata. In Byzantium appears the great clock of Gaza (ca. A.D. 500) described as having a moving sun-god and hourly appearances of Hercules pursuing his labors (reputedly illuminated by night). For the Islamic period, a key text is that of Al Jazari (A.D. 1205) whose automata range from solitary mechanical figures to whole boatloads of musicians. In Europe the most admirable automata are associated with fountains and clocks such as the great clocks of Cluny and Strasbourg (created in the second half of the fourteenth century).[6]

Automata in Renaissance Europe. From our point of view, special interest attaches to later sixteenth-century developments—those, namely, that occurred on the eve of the Galilean revolution. It was in 1574, for example, that the Halbrechts rebuilt the clock at Strasbourg using designs of Dasypodius. By this time, on innumerable clocktowers throughout Europe, mechanized saints, devils, knights, and jacquemarts of every variety struck the hour. And in certain churches, altar figures were mechanized. In the meantime, there occurred an almost explosive production of water-driven mechanical fountains including such marvels of complexity as those created for the Duke of Florence in Tuscany, for the Dukes of Burgundy at Hesdin, for Count Hohenems at Salzburg, and for the King of France at Saint-Germain-en-Laye.[7] One of the fountains at Saint Germain appears, incidentally, to be that described by Descartes in a famous passage in his *Treatise of Man*.[8] The visitor to Saint Germain mounted a three-tiered terrace where in grottoes Orpheus bewitched the animals with actual music, and—as described by Descartes—an invitingly mobile Venus was protected by a menacing, trident-wielding Neptune. At the very beginning of his *Treatise of Man* (see chapter 18), Descartes announces as his subject a hypothetical "machine" or "earthen statue."

An interesting point of scholarship is the extent to which mechanistically oriented physiologists were influenced, in constructing a new image of man, by the automated imitations of vital action familiar, as we have just noted, since earliest times. One recent careful student of this question, D. J. de Solla Price, considers the connec-

---

[6] For these examples, we have depended on Alfred Chapuis et Édouard Gélis, *Le monde des automates* (Paris, 1928). See also Conrad W. Cook, *Automata Old and New.*

[7] Part of the mechanism of this is described by Salomon de Caus, *Les raisons des forces mouvantes*, especially, bk. 1, pl. 35 (Descartes' grotto) and bk. 2 "ou sont Desseignées Plusieurs Grotes et Fontaines propre pour l'Ornement des Palais [,] Maisons de Plaisances et Jardines."

[8] René Descartes "Traité de l'homme" (first publ. Amsterdam, 1662), *Oeuvres de Descartes* (Paris, 1909), 11:131.

tion a real and important one. Thus when a serious study of the origins of mechanistic physiology is written it must pay attention to automata as an important source along with other aspects of applied mechanics.[9]

Theoretical Mechanics

Simultaneously with advances in technology there developed in the sixteenth and especially the seventeenth century a partly new science of abstract or theoretical mechanics.[10] As a comprehensive theory, or evolving pattern of theories, it was distinguished in two special ways from other world views. First, it tended, initially in cosmology and presently in biology, to do away with putative psychic agents as motive causes, to depart from the Platonic idea of a cosmic soul—and from the related patristic idea of angelic beings—as causes of celestial movement and to dispense with the Platonic-and-Aristotelian physiological life-soul as the mover of the living microcosm (the organism). Second, it had, as a major goal, to reduce the interactions of bodies, in the heavens and on earth to mathematical rules.

In physics and cosmology, Galileo is usually and not unjustly regarded as the major prophet of the new general movement. Mechanics, however, did not begin suddenly with Galileo. His endeavor was embodied in a partly antecedent, partly concomitant transformation in mathematics and physics due to such men as Cardano (1545), Targtaglia (ca. 1556), Benedetti (1580), Stevinus (1586), and an even more profound revolution in astronomy due to Copernicus (1543), Brahe (1573), and Kepler (1627)—though Galileo neglected Kepler's contributions. Science historians, among them René Dugas and Marshall Clagett, have traced the contributions to mechanics of such still earlier students of cosmology and physics as Thomas Bradwardine (d. 1349), John Buridan (d. 1358?), Nicolas Oresme (d. 1382), and Albert of Saxony (d. 1390)—who must be

---

[9] D. J. de Solla Price, "Automata and the Origins of Mechanism and Mechanistic Philosophy," *Technology and Culture*, 5 (1964):9–23.

[10] For the steps in the evolution of Aristotelian into Newtonian physics see Herbert Butterfield, *The Origins of Modern Science* (London, 1949, 1957), chaps. 1, 4, 7, 8; also Stephen F. Mason, *Main Currents of Scientific Thought* (London, New York, 1953, etc.), chaps. 11 and 15; and Marshall Clagett, *The Science of Mechanics in the Middle Ages* (Madison, 1959). One of the clearest modern interpretations of Aristotelian motion theory is that of E. J. Dijksterhuis, *De Mechanisering van het Wereldbeeld* (Amsterdam, 1950), trans. C. Dikshoorn, *The Mechanization of the World Picture* (Oxford, 1961, 1964), pp. 17–42.

accounted pre-Renaissance inceptors of the new philosophy.[11] These thinkers built in turn on inchoate leanings in early Christian thought toward the substitution of physical force (*impetus*) for psychistic and angelic causes of celestial movement. The story of this "depsychization" of the cosmos has been often retold.

Yet, though Galileo's revolution thus had roots in earlier developments, it was not for that any less a revolution. A newly productive phase in mechanics commenced with the great Italian's studies of acceleration, pendular motion, trajectories, virtual velocities, acoustical resonance, and magnetism. This new phase was to culminate in the grand mechanical synthesis of Isaac Newton. Its major interim heroes included Toricelli (gas mechanics), Pascal (hydrostatics and pneumatics), Huyghens (oscillations, centrifugal forces, impact), and two others, Descartes and Boyle, to whom we shall return because they were biological as well as physical thinkers.

Mechanistic Interpretation of Nature

Three sciences of mechanics. Our intent in summarizing, even thus inadequately, the status of mechanics in the late sixteenth and early seventeenth centuries is to suggest something of the climate in which the new biology developed. It was an atmosphere in which all physical objects appeared to certain scientists as material masses reciprocally determining each other's motions or positions.

This view, which we term mechanistic, developed with respect not only (*a*) to familiar and more or less manipulable objects (such as a clock or a lever or part or all of a plant or an animal) but also (*b*) to very large things (such as the solar system, or the cosmos), as well as (*c*) to very small ones (from microscopic tissue components downward to individual atoms). One may thus properly think not of one but of three mechanical sciences undergoing simultaneous development, viz. a macro-, a meso-, and a micromechanics, whose basic principles were more or less transferable from one to the other.

The idea of a mechanical cosmos belonged, incidentally, to a very old tradition. It may seem at first to stand at the exact antipodes of the cosmic animal conceived by the Pythagoreans and Plato. Yet

[11] See Ernst Mach, *Die Mechanik in ihrer Entwicklung* (Leipzig, 1879), trans. T. J. MacCormack, *The Science of Mechanics* (London, 1893; Chicago, 1902, 1909); René Dugas, *Histoire de la Mécanique* (Neuchatel, 1950), trans. J. R. Maddox, *A History of Mechanics* (New York, 1955); and Marshall Clagett, *The Science of Mechanics in the Middle Ages* (Madison, 1961).

Plato's model of the cosmic *psyche* with separate circular bands for propelling the planets is suggestive of an armillary sphere, and some authorities suppose that Plato possessed such an instrument. Besides, in antiquity, the two ideas—animal and machine—were not mutually exclusive. Plato's cosmos partook of both the animal and the mechanical. Does it seem to us that Pythagoras conceived the cosmos too much in the image of Pythagoras, reasoning outward from the living microcosm to the cosmos at large? To Pythagoras modern mechanistic biology would perhaps seem to commit the opposite error, that of robbing the world of its soul and of conceiving of microcosmic man in the image of the soulless cosmos.

The other two subsciences of mechanics dealing with its meso- and micro-aspects of nature (meso-and micromechanics), will occupy us—in their biological applications—in the next ten chapters.

Nontechnical aspects

Before considering mechanistic biology more specifically, we should note another expression of the mechanistic outlook, namely the literary and artistic.

It is a truism that we ordinarily understand by *the Renaissance* not separate and distinct revolutions in art, technology, and science but rather a simultaneous rebirth of the whole human spirit. In the case of a Leonardo, for example, it is often impossible to classify his interest in a problem as simply scientific or simply artistic. A series of anatomical sketches showing carefully measured proportions will culminate in a freehand rendering of unsurpassed realism and beauty. It is as though the mathematical studies were a disciplinary prelude to a final free act of esthetic intuition. One finds, in the notebooks, hundreds of entries like the following on the flight of birds. "The imperceptible fluttering of the wings without any actual strokes keeps the birds poised and motionless amid the moving air."[12] Or, "A bird descending makes itself so much the swifter as it contracts its wings and tail."[13] In such statements one feels the artist behind the scientist and vice versa. In art and science mechanism becomes a single way of exploring the exact reality of things.

In literature, two figurative uses of *machine* are of special interest

[12] Edward MacCurdy, *The Notebooks of Leonardo da Vinci,* 2 vols. (New York, 1938). 1:455.
[13] Ibid., p. 464.

to science, namely the role of this term as a metaphor for *the world* and, equally important, its role as a metaphor for *man*. When a proper history of mechanicism is written it will concern itself with— among other things—the way in which these metaphors move back and forth between the realms of literature and of science.

Thus, when Lucretius sees ruin in store for the mass and machine of the world (*moles et machina mundi*) does he use *machine* in a technical sense or by a sort of poetic inadvertance? Again, how happens a sixteenth-century hymnist to be praising "this machine round, this vniverse, this vther world he [God] wrochte"? Or through what provenience comes the Prince of Denmark to offer to Ophelia the "machine" that is the physical Hamlet? Merely to ask these questions (to answer them is outside the scope of our venture) is to emphasize that the technical uses of *machine*—as concrete fact, and as concept —are embedded in a web of less technical meanings, some traceable to classical beginnings. Thus *mechane,* for the Greeks, was an engine of war; a scaffold or place where slaves could be shown; a derrick or hoist; a theatrical contraption for bringing gods to the stage; or more generally any contrivance or scheme or way of accomplishing something. And *machina* had, for its Latin users, a similar spectrum of meanings. The evolving contents of these words for machine, of their later derivatives, and of the cognate expressions, *mechanism* and *mechanics,* are more than philological matters: they form the substance of an important and largely unwritten chapter of intellectual history.

Early Biomechanics

The early mechanists of the seventeenth century transferred their concepts with notable ease from one level on the scale of magnitudes to another, and back and forth between the nonliving realm and the living. Their disposition to extrapolate in this fashion from one system of objects to another is perhaps less surprising in that these were scientists whose interests ranged very widely, whose knowledge was comprehensive, and whose own creativity transcended the barriers of a single field.

Thus Galileo was to some extent a macro- as well as a mesomechanist and in addition saw the need for a science of micromechanics. Descartes and Newton conceptualized on all three of these levels. Harvey and Borelli carried the mechanical viewpoint from physics to physiology. Indeed it was only through this ease of movement from realm to realm and level to level that mechanical analysis spread its web and achieved the status of a generalized system.

To the existence of a mechanistic world system physiology responded by undergoing a radical transformation. We shall discover, however, that that transformation was anything but simple or orderly. To suggest the first steps in its occurrence we shall trace the theories of the new biology's two principal founders, Descartes and Harvey, giving special attention to their views of life and matter and their connections. We shall pause likewise to become acquainted with Francis Bacon. Although Bacon stood so early in the sequence of events as not to be personally much influenced by the new mechanical viewpoint, he nevertheless paved the way for its spread from physics to physiology.

Descartes and Harvey were both very much caught up in what Dijksterhuis has aptly termed the mechanization of the world picture.[14] Yet between the two we notice a fundamental difference in their use of mechanical models. To grasp this difference we need to realize that biology became mechanistic simultaneously on two levels, the meso- and micromechanical. At the *meso*mechanical level, the whole organism and its main working parts were viewed as machines. For Descartes (1635) the whole animal was an automaton and so were such working components as the heart and the various pairs of reciprocating muscles. Borelli (1685) had a comparable attitude toward the bird and its wings; the whole bird was a flying machine of which the wings are necessary working parts. Harvey, writing earlier than Descartes, compared the propulsion of blood through the heart to that of water through a water bellows.

At the *micro*mechanical level we encounter numerous devices, of which the most elaborate are, perhaps, those which Descartes imagined in the brain, apertures which, by opening and shutting and by being movable so as to face in various directions, control the flow of spirits between the ventricles and the nerves. In a somewhat more objective but still speculative spirit, Hooke (1665) suspected that his cells might be equipped with small mechanical contrivances. Thus "though I have with great diligence endeavoured to find whether there be any such thing in those *Microscopical* pores [cells] of Wood or Piths as the Valves in the heart, veins, and other passages of Ani-

---

14 E. J. Dijksterhuis, *De Mechanisering der Wereldbeeld*, see above n. 9.

mals, that open and give passage to the contain'd fluid juices one way, and shut themselves, and impede the passage of such liquors back again, yet I have not hitherto been able to say anything positive in it; though, me thinks, it seems very probable, that Nature has in these passages as well as in those of Animal bodies, very many appropriated Instruments and contrivances, whereby to bring her designs and ends to pass, which 'tis not improbable, but that some diligent Observer, if help'd with better Microscopes, may in time detect."[15]

Some theorists carried the mechanical analogy even beyond the level of things potentially visible through a microscope. Thus both Descartes and Giovanni Borelli (to whom we shall return in later chapters) extended mechanical models all the way to the unit particle. It is at this microlevel that we encounter a striking difference between Harvey and Descartes. Descartes maintained a mechanistic viewpoint with respect to the whole spectrum of natural objects, from stars to particles. Harvey—curiously, in view of his immense successes in mesomechanics—abandoned his mechanistic viewpoint when it came to problems of microdynamics. Even his early masterpiece on the circulation reflects the Aristotelianism of his Paduan education. In his study of generation which ripened through the years and was published decades later, he departed completely from his own implicit canons of scientific austerity and experimental objectivity and proceeded in a manner which was not only non-mechanistic but outrightly Aristotelian.

The picture we obtain from a study of Renaissance and post-Renaissance biology is, then, not a picture of sudden emancipation but rather one of slow and laborious struggle to escape the enslavement of old axioms and procedures. A new scientific weapon had been forged, the method of mechanical interpretation, but its use was mastered only slowly.

[15] Robert Hooke, *Micrographia* (London, 1665), p. 116.

First Efforts toward Modern Theory

FRANCIS BACON [1561 · 1626]

We quickly discover in Francis Bacon consistent—or at least re-concilable—concepts of life and matter and their interrelations. Two questions naturally suggest themselves about his physiological ideas. First, how well does Bacon follow, in studying living nature, his own famous precepts for objective inquiry? Second, in what way do his conclusions relate to earlier, and later, physiological theory? The answers to these questions, it must be acknowledged, do not reflect fully favorably on Bacon's endeavor. His physics, where matter is concerned, is a reasonably well-ordered synthesis of alchemy and Stoic pneumaticism—with added elements of something like the hylozoism and the corpuscularism of ancient Greece. The skepticism and inductivism he elsewhere recommends are scarcely apparent in Bacon's theory of matter.

His physiology follows the investigative pattern laid down by Greek interpretive science: he treats classic questions; he treats them both phenomenally and cryptomenally; he adapts his own thinking to broad ordering ideas or themes; and he brings cosmology and biology together in a single general scheme. In Bacon's behalf, it is arguable that his conclusions seem less credulously Greek than Fernel's and less mixed with fantasy than van Helmont's, his orientation being, in these respects, toward the future. He is conscientiously and systematically reductive in his instincts but unfortunately, the physics to which the vital phenomena are reduced in his system is mostly archaic. On the life-matter problem, he is within these short-comings discerning. He sees the need for a conceptual micromodel—partly structural and partly dynamic—capable of explaining the patent expressions of life-as-action.

Bacon on Matter

Bacon's world is, first of all, dual—but his is not the mind-matter duality that we associate with a Plato (or in a quite different way, as we shall see, with Descartes). The question here is one, rather, of two different sorts of matter—one tangible and corporeal, the other so finely subdivided and rapidly mobile as to be completely intangible.[1] There is, thus, in every object both a corporeal component and, in addition, its own intangible (but material) "spirit."[2] This somatopneumatic (as opposed to somatopsychic) dualism is a modified derivative of alchemical and Stoic antecedents. It applies to living as well as nonliving things but also offers, as we shall note, a basis for distinguishing between them.

Having adopted a partly Stoic, partly alchemical dualism, Bacon uses it to explain a full sampling of natural phenomena. These range from various properties of brute matter to virtually every vital process accessible to scientific interpretation, for instance, organic movement, nutrition, generation, perception, and even emotion and imagination.

As between tangible and intangible matter it is as much the latter as the former that is used by Bacon as an instrument of explication. Indeed, we learn that "spirits are the agents and workmen that produce all the effects in the [tangible] body."[3] Hence, we are told, every effort should be made to understand them.

In addition, Bacon urges, science should discover what it can about the *minute parts* of tangible matter. These particles may indeed be "invisible, and incurre not to the Eye; but yet they are to be deprehended by Experience. . . ."[4] Their presence is indicated, for example, by the extreme solubility of dyes and by the diffusibility of odors. Early thinkers are belabored by Bacon for having attributed too much to size and shape, to strife and friendship, to symbolic qualities, to sympathies and antipathies, to virtues and properties, to fate, fortune, and necessity—and not enough to motion. On this

---

[1] *Francis Bacon, Sylva Sylvarum or A Naturall Historie in Ten Centuries Written by the Right Honourable Francis Lord Verulam Viscount St. Alvan. Published after the Author's Death, by William Rawley* Doctor of Divinitie, late his *Lordship's* Chaplaine (London, 1627); J. Spedding, R. L. Ellis, D. D. Heath, eds., *The Works of Francis Bacon* (Boston, 1862) 4:219, 5:83.

[2] *Francisci Baconis de Verulamo . . . Historia Vitae et Mortis. Sive, Titulus Secundus in Historiâ Naturali et Experimentali ad Codendam Philosophicum: quae est Instaurationis Magnae Pars Tertia* (London, 1623), Spedding, Ellis, Heath, eds., "The History of Life and Death," *Works*, 10:156–57.

[3] Ibid., p. 83.

[4] ". . . as *Democritus* said well, when they charged him to hold, that the World was made of such little Moats, as were seen in the Sunne; *Atomus* (saith he) *necessitate Rationis & Experientiae esse convincitur. . . .*" "Sylva," 4:220.

premise, Bacon speculates, with variable success, on consistency (solidity, fluidity), solubility, cohesion, capillarity, the movements, of projectiles, explosions, earthquakes, comets, and other phenomena.[5]

Equally needful of study, he believes, is the "passage of Effects" between the tangible parts and the spirits. It is the distribution and interaction of these two that account for the characteristics of natural objects, including animate ones.

Tangible matter. It is in its treatment of tangible matter that Bacon's theory mixes alchemical with Ionian (hylozoic) suppositions. The Paracelsian *tria prima* are reduced to the alchemical two "Great Families of Things," namely mercury and sulphur. Paracelsus' supposed third member, salt, is viewed by Bacon as "mixed of both [the others] connexed into one." Sulphur is familiarly inflammable, oily, and elaborate; mercury, noninflammable, watery, and crude. Mercury and sulphur are the chief tangible constituents of minerals; water and oil, of animate beings. One may abstract from Bacon's several treatments of this topic the following tabulation of the world's elementary constituents:[6]

TABLE 9

TABLE OF ELEMENTS

CHARACTERIZATION	PRINCIPLES	
Tangible		
Subterraneal (mineral)	Sulphur	Mercury
Vegetable and Animal	Oyle	Water
Intangible or Pneumaticall		
Inferior Spiritual	Flame*	Aire*
(or Pneumaticall)		
Superior Spiritual	Body of Starre	Pure Skie
(or Pneumaticall)		

* The running text on which this table is based suggests a reversal of the position of flame and air but the arrangement shown here seems more consonant with Bacon's theory, and is what he probably had in mind.

The most interesting attribute of Bacon's matter is a certain property of "Perception," that is, an "Election to embrace that which is Agreeable and to expel that which is Ingrate." The elective act, and this point is important, is seen as *preceding* the act of embracing or

[5] Francis Bacon, *Cogitationes de natura rerum* (London, 1605), trans. "Thoughts on the Nature of Things," *Works,* 10:287.
[6] Ibid., p. 331; see also, "The History of Sulphur, Mercury, and Salt. Introduction," *Works,* 9:472.

expelling; and this fact makes perception something separate and distinct from the later, overt reaction. Perception, according to Bacon, is similar to but more acute than ordinary sensation. It is, or permits, a kind of "Discovery" and "Divination."[7]

Bacon's evidence for percipient matter seems at first somewhat strained—his examples ill-selected, and not clearly relevant to the theory. A feather or flame "perceives" a subtler wind than man's dull sense can perceive. *"Wormes* in the *Oake-Apple," "Plenty of Frogs,"* or *"Early Heats* in the *Spring"* are all prognostic [!] of *"Pestilentiall* and *Vnwholsome Yeeres."* And (delightfully, one must admit) "the *Lineaments* of the *Body* will discover those *Naturall Inclinations* of the *Minde,* which *Dissimulation* will conceale, or *Discipline* will suppresse." These instances of discovery and divination are loosely connected by Bacon with instances of actual sensory perception. Thus *"Creatures* that Live in the *Open Aire* (Sub Diô) must needs have a Quicker *Impression* from the Aire, than *Men* that live most within *Doores;* And especially *Birds,* who live in the *Aire,* freest, and clearest; And are aptest by their *Voyce* to tell Tales, what they finde; . . . ."[8]

One's first impression from this odd compilation is that Bacon has—in an *éclat* of inductive enthusiasm—thrown a hodge-podge of observations together. His lumping of these things seems, and to some extent is, artificial. Yet this passage, later applauded by Whitehead,[9] is pregnant with the antecedents of important scientific developments. It approaches, albeit awkwardly, the crucial questions of attraction and repulsion (Bacon himself handles these more satisfactorily elsewhere).[10] It emphasizes, validly, the inferior sensitivity of human as contrasted with other sensors and detectors and the desirability of augmenting man's sensory equipment by proper instrumentation. It raises real questions about time relations of responses, both physical and physiological.

Most importantly, this passage sustains—although in modified form—the Ionian theory of sentience viewed as immanent in matter. In doing this, it helps to bring the mind-matter problem to the forefront of Renaissance science.

[7] "Sylva," 5:63.
[8] Ibid., p. 71.
[9] Alfred N. Whitehead, *Science and the Modern World* (New York, 1925), pp. 57–79.
[10] E.g. at "Sylva," 4:300.

Intangible matter (spirits). Of "Pneumaticalls" (a term Bacon uses more or less interchangeably with spirits) we are told that two, or two orders, exist. Of these the inferior (presumably terrestrial) order comprises air and flame. The superior Pneumatical (celestial) includes "Pure Skie" and "body of the Starre."[11]

The "superficiall speculations" of earlier theorists have confused spirits with vacua, air, fire, natùral heat, qualities, virtues, and other such "idle matters."[12] They are none of these, according to Bacon, but are "infinitely materiall in *Nature*." A spirit is "nothing else but a *Naturall Body,* rarefied to a Proportion, and included in the *Tangible Parts* of *Bodies,* as in an Integument. . . ." We are told that ". . . from them, and their *Motions,* principally proceed *Arefaction* [drying], *Colliquation, Concoction, Maturation, Putrefaction, Vivification* [!] and most of the Effects of Nature; . . ." including such properties as density, hardness, toughness, flexibility, liquefiability, and others.[13] In general, "*Tangible Parts* in *Bodies* are Stupide things; And the Spirits doe (in effect) all."[14] Later we shall learn that the "tangible integument" imposes certain restrictions on what the spirits can do.

Animate vs. Inanimate

Bacon breaks down the world of objects as follows (plants are not strictly living):

Inanimate Bodies
Animate Bodies
    Plants
    Living creatures (animals)
        Insects (animals sometimes produced
            through putrefaction)
        Perfect animals (produced only through
            copulation)

The difference between animate and inanimate things—in the familiar sense of life-as-action—is raised by Bacon directly. The distinction resides in the constitution and spatial distribution of their intangible and tangible components. The living body is, like all

---

11 Ibid., p. 331; and see n. 6.
12 Ibid., p. 291; and "Life and Death," *Works,* 10:157.
13 "Sylva," 5:81.
14 Ibid., 4:220.

bodies, a dual (tangible-intangible) system. Its tangible phase is composed of a mixture of water and oil; its intangible phase, a mixture of air and flame. The intangible (air and flame) phase forms a completely interconnected network of channels with the tangible (water and oil) phase interrupted by them.[15]

Applications of the Theory

Bacon both derives his theory from, and applies it to, such observable acts of living things as nutrition, senescence (and death), "spermatick" generation (the kind we recognize today), and putrefactive vivification (spontaneous generation in decaying materials).

Nutrition. As to nutrition, the organism is a scene of simultaneous and opposite acts of dissolution and repair. The two sets of processes keep pace until old age begins. Then certain organs ("membranes, Tunicles, Nerves, Arteries, Veins, Bones, Cartilages, and most part of the Inwards") lose their power of reparation. Soon they spread their incapacitation to other, ordinarily more reparable, parts ("Spirits, Bloud, flesh, and Fat"). Finally the whole system falls into ruin.[16] The "way of death" is that "the more reparable parts perish in the embrace of the less reparable."[17]

The consumptive or depradative phase of this activity is promoted by air,[18] but air is a spirit. Spirits then are both necessary and, potentially, dangerous. About these ideas, more in a moment. The important point to be noted here is Bacon's insistence upon a balanced equilibrium—between Consumption and Reparation, between Depradation and Refection. The notion is in the tradition established two thousand years earlier by Greek students of nutrition. It is also in the tradition still surviving three hundred years later, when certain biologists will think of life as the sum of linked "anabolic" and "catabolic" molecular activities.

Life-flame. Bacon calls attention to the flamelike quality of "that which may be repaired by degrees without destruction." But life, like a flame, entails eventual extinction. We noted earlier the preoccupation in Renaissance times with factors determining longevity. With Bacon, significantly, the "cause of the Period [life span] is

---

15 "Life and Death," 10:159; "Sylva," 4:149, 331.
16 Ibid., p. 195; "Life and Death," 10:12–13.
17 "Life and Death," 10:173.
18 Ibid., pp. 12, 84.

because the spirit [that is, the spiritual or intangible part of the body] preying always like a still and gentle Flame, the external air which also sucks and dries up the bodies conspiring with it, at the last ruins the frame of the body and organs, and makes them unable to performe the act of Reparation."[19]

The flamelike phenomenon resides in the body's own spirit which "has in it a degree of inflammation, and is like a breath compounded of flame and air, as the juices of animals contain both oil and water. . . . But the inflammation of the vital spirits is gentler by many degrees than the softest flame whether of spirit of wine or other; and besides, it is largely mixed with an aerial substance, so as to be a mysterious combination of flammeous and aerial nature."[20] We shall return to the flamy and aerial vital spirit in a moment.

Air. On the biology of air Bacon's ideas are prescient, if not always easily fitted together. Air, for him, is interconvertible with water.[21] Aerial plants feed themselves by converting air into their substance; whence air is nutritious.[22]

In living bodies, air mingles with flame in the vital (as contrasted with the inanimate) spirit just as oil and water mingle in the tangible part of the body."[23] Thus ". . . the Spirits of *Animate Bodies,* are all in some degree, (more or lesse,) kindled and inflamed; And have a *Commixture* of *Flame,* and an *Aeriall Substance.*"[24] "It is no marvell therefore, that a small *Quantity* of *Spirits,* in the Cells [ventricles] of the Braine and Canales of the Sinewes, are able to move the whole Body, (which is of great Masse) both with so great Force, as in Wrestling, Leaping; and with so great Swiftnesse, As in playing Division vpon the *Lute.* Such is the force of these two Natures, *Aire* and *Flame,* when they incorporate."[25]

In passages somewhat at variance with the foregoing we are warned of the *danger* of external air and the need for preventing its too ready access especially to the surface of the body. "Exclusion of air con-

[19] Ibid., p. 13.
[20] Ibid., p. 160.
[21] At least, "It seems that there be these ways (in likelihood) of version of vapors, or air, into water and moisture." "Sylva," 4:172.
[22] Ibid., pp. 175–76.
[23] Ibid., p. 331.
[24] Ibid., p. 149. The idea of spirit as a fire-air combination had been put forward by the Stoics. See Samuel Sambursky, *The Physical World of the Greeks,* trans. M. Dagut (London, 1956), p. 153.
[25] "Sylva," 4:177.

tributes to longevity, if you guard against other inconveniences" of doing without it![26]

Procreation. Turning now to procreation (of either sort, whether by copulation or putrefaction) the requisite conditions are as follows: The first essential is a *"Gentle* and *Proportionable Heat."* This heat "doth bring forth the spirit" in a substance that must be properly *"Glutinous"* yet *"Yeelding"*—glutinous to detain the spirit and yielding enough to respond to its movements. "Therefore all *Sperme,* all *Menstruous Substance,* all *Matter* whereof *Creatures* are produced by *Putrefaction,* have evermore a *Closenesse, Lentour,* and *Sequasity,"*[27] qualifications without which no generation would occur. In sum, the "great *Axiome* of *Vivification* is, that there must be *Heat* to dilate the *Spirit* of the *Body;* an *Active Spirit* to be dilated; *Matter, Viscous or Tenacious,* to hold the Spirit; and that Matter to be *put forth,* and *Figured."*[28]

With respect to the requisite matter, this must be "proportionate," and must be something like Chalcites (copper sulfate) "which hath a Spirit, that will Put Forth and germinate, as we see in Chemicall Trialls." Bacon responds here to the frequently invoked comparison of living bodies with (in some cases, branching) crystals. A number of observers after Bacon will inquire into the possible analogy between organic growth and the growth of crystals (see Hooke, Buffon, Maupertuis, Schwann, Haeckel, and Virchow).

A central question is whether the cause of differentiation lies in the tangible or the intangible component. To this question Bacon gives a somewhat ambivalent answer. "Vivification therefore always takes place in a matter tenacious and viscous, but at the same time soft and yielding, that there may be at once both a detention of the spirit, and a gentle yielding of the parts, as the spirit moulds them."[29] On the other hand, we hear that the "actions or functions[30] of the individual members follow the nature of the members themselves; . . . but yet none of these actions would ever be set in motion without the vigour, presence, and heat of the vital spirit."

Putrefactive as contrasted with copulative procreation seems to

[26] "Life and Death," 10:104, 168.
[27] "Sylva," 5:115–16.
[28] Ibid., 4:473.
[29] "Life and Death," 10:158.
[30] Ibid., p. 161. Among those enumerated are "attraction, retention, digestion, assimilation, separation, excretion, perspiration . . . ," and sensation.

Bacon a specially fruitful subject for scientific exploration since one may hope to discover thereby the "originall of vivification" and of "figuration." We may also learn much, from such study, about Perfect Creatures (those which arise invariably by copulation and never by putrefaction), and we may find out how to "work effects" upon such creatures.[31] There is, Bacon believes, an essential difference between copulative and putrefactive procreation. The former occurs in a leisurely fashion and only in matter "exactly prepared according to the species" and in a place (the womb) *where the spirit can be prevented from escaping.* Putrefactive procreation, by contrast, occurs rapidly and in matter which is less definitely prefigured.

Vital spirit. We have seen that "protoplasm" is, for Francis Bacon, a two-part system consisting of tangible and intangible components. Both of these are material. Each has a counterpart in inanimate nature from which it differs in a specified manner. The tangible component is distinguished, as we have seen, by converting *watry iuyce* into *oily iuyce* and maintaining these in an equilibrium not found in inanimate objects. We need, finally, to establish in what way the intangible vital or living spirit differs from the lifeless. We noted earlier that, as the tangible component mingles water and oil, so the intangible mingles air and flame.

We hear, further, that the vital or living (air-flame) kind of spirit "governs" and has "some agreement" with the lifeless but "differs in being (essence)" from it. Living spirit differs in that it is "integral" and "self-subsisting" (neither of which Bacon explains very clearly). Living also differs from lifeless spirit in other important ways. Lifeless spirit "desires" to "multiply itself" and to "go forth and congregate with its connaturals," (the Greek idea of like to like). Living spirit likewise multiplies but "has a special abhorrence of leaving the body seeing it has no connaturals near at hand."[32] The two sorts, living and lifeless, should, incidentally, be understood as coexisting in every animate object.

As to the source of vital spirit we learn that it is "repaired from the fresh and lively blood of the small arteries which are inserted into the brain, but this repair takes place according to its own manner, whereof I am not now speaking."[33] This statement is an unelaborated

---

[31] "Sylva," 4:470–71.
[32] Life and Death," 10:162.
[33] Ibid., p. 176.

(and unattributed) allusion to the derivation of spirits from the blood plexuses, a characteristic sixteenth-century modification of Galen's ideas.

Of very great importance is the action of the vital spirit on the body's tangible component. Vital spirit has good and bad effects and must exist in just sufficient abundance as to be "robust" without being "eager," "operative" without being "predatory" or consumptive.[34] The consumption consists in making more of itself at the expense of the tangible component. If present in proper abundance, vital spirit will likewise perform properly its function of "inteneration." That is, it will moisten and soften the hard parts without drying up or dissipating the soft ones.[35]

In individual development, vital spirit plays a role of fundamental importance in exerting a "moulding" effect. Tangible matter "follows the motion of the spirit, which as it were inflates and thrusts . . . [it] . . . out into various figures; whence proceeds . . . generation and organization."[36] Bacon's thought here is weak from the explicative point of view, but adequate explanatory models were fated to come later in embryology than in almost any other branch of biological thought.

RECAPITULATION

In sum, Bacon would limit animateness to a properly organized tangible-intangible material system. Each of these two components —the tangible and intangible—differs from its inanimate counterpart and each is enabled, by its special constitution, to alter and be altered by the other. Of the two, the intangible (spirit) seems in a way more important since it moves, moulds, and maintains the proper consistency of the other. To accomplish these effects, however, the spirit must be self-continuously distributed (in channels) in tangible matter of the right composition, that is, of a composition that inclines it to acquire a definite organization.

Bacon's biology is notable, but not original, in setting forth the main themes and issues which others will develop and debate almost

[34] Ibid., p. 164.

[35] Ibid., p. 163.

[36] Ibid., pp. 84, 169. It is in these passages that Bacon uses one of his few biomechanical metaphors. "If it were possible for young spirits to be put into an old body, it is probable that this great wheel might put the lesser wheels in motion. . . . The nature of the spirits is as it were the masterwheel. . . ."

down to the present period. Is life immanent or emergent? What are the material conditions of its expression? What is the requisite micro-configuration? How does this account for motion, nutrition, genera-tion, imagination, reason? Whence the impetus (later, force or energy) for vital motions? What is the character of displacement and replacement? How do heat and air figure in life's preservation and varied manifestations? What deviations are responsible for disease, senescence, and death and how can these be prevented?

In answering these questions, Bacon proceeds in a highly deduc-tive fashion and bases his deductions upon a wealth of unexamined assumptions, some at least of which are medieval and classical in origin. In other words, he departs conspicuously from his noble ad-monitions concerning the proper method for the advancement of learning.

Students of Bacon have disagreed about the significance of his contributions to science and scientific method. His intentions were, at any rate, impeccable. "His Lordship's course," in the words of his chaplain, editor, and biographer W. Rawley, "is to make Wonders Plaine, and not Plaine things Wonders."[37] What Bacon failed to achieve—in his biology as in the other departments of his scientific effort—was any concrete discovery on which the future might build. It was not in his errors that his scientific shortcomings lay, but in the fact that his errors were nonproductive. They lacked that special heuristic quality of demanding to be corrected. And this fact is iron-ical since it was exactly in biology that Baconian methods of tabula-tion and description were to have their most fruitful application, namely in the comparative and taxonomic studies that ultimately produced the theory of evolution.

[37] From editor "To the Reader," "Sylva."

· 17 ·

The Blood as Locus of Life

WILLIAM HARVEY [1578 · 1657]

William Harvey is rightly regarded as a prime founder of mechanistic biology, but this point requires a certain qualification. We noted earlier that mechanistic science advanced during the seventeenth century, on three distinguishable levels. There was a mechanics that dealt with celestial phenomena; another that dealt with terrestrial "machines" such as windmills, pumps, plants, animals, and people; and still another that dealt with the imperceptibly small constituents that make up the preceding two. Descartes' world, we discovered, was mechanical on all of these levels. With Harvey we encounter a quite different orientation. The point is especially apparent when we turn from his early work on the circulation (1628) to the much later and more compendious study of reproduction (1651).[1]

Matter and Living Matter

Significantly, the mature Harvey has strong antimaterialist convictions. "Neither Aristotle himself," Harvey reminds us, "nor anyone else has ever demonstrated the separate existence of the elements in the nature of things, or that they were the principles of 'similar' bodies [tissues]."[2] Reductive materialism seems, to Harvey, a mere exercise in logical deduction. Things are resolved into elements "according to reason rather than in fact." The four elements? The atoms of Democritus? The *tria prima?* If elements exist they exist in *posse*

[1] William Harvey, *Exercitatio de motu cordis et sanguinis in animalibus* (Frankfurt, 1628). *Exercitationes de generatione animalium, etc.* (Amsterdam, 1651).

For discussions of the comparative merits of the *De motu* and *De generatione,* see Arthur W. Meyer, *An Analysis of the De Generatione Animalium of William Harvey* (Stanford, 1936), pp. 1–10; also Joseph Needham, *A History of Embryology,* 2d ed. (Cambridge, 1959), pp. 133–53; also H. P. Bayon, "William Harvey, Physician and Biologist, etc." *Annals of Science,* 3 (1938):59–118.

[2] William Harvey, "On Generation," *The Works of William Harvey, M.D.,* trans. Robert Willis (London, 1847), p. 517. Quoted passages are from this translation unless specified.

rather than in *esse*. If they are to be found at all in nature, "The so-called elements, . . . are not prior to those things that are engendered, or that originate, but are posterior rather—they are relics or remainders rather than principles."[3]

Living matter is not an organization of elements, in Harvey's view. Nor has he "been able to trace any 'similar' parts, such as membranes, flesh, fibres, cartilage, bone, etc., . . . such . . . that from these, as the elements of animal bodies, conjoined organs or limbs, and finally, the entire animal, should be compounded."[4] How then, will Harvey account for vital organization?

One gets an inkling of an answer from his famous formulation of an epigenetic theory of development, development in which the complex organism is supposed to arise by progressive differentiation from a relatively undifferentiated starting substance. Harvey's extensive observations convince him that "the first rudiment of the body is a mere homogeneous and pulpy jelly, not unlike a concrete mass of spermatic fluid; . . ."[5]

Out of this "at the same time split or multifariously divided, as by a divine fiat, from an inorganic an organic mass results; . . ." We hear that "this [part of the mass] is made bone, this [part becomes] muscle or nerve, this a receptacle for excrementitious matter, etc. . . ." The central point is that "from a similar a dissimilar is produced; out of one thing of the same nature several of diverse and contrary natures . . ." arise. The whole process involves "the segregation of homogeneous rather than the union of heterogeneous particles"[6]—but these "particles" are not atoms or Platonic primary bodies.

Harvey's epigenesis is offered not merely as an account of the way the embryo develops. He makes it part, rather, of a wider concept of physics and the cosmos. Epigenesis is the logical procedure for all coming-into-being in a world where there were no principles to aggregate as alleged by the atomists (or other kinds of reductive theorists). Harvey believes "that the same thing (differentiation) takes place in all generation, so that similar bodies have no mixed elements prior to themselves, but rather exist before their elements. . . . Gen-

3 Ibid., p. 517.
4 Ibid., p. 516.
5 Ibid., p. 517.
6 Ibid., p. 513.

eration, . . . is the distribution of one similar thing having undergone change into several others."[7]

Harvey here voices one of the oldest concepts of the nature of generation. Out of a homogeneous ground-substance, a primal chaos, an *arche,* there arises, through *ekkrisis* or *diakrisis,* an organization. Greek thinkers applied this idea not only to men and animals, but also to the cosmos, and the idea reaches back beyond Greek science into mythology (though we do not mean to suggest that there was homogeneity in Greek thought on this subject).

The Locus of Life

The locus of life lies, in Harvey's conception, in two different fluids, each with very special properties; first, in that "crystalline colliquament, from which the foetus and its parts primarily and immediately arise . . ."; second, in that "fountain of life" that is blood.

The primigenial moisture. The crystalline colliquament with which individual life starts is also designated, by Harvey, "the radical and primigenial moisture." The phrase will remind us of Fernel's *"l'humide radical, ou l'humeur la première née"* (with its numerous medieval predecessors) and of the *liquor vitae* of Paracelsus. In the chick's egg the primigenial moisture "presents itself, after a brief period of incubation, as the first work of the egg's inherent fecundity and reproductive power. This fluid is, also the most simple, pure, and unadulterated body, in which all the parts of the pullet are present potentially, though none of them are there actually."[8]

Scientifically, this proposition makes sense if Harvey means to suggest by potentiality some sort of material organization predisposing to later organ formation. This, however, he does not mean. Although he calls upon science to limit itself to efficient causes, no one is more openly Aristotelian, in respect to development, than Harvey. He says of the primigenial moisture, "It appears that nature has conceded to it the same qualities which are usually ascribed to first matter common to all things, viz. that potentially it be capable of assuming all forms, but have itself no form in fact." Later on we shall even hear the primigenial moisture compared, like Aristotle's seminal *pneuma,* to the substance of the stars. Its potentialities exist meta-

[7] Ibid.
[8] Ibid., p. 513.

physically, not materially. How nonmaterial they are becomes clear in the following polemic: "When I see, therefore, all the parts formed and increasing from this one moisture, as 'Matter,' and from a primitive root, . . . I can scarcely refrain from taunting and pushing to extremity the followers of Empedocles and Hippocrates, who believed all similar bodies [tissues] to be engendered as mixtures by association of the four contrary elements, and to become corrupted by their disjunction; nor should I less spare Democritus and the Epicurean school that succeeded him, who compose all things of congregations of atoms of diverse figure."[9]

Blood. The first and principal differentiation arising out of the primigenial moisture is the blood.

According to certain historians, pre-Harveian opinion placed the motive impetus of the blood in the blood itself. Nordenskiöld sees Harvey as reversing this idea by proving that "the movement of the blood is due to the purely mechanical function of the heart. . . ."[10] Actually, the problem, as Harvey sees it, is not nearly so simple.

The cause of motion turns out to be different for blood entering and leaving the heart. Harvey does not "imagine that the blood has powers, properties, motion, or heat, as the gift of the heart; . . . neither do I admit that the cause of the systole and contraction is the same as that of the diastole or dilatation, whether in the arteries, auricles, or ventricles; for . . . I hold that the innate heat is the first cause of dilatation, and that the primary dilatation is in the blood itself, after the manner of bodies in a state of fermentation, gradually attenuated and swelling. . . ."[11]

Harvey makes the foregoing point, that the blood enters and dilates the auricles under the impetus of its own innate heat, for a very special reason. He means to establish the priority, the supremacy, of the blood as compared with the heart and every other part of the body. He wishes to establish the blood as the "primary cause of life." His insistence upon this point that blood is the physical vector of life is emphatic and unequivocal. Herbert Butterfield correctly asserts that Harvey "waxes so lyrical sometimes [about the heart] that he

[9] Ibid., p. 516.

[10] See, e.g., Erik Nordenskiöld, *The History of Biology,* trans. Leonard Bucknall Eyre (New York, 1928), p. 114–18.

[11] "A Second Disquisition to John Riolan, Jun., in which Many Objections to the Circulation of the Blood are Refuted," *Works,* p. 137.

half-reminds us of Copernicus on the subject of the sun."[12] Harvey does so, in the *De Motu,* but in the *De Generatione* he waxes equally lyrical on the subject of the blood.

Long passages in the *De Generatione* are, in effect, a sort of apotheosis of blood. Blood is the first thing to differentiate out of the primigenial moisture.[13] In chick eggs it appears as a pulsating particle in which "there is a primary vital principle [*anima*] inherent, which is the author and original of sense and motion, and every manifestation of life."[14]

This pulsating point is likewise the first locus of that innate heat (*calidum innatum*) which is the invariant concomitant of life. The blood vessels are constructed later "for the sake of" (we seem to hear Aristotle speaking) the already-formed blood. The pulsation of the blood is due to something inherent in itself. "The diastole, I say, takes place from the blood swelling, as it were, in consequence of containing an inherent spirit. . . ."[15]

Even as early as in the *De Motu* Harvey is already asking

Has not the blood itself or spirit an obscure palpitation inherent in it, which it has even appeared to me to retain after death? and it seems very questionable whether or not we are to say that life begins with the palpitation or beating of the heart. The seminal fluid of all animals—the prolific spirit, as Aristotle observed, leaves their body with a bound and like a living thing; and nature in death, as Aristotle further remarks, retracing her steps, reverts to whence she had set out, returns at the end of her course to the goal whence she had started; and as animal generation proceeds from that which is not animal, entity from non-entity, so, by a retrograde course, entity, by corruption is resolved into non-entity; whence that in animals, which was last created, fails first; and that which was first [i.e. the blood, not the heart], fails last.[16]

In eclectic fashion—borrowing ideas from Plato, Aristotle and

[12] Herbert Butterfield, *The Origins of Modern Science, 1300–1800* (New York, 1951) p. 31.
[13] *Works,* p. 373.
[14] Ibid., p. 373.
[15] Ibid., p. 375.
[16] Ibid., p. 29.

Galen—Harvey makes blood now the cause, now the instrument, now the locus, now the material equivalent of the vegetative soul. "The life, therefore," to cite only one of numerous allusions, "resides in the blood (as we are also informed in our sacred writings),[17] because in it life and the soul first show themselves and last become extinct."

From all which considerations "it clearly appears that the blood is the generative part, the fountain of life, the first to live, the last to die, and the primary seat of the soul; the element in which, as in a fountain head, the heat first and most abounds and flourishes; from whose influxive heat all the other parts of the body are cherished, and obtain their life; . . . ."[18] Harvey agrees with Aristotle when the latter says that "the blood, moreover, is that alone which lives and is possessed of heat whilst life continues."[19]

The blood, in fine, is "the cause not only of life in general—inasmuch as there is no other inherent or influxive heat that may be the immediate instrument of the living principle except the blood —but also of longer or shorter life, of sleep and watching, of genius or aptitude, strength, &c."[20]

As for the *pneumata* of Galenic medicine, Harvey sees no need for considering them as anything distinct from the blood. He does admit a (more Aristotelian sort of) pneuma in the blood and seems, as we shall note presently, to equate this with soul. To his mind, "the blood and these spirits signify one and the same thing, though different—like generous wine and its spirit. . . ."[21]

Harvey returns repeatedly to his attack on the *pneumata*. We hear that "There is, in fact, no occasion for searching after spirits foreign to, or distinct from, the blood; to evoke heat from another source; to bring gods upon the scene, and to encumber philosophy with any fanciful conceits; what we are wont to derive from the stars is in truth produced at home: the blood is the only *calidum innatum,* or first engendered animal heat. . . ."[22] Note how Harvey literally materializes the innate heat. It *is* the blood.

> Scaliger, Fernelius, and others, giving less regard to the admirable qualities of the blood, have imagined other spirits

[17] Lev. 17:11, 14.
[18] *Works,* p. 377.
[19] Ibid., p. 379. Aristotle *Historia Animalium* bk. 3, chap. 19.
[20] *Works,* p. 380.
[21] Ibid., p. 117.
[22] Ibid., pp. 136–7, 502.

of an aerial or ethereal nature, or composed of an ethereal or elementary matter, a something more excellent and divine than the innate heat, the immediate instrument of the soul, fitted for all the highest duties. . . . They have, therefore, feigned or imagined a spirit, different from the ingenerate heat, of celestial origin and nature; a body of perfect simplicity, most subtile, attenuated, mobile, rapid, lucid, ethereal, participant in the qualities of the quintessence. They have not, however anywhere demonstrated the actual existence of such a spirit, or that it was superior to the elements in its power of action, or indeed that it could accomplish more than the blood by itself. We, for our own part, who use our simple senses in studying natural things, have been unable anywhere to find anything of the sort.[23]

Harvey, however, fails to live up to the standard of secularism and simplicity he sets for others. His hymn to the blood leads him to associate something in it with the substance of the stars. With Aristotle he accords to semen a vital something "bearing a proportion to the element of the stars."[24] Speaking for himself he says,

Now I maintain the same things of the innate heat and the blood [as Aristotle does of semen]; I say that they are not fire, and neither do they derive their origin from fire. They rather share the nature of some other, and that a more divine body or substance. They act by no faculty or property of the elements; but as there is a something inherent in semen which makes it prolific, and as, in producing an animal, it surpasses the power of the elements—as it is a spirit, namely, and the inherent nature of that spirit corresponds to the essence of the stars— so is there a spirit, or a certain force, inherent in the blood, acting superiorly to the powers of the elements, very conspicuously displayed in the nutrition and preservation of the several parts of the animal body; and the nature, yea, the soul in this spirit and blood (*natura imo anima in eo spiritu & sanguine*) is identical with the essence of the stars (*respondens elemento stellarum*).

And yet Harvey sees this spirit not as something separate from blood; for him, blood and spirit are one: "Sanguis et aquae est

23 Ibid., pp. 502–3.
24 Ibid., p. 314.

spiritus . . ." Blood qua spirit "is possessed of plastic power (*virtutis plasticae*), and endowed with the gift of the vegetative soul, [and] is the primordial and innate heat, and the immediate and competent instrument of life (*vitaeque instrumentum immediatum atque idoneum*)."[25]

How far removed all this seems from the careful empiricism and restraint that were, by and large, the earmarks of Harvey the experimentalist! It would be wrong to think of the great Englishman as suddenly separating biology (as we understand it) from its traditional involvements with Greek thought, with the doctrines of the church, and with the science and pseudoscience of the Middle Ages. The separative process was to be gradual, Harvey's contribution to it signal. That contribution lay not in the area of reductive explanation but in the transfer from physics to biology of quantification and experimentalism (apparent in his study of the circulation but not in his theory of the material basis of life).

Life and Matter

Harvey's "biochemical" theories are perhaps chiefly interesting as illustrating a dichotomy of physical concepts whose origins belong to antiquity. Galen puts it clearly when he says that "one class (of philosophers) assumes that all substance that is subject to generation and destruction is at once *continuous* and susceptible of *alteration*. The other school assumes substance to be unchangeable, unalterable, and subdivided into fine particles, which are separated from one another by (and move about in) empty spaces."[26]

Elsewhere we have termed these two viewpoints dynamic and dimensional respectively. The two are not really mutually exclusive. Plato achieved a fruitful synthesis of them, and so, in our day, has modern physics. But Harvey takes sides firmly against the dimensional and rules out elements altogether. Life manifests itself first in a primigenial jelly whose differentiative capacities are purely potential, and manifests itself next in blood or in a blood-associated spirit that is somehow "analogous to heaven."[27]

In sum, Harvey is not prepared to extend to matter the mechanistic

[25] Ibid., pp. 505–6.
[26] Galen, *On the Natural Faculties,* bk. 1, chap. 12 (trans. Brock).
[27] *Works,* p. 507.

thought that we so admire in his analysis of the circulation. It is this absence of micromechanics from his outlook that distinguishes him sharply from his contemporary Descartes. And, significantly, Descartes the mathematician leaves numbers behind him when he turns to biology and studies life, as we shall see, in a nonquantitative way.

· 18 ·

Microbiomechanics

RENÉ DESCARTES [1596 · 1650]

The factors that affected Descartes' scientific endeavor—his physiology and his physics—are familiar enough. Chief among them was
the increased attention on every hand to mechanism, materialism, and
mathematics as conceptual tools for the interpretation of nature. The
most powerful innovators in the use of these tools were not biologists
but physicists among whom Galileo shone with special brilliance.

Descartes agreed with Galileo that mathematics was the key to
understanding the physical universe and held that "no other means
exists for finding the truth."[1] Beyond this, however, he was critical,
even contemptuous, of the great Italian, acknowledging no debt to
him nor to any other famous contemporary. Created by and helping
to create a scientific revolution, Descartes dissociated himself, as
far as possible, from his fellow rebels. He likewise declared his
independence of the authors of the past. If he agreed with their ideas,
he preferred that it be through a painstaking process of personal
rediscovery. To begin to think, he felt that he must wipe the slate
clean. How close he actually came to the procedural goal he established for himself is a question to which we will return.

Interestingly, for the chief inventor of analytical geometry, Descartes did not build his physiology on quantitative lines. Measure-
ment was virtually absent from it. His biology does resemble his
mathematics but only in its use of the deductive rationalistic method
in reaching solutions to problems. This method partly agreed with
and partly differed from that proposed slightly earlier by Francis
Bacon. Descartes shared with Bacon an antipathy toward ill-
grounded preconceptions. Beyond this point the two moved in opposite directions. Bacon would have had science generalize from the

---

[1] René Descartes, Lettre à Mersenne, Oct. 11, 1638, *Oeuvres de Descartes* (ed.
Charles Adam and Paul Tannery, 1897), 2:380.

largest possible number of new, direct observations and verify by still further observations obtained from experimentation. Descartes' plan was to begin with the smallest number of self-evident axioms or assumptions and to create a conceptual structure explaining the world as we see it.

Matter

Atomic theory, of the sophisticated variety that developed in Ancient Greece, was reintroduced to Europe in varied formulations by persons slightly antecedent to or contemporary with Descartes, among them Francis Bacon, Daniel Sennert (from 1619), Sebastian Basso (1621, 1630), J. C. Magnenus, Descartes' special rival and critic (esp. 1646 and 1648), and Pierre Gassendi.[2] The works of these theorists were either known directly to Descartes or brought to his attention by his friend, Isaac Beeckman, who, in close contact with the scientists of the time, acted as an early mentor, almost as an older brother, to him. As for Gassendi, he wrote a critique of Descartes' *Méditations* to which Descartes replied in detail. In a letter to Beeckman, April 23, 1619, Descartes writes: "I slept—and you awakened me."[3] Between 1619 and 1629, Descartes developed his own particulate theory, which follows the ancient atomists with respect to the corpuscularity of matter but not with respect to its supposed ultimate atomicity (indivisibility). Descartes was a corpuscularist but not an atomist. Like most of his doctrines, his corpuscularism is partly derivative and partly markedly his own.

For Descartes, matter has no property other than extension *"in longum, latum e profundem."*[4] Matter is indistinguishable from space.

[2] Francis Bacon, *Cogitationes de natura rerum* (1605), *Novum organum* (1620); David Sennert, *De Chymicorum cum Aristotelicis et Galenicis consensu ac dissensu* (1619), *Hypomnemata physica* (1636); Sebastian Basso, *Phylosophia naturalis adversus Aristotelium* (1621); J. C. Magnenus, *Democritus reviviscens sive de atomis* (1646); Pierre Gassend (or Gassendi), *De vita et moribus Epicuri* (1647).

Descartes developed his particle physics in *Le Monde,* published posthumously in 1664, written about 1633, and in his *Principia philosophiae* (see below, n. 4) published in 1644. For the role of Descartes in the revival of atomism, see J. R. Partington, "The Origins of the Atomic Theory," *Annals of Science,* 4 (1939): 245; C. Adam, "Vie de Descartes," *Oeuvres,* 12:44; "Descartes et Beeckman," *Oeuvres,* 10:17; Marie Boas, "The Establishment of the Mechanical Philosophy," *Osiris,* 10 (1952):422.

[3] C. Adam, "Vie de Descartes," *Oeuvres,* 12:45; Lettre à Beeckman, April 23, 1619, *Oeuvres,* 10:162.

[4] René Descartes, *Principia philosophiae* (first publ. in Latin, Amsterdam, 1644; trans. into French by Claude Picot, reviewed by Descartes and publ. in Paris, 1647), part 2, sec. 4.

The universe is full of matter, is a *plenum.* "Since it is in all ways repugnant that *nothing* should possess extension," he says, "so concerning space which is supposed empty, it must be concluded that since there is extension in it so there is in it, necessarily, substance also."[5] There is no such thing as a vacuum.

Descartes' matter is, then, minutely corpuscular but not in the sense of being ultimately indivisible.[6] When a material body moved, he supposed that—there being no vacuum—its place must be taken by another body. Hence, "we must conclude that there must necessarily always be a circle of matter or ring of bodies together moving at the same time."[7] Descartes becomes intoxicated with the idea of circular motions or vortices (*tourbillons*); individual masses spin or revolve; concatenations of masses move round in such a way that no space will be left vacant.

Three primary elements make up the visible world (*mundus adspectabilis*). All three are derived cosmogonically from a single sort of original particle.[8] The originals had been of medium caliber; and, at medium speed, had (*a*) rotated round their own axes and (*b*) circled round what are now the axes of all the fixed stars. The motions of these original particles were mutually abrasive causing them to tend to round off. They could not have started as round particles, incidentally, since "several balls joined together did not compose an entirely solid and continuous body such as this universe is. . . ."[9]

The first element (comparable to fire) is a dust of innumerable minute particles of indeterminate shape and extreme mobility chipped off in the abrasive process. The second element (celestial or subtle matter corresponding to ether) comprises particles well rounded by abrasion. The particles of the third element (terrestrial matter corresponding to earth) are larger, less thoroughly rounded, and less easily moved. The first element composes the bodies of the sun and the fixed stars. The second element composes the interstellar substance or "heavens." The third composes the earth (along with its water and air) as well as the planets, and comets. The sun and fixed stars emit, the heavens transmit, the earth absorbs and reflects light.

[5] Ibid., 2.16 (cited by part and section).
[6] Ibid., 2.20.
[7] Ibid., 2.33. See also, "Traité de la Lumière," *Oeuvres* 11:19–20.
[8] *Principia,* 2.23.
[9] Ibid., 3.48.

"So I have reason to avail myself of three differences: 1) to be luminous 2) to be transparent 3) to be opaque; which are the principal differences which can be seen to distinguish the three elements of the visible world."[10]

As all three of Descartes' elements are presumed to be derived from the same kind of matter, differentiated only by the size, shape, and motion of their constitutent particles, their common derivation likewise renders them readily intertransmutable (unlike the atoms of Democritus, Epicurus, and Lucretius).

In general, the interstices between the variously shaped particles of the third element are occupied by the rounded particles of the (subtle) second element whose interstices are, in turn, completely filled by the exceedingly swift, small particles of the (still subtler) first element. The minuteness, mobility, and changeability of the particles of the latter enable it to take the shape of whatever space it occupies and thus to prevent a vacuum.[11] All three elements are to be found in all living bodies.

Descartes derives many sensible and chemical properties from particle shape and movement. Air consists of slender detached particles of the third element, obeying each within its own spherical field of movement, impulses imparted by the intervening particles of the other two elements.[12] Water particles are long, slippery, "like little eels" so that they can become interlaced without becoming entangled.[13] Finally, earth particles are "of all sorts of shapes," and large enough so that the first two elements cannot carry earth along with them as they can air and water.[14] In general, earth particles are either *branched,* compact (spheres or cubes), or attenuated.[15] On their shapes, and to an even greater degree on their varying degrees of rest and mobility, the sensible properties of matter depend. We shall return to some of these when we consider what Descartes has to say about the microorganization of living bodies.

---

[10] Ibid., 3.52. (Note: In his "Traité de la Lumière," written slightly earlier but published posthumously, Descartes called the three elements Fire, Air, and Earth. See *Oeuvres,* 11:24–5.

[11] Lettre à Mersenne, Jan. 9, 1639, *Oeuvres,* 2:483.

[12] *Principia,* 4.45.

[13] René Descartes, "Les météores" in "Discours de la methode, plus la dioptrique, les météores et la géometrie, qui sont des essais de cete methode" (first publ. anonymously, Leyden, 1637), *Oeuvres,* 6:233.

[14] *Principia,* 4.57.

[15] Ibid., 4.33.

Position of Descartes as a Physiologist

The two principal physiological treatises of Descartes have been largely neglected by Cartesian scholars, and generally misrepresented by historians of science. The one penetrating analysis of them, by Auguste Georges-Berthier (1914, 1920 posth.),[16] is ultimately negative in its assessment, primarily on the grounds that Descartes was responsible for no solid additions to scientific knowledge of the sort that, say, Vesalius made in anatomy and Harvey in one aspect of physiology.

It is arguable, however, that Descartes' importance as a biologist must be sought in a different direction. A reexamination of his sources has recently been undertaken by the present author, and, although the search has not been completed, it justifies the following judgment of the character and significance of his effort. It is true that Descartes makes no important new anatomical or physiological discoveries. His undertaking amounts, rather, to a reductive reanalysis of primarily borrowed information. The sources of this information (some of it misinformation) are: first, the works of the major somewhat earlier medical biologists (Fernel, Colombo, du Laurens, Bauhin, Caspar Bartholin, Piccolomini, and Jean Riolan); second, a spectrum of Scholastic texts on physiological subjects—texts that were used, as Etienne Gilson points out, in the Jesuit college that Descartes attended; and, third and especially, the major works of Galen (with which Descartes was directly or indirectly familiar), as well as those of Aristotle and Plato.

In effect, Descartes developed a synthesis based on all these authors, on Galen more than any other, and then systematically *explained* this synthesis in terms of his own corpuscular physics. More thoroughly than any other author he shows what it means to reduce the patent phenomena of biology to the latent cryptomena of physics. If the physics in question proves for the most part ultimately defective, Descartes stands nevertheless as a pioneer of reductive procedure, and this fact alone establishes him as a major founder of the physiology of the future. It is easy, and partly correct, to think of Descartes as philosophizing while Harvey experimented. But of the two, Descartes was the more committedly mechanistic and materialistic.

---

[16] A. Georges-Berthier, "Le mécanisme cartésien et la physiologie au XVIIᵉ siècle," *Isis,* 2 (1914):37–89; 3 (1920):21–58.

Harvey's glory was to have illustrated in the most monumental and productive way the importance in biology of quantification and experimentation. But his mechanicism was one dimension in a paradigmatic amalgam, and his total system was at least three-dimensional in the sense of being at once mechanical, teleological, and animistic. Descartes was not the first of his era to reason nonpsychistically (Sanctorius, especially, preceded him in this). Yet Descartes, more than Harvey or anyone else, understood the need for the substitution of physical for psychistic explanations of vital activity in general. While Harvey never repudiated the Greek physiological soul, Descartes dealt it a blow from which it was never to recover. Therein lies, especially, his contribution to the evolution of physiological theory, notwithstanding the prompt rejection of most of the explanations he offered.

Descartes' Physiological Program

It would be difficult to improve on the outline of intellectual intentions that Descartes inserts prefatorially in his treatise *The Description of the Human Body,* which he wrote near the end of his career.[17]

He begins by making the heart's heat the prime cause of vital movements: "So as to have first a general notion of the entire machine which I have to describe, I shall say here that it is the heat which it has in its heart which is like a mainspring and principle of all the movements that are in it." We shall return to heat's role in a moment.

Next, Descartes brings out the tubular structure of veins, arteries, and the intestines and their respective roles in conduction. He tells us that "the veins are tubes that conduct blood from all parts of the body to the heart where it serves to nourish the heat that is there." Descartes is thus in agreement with a certain "English physician" (Harvey is meant) who sees blood as moving unidirectionally through the veins. The "stomach and bowels are another, greater tube, studded with many little holes whereby the alimentary juice flows into the veins which carry it right to the heart." Next, our attention is directed to the arteries and we learn that these "are still other tubes whereby the blood, warmed and rarefied in the heart,

[17] René Descartes, "La description du corps humain" (alternate title, "De la formation du foetus") (first publ. jointly with "L'homme de René Descartes," posth., Paris, 1664).

passes thence to all other parts of the body to which it carries heat and material to nourish them."

Mention of the arteries provides a transition to nervous and neuromuscular function. Thus, food and nourishment are not the only things transported by the blood, since its "most agitated and most lively particles, being carried to the brain through the arteries that come straightest from the heart, compose something like air or a very subtle wind, which we call the animal spirits which, dilating the brain, render it fit to receive impressions of external objects, as well as impressions from the soul, i.e., to be the organ or seat of common sense, of imagination, and of memory." Moreover, and finally, "this same air or these same spirits flow from the brain through the nerves into all the muscles, by which means they dispose these nerves to serve the organs of external sense, and, differently inflating the muscles, to give movement to all the members.

"Here, in sum, are all the things I have to describe, so that knowing distinctly what, in each of our actions, depends only on the body and what on the soul, we may make better use of both the former and the latter, and may better cure or prevent their maladies."[18]

Having promised to discuss all these subjects, Descartes does so faithfully. Many of them he has already detailed in the preceding *Treatise of Man.* He also has much to say about the origins of the body parts and functions.

Notes on the Nature of the Cartesian Program

The decisive assumption underlying Descartes' physiology is the clearcut distinction it draws between life (which Descartes equates with motion and especially with heat) and soul (which he equates with mind and thought and admits in man alone among all living things). He denies animals a soul—though he agrees that as automata that move without power of judgment they may be more efficient than man and in this way superior to him.[19]

The difference between a living body and a dead body is the dif-

---

[18] Ibid., 1.7 (cited by part and section), *Oeuvres,* 11:226.

[19] Lettre à Morus, Feb. 5, 1649, *Oeuvres,* 5:267, "Life I deny to no animal . . . except insofar as I lay it down that it consists simply of the warmth of the heart." See also Lettre au Marquis de Newcastle, Nov. 23, 1646, *Oeuvres,* 4:573. See also Henry A. P. Torrey, *Philosophy of Descartes* (1892), p. 281, on "Automatism of Brutes." Note: Rarely, Descartes departs from his insistence on the soullessness of animals and grants them a material, as opposed to man's immaterial, soul (see Descartes to Voëtius, *Oeuvres,* 10:30).

ference between a watch that is wound up and a watch that is broken and whose movements cease.[20] It is wrong to believe that soul gives the body its movement and heat.

> The error is that, from observing how all dead bodies are devoid of heat, and consequently of movement, it has been thought that it is the absence of the soul which has caused these movements and this heat to cease; and thereby, without reason we have come to believe that our natural heat and all the movements of our body depend on the soul. What, on the contrary, we ought to hold is that the reason why soul absents itself on death is that this heat ceases and that the organs which operate in moving the limbs disintegrate.[21]

Descartes' separation of life from soul signals the close of a long period of conceptual and semantic cross-connections between the two beginning in Greece where one word, *psyche,* meant both. Not everyone immediately accepted the dichotomy drawn by Descartes, but the distinction was fundamental in all later physiological inquiry and fully entitles Descartes to be considered a main founder of modern scientific biology. Animism was not destined to die with Descartes; but, after Stahl's heroic attempt to fan it, (from 1688), the spark was to glow more and more dimly. The nature of the world— and man—would not seem the same.

Life-as-Action

Descartes' physiology is in essence an attempted restatement of gross form and function in terms of corpuscular composition and action. To be sure, he also drew analogies as between the body and a fountain, and a "clock or other automaton," and he used machine-like models (springs, counterweights, wheels) to explain certain functions. But he was at least as interested in *petites parties* as he was in *ressorts et roues.* He was more a micro- than a mesomechanist.

Heat. The importance of heat is made plain both in the *Treatise of Man,* and in *The Passions of the Soul* where we learn that "there is a continual heat in our heart that is a sort of fire sustained there by the blood from the veins and that this fire is the corporeal principle

---

[20] René Descartes, *Les passions de l'ame* (first publ. Amsterdam, 1649), part 1, art. 6.

[21] Ibid., 1.5 (cited by part and article).

of all the movements of our members."[22] In nature generally, we encounter: light with heat, light without heat, and heat without light.[23] The heart's fire is *"un de ces feux sans lumière."*[24] More precisely, the heart's heat, "is a motion of the particles of the third element which is so brisk that it becomes sensible."[25] Aside from this "there is no need to imagine it to be different in nature from all such caused by any mixture of liquids or ferments that causes the body they are in to expand."[26]

Blood is a readily expansible fluid. The heat of the heart tends to subtilize and rarefy blood that comes in from the veins intermittently *("goutte à goutte")*,[27] and the rarefied blood dilates the heart and passes, *"avec effort,"* [28] into the arteries.

The beat of the heart and the arterial pulse are due solely to this heat-induced expansion of the blood which, moved in this way, carries natural heat, along with nourishment, to all the parts of the body. Later we shall see that in rarefying the blood the heat of the heart also alters some of its particles and so prepares them to enter into the position of living parts of the body.

Spirits. Descartes' tenacious cleavage to a mechanist-materialist view comes out especially clearly in this treatment of animal spirits which he adapts to his own way of looking at things. Animal spirits "have no other property than that they are very small bodies that move very fast . . . and do not stop in any one place."[29] They are like "a certain very subtle wind, or rather a very live and very pure flame."[30]

What the ancients called "vital (as opposed to animal) spirits" are not different, according to Descartes, from arterial blood itself.[31] Animal spirits, too, are already present as the "liveliest, strongest, subtlest" particles in the blood reaching the brain. These particles "climb straight up to" the brain since, being the liveliest they are

22 Ibid., 1.8.

23 J. R. Partington, *A History of Chemistry* (London, 1961), 2:436.

24 René Descartes, "L'homme" (first publ. posth., Paris, 1664, under the title "L'homme de René Descartes et un traitté de la formation du foetus du mesme auteur avec les remarques de Louys de la Forge, etc.") *Oeuvres,* 11:123.

25 Partington, *History of Chemistry,* 2:436.

26 "Du corps humain," (see n. 17), 2.8, *Oeuvres,* 11:228.

27 "L'homme," *Oeuvres,* 11:123.

28 "Du corps humain," 2.10, *Oeuvres,* 11:231.

29 "Les passions de l'ame," 1.10, *Oeuvres,* 11:334.

30 "L'homme," *Oeuvres,* 11:129.

31 Lettre à un Seigneur (Marquis de Newcastle) April (?) 1645, *Oeuvres,* 4:191.

most inclined, by their momentum, to travel in a straight line, and the carotid arteries give them a better chance to do this than do any other blood vessels carrying blood from the heart. On their arrival, specifically in the blood vessels surrounding the conarium or pineal gland, they separate from the coarser blood particles and leaving the bloodstream pass first into the pineal gland and then to the brain ventricles.[32]

Animal spirits moving out from the ventricles via tubular nerves cause muscular movement. "Not that the spirits that come immediately from the brain suffice by themselves to move the muscles, but they determine other spirits already in these two muscles, i.e., the flexor and extensor to leave one of these very promptly and enter the other."[33] Descartes here develops an ingenious (but erroneous) guess about the mechanism of reciprocal muscle action.[34]

Nerves likewise contain a central marrow of fibrils surrounded by spirits and extending to each sensory end-organ from a special part of the brain. This happens ". . . in such a way that the least thing that moves the body part or extremity causes movement of the part of the brain from which it [i.e., its fibril] comes just as when one pulls one end of a cord one thereby moves the other."[35] The motion occasioned in the brain somehow acts through the spirits contained in the ventricle to impinge on and move the pineal gland whose movement acts, in turn, on the soul.

Nutrition and Microdynamics

In discussing nutrition, Descartes comes to grips with problems of microstructure and microdynamics. Adopting a tradition that appears to be Alexandrian or older in origin, he gives the solid parts a fibrillar constitution. "All the solid parts are composed entirely of differently extended, folded and sometimes interlaced fibrils (*petits filets*) each of which arises at some point on an arterial branch. . . ."[36]

The central feature of nutrition—here he follows an equally an-

---

[32] "L'homme," *Oeuvres*, 11:129. In an excess of mechanistic deductivism, Descartes rushes on to note that the second liveliest particles being forced to turn downwards with the descending aorta still tend to go straight, heading, in this case, directly for the generative organs!

[33] "Les passions de l'ame," 1.11, *Oeuvres*; 11:335.

[34] For a more extensive development of the same subject, see "L'homme," *Oeuvres*, 11:132.

[35] "Les passions de l'ame," 1.12, *Oeuvres*, 11:337.

[36] "Du corps humain," 3.19, *Oeuvres*, 11:246.

cient tradition—is the continual renewal of the body. The problem, then, is to decide how the basic fibril is replaced. Descartes solves this by having the fibrils secreted continuously by the arteries. With each pulsebeat, arterial pores open slightly and eject blood particles which "strike the roots of the fibrils." When the pulse recedes and the pores shut down, the extruded particles remain attached to the tissue fibrils and "thus enter into the composition of the body."[37]

Meanwhile the whole fiber moves outward and away from the artery which secretes it. For a mechanist, a major concern must be the cause of this movement. Fibers are encouraged to move, we are told, by currents in the fluids that surround them. These "fluid parts" as opposed to the "solid parts" are made up of "humors and spirits" which flow alongside the fibers in an "infinity of little rivulets."[38]

For the source of this movement Descartes turns to the "subtle matter" of the first and second elements. "In addition to the pores through which the humors flow there are quantities of other smaller ones through which there continually passes some of the matter of the first two elements which I described in my *Principles;* and just as the agitation of the matter of the first two elements sustains that of the humors and spirits, so the humors and spirits while flowing alongside the fibrils . . . cause them continually to advance a little. . . ."[39] Moreover "this subtle matter which is in our body does not stay there a single moment, but is constantly going out and coming back in."[40]

There is, thus, a continual process of displacement and replacement. "As fast as any particle is detached at the extremity of each fibril, another is attached at its root in the manner of which I have already spoken. And that which is detached evaporates into the air if it is from the external skin that it leaves; but if it is from the surface of a muscle or some other internal part, it mixes with the fluid parts, and flows with them wherever they go, i.e., sometimes outside the body and sometimes via the veins toward the heart which they often happen to re-enter."[41]

Descartes thus properly envisions a flow of material into and out of each organ. He makes a guess at the nature of the flow and por-

37 Ibid., 5.61, *Oeuvres,* 11:274.
38 Ibid., 3.19, *Oeuvres,* 11:246.
39 Ibid., 3.20, *Oeuvres,* 11:247.
40 Lettre à Mersenne, Dec. 25, 1639, *Oeuvres,* 8:186.
41 "Du corps humain," 5.61, *Oeuvres,* 11:274.

trays it in a manner which is consistent with his biophysics but—as the microscope will ultimately demonstrate—erroneous. Even the microscope, however, will by no means immediately repudiate the fiber as the architectural unit of living systems. During the eighteenth century, the fiber will be widely accepted as the basic building block (see Boerhaave, Haller, Diderot). The achievement of Schwann in his general cell theory (1838–39) will be to show that all supposed building blocks—fibers, granules, globules—are cellular in origin (though not necessarily in adult composition); hence the true unit— that from which everything arises—is not the fiber but the cell.

CONCLUSION

Descartes had, in two ways, a profound and ultimately salutary effect on the future of physiology. First, he asked many sound questions about the body functions and thus laid down a detailed program of investigation that is still being followed, in part, more than three hundred years later. What is blood? How is it moved? What factors assure that the particular blood components enter the very tissues that need them? The questions he asked about brain, nerve, and muscle—though we have paid less attention to them here—were equally perspicacious. Second, Descartes insisted on nonvitalistic, material-mechanical answers to the questions he asked, thus establishing a procedural canon for all fruitful later investigation. That his own material-mechanical answers were often wide of the mark we can attribute to his excessively deductive and insufficiently experimental method.

Where Descartes failed, other material-mechanists, proceeding inductively and experimentally, succeeded. His way had the advantage, to be sure, of being quicker than the experimental way. Three hundred years later, experimentalists would still be struggling for adequate solutions to questions Descartes "solved" by the prompt and almost automatic process of logical deduction. The trouble is just that Descartes' quick deductive method often led him to false conclusions.

Descartes preserved a curious detachment, even ambiguity, in presenting his own scientific conclusions. This is not surprising when we consider the climate of the times. In 1624, the Parliament of Paris set death as the penalty for teaching, or even holding, ideas in opposition to Aristotelian doctrine, and Aristotle held the doctrine of corpuscularity in contempt.[42] In 1633, just as Descartes was about to publish his treatise on *The World,* Galileo was condemned by the Inquisition in Rome. Since the movement of the earth was an integral part of Descartes' physics, he wrote to Mersenne that he would suppress his treatise entirely ("I do not want ever to publish it").[43] This attitude, according to his biographer, Adam, was due not to personal fear, but rather to Descartes' desire for "the triumph of his ideas."[44] Why condemn them by publication when one's ultimate ambition is to have them accepted and taught?

The suppressed work on the world, written between 1629 and 1632, contained two parts, a *Treatise of Light* and a *Treatise of Man.* In the opening sentence of the latter, Descartes tells us that his subject is not *Man* but a *Machine;* he adds that this is not a real but only a hypothetical machine such as God might have made had he wanted the result to resemble a man. What we are given is an ambiguous robot, and the ambiguity is maintained to the very end of the *Treatise,* as witness the closing paragraph:

> I desire that you should consider that all the functions that I
> have attributed to this machine . . . [he here enumerates the
> functions] . . . *imitate* as perfectly as possible those of a *real* man;
> I desire, I say, that you should consider that all these functions
> follow quite naturally solely from the arrangement of its parts
> neither more nor less than do the movements of a clock or other
> automaton from that of its counterweights and wheels; so
> that it is not necessary to conceive in it any other principle of
> movement or of life than blood and spirits set in motion by
> the heat of the fire that burns continually in its heart and is of
> no other nature than all those fires that are in inanimate bodies.

Here then is a hypothetical machine very much *like* a true man. As to man himself, the reader is left to draw his own conclusions. Of the origin of the universe we are given the same sort of equivocal explanation:

> Though I should wish no one to be persuaded that bodies composing the world really were produced in the manner described,

---

[42] J. R. Partington, "The Origins of the Atomic Theory," *Annals of Science,* 4 (1939):263.

[43] Lettre à Mersenne (fin Novembre, 1633), *Oeuvres,* 1:271.

[44] C. Adam, "Vie de Descartes," *Oeuvres,* 12:168.

... if one can by this means and no other give intelligible explanations of all that one sees in the world one must conclude that—even though the world was not originally made in this way but was made directly by God—all things in it are such as would have been produced in the indicated fashion.[45]

God, in other words, produced the same cosmic results that would have been produced by a Godless cosmic machine! Such ostensibly noncommittal statements were meant to clothe heretical, or at least ambiguous, thought in reassuring terms. Not all contemporary authorities *were* reassured, but that is another story.

In summarizing Descartes' ideas of life and matter, we have emphasized his interest in nutritive assimilation—a central phenomenon, certainly, but only one of his many concerns. In his physiological treatises *Of Man* and *Description of the Body,* in his *Dioptrics,* his *Principles,* and his *Passions of the Soul,* and in posthumous fragments, letters, and notes, Descartes developed an integrated (but evolving) analysis of digestion, secretion, absorption, circulation, respiration, nervous and muscular action, and (in his later works) generation. His attention to these subjects was unequal; the sense organs, nerves, and brain, receive special emphasis in *Of Man;* embryonic development, in *Description of the Body.* In his picture of generation, separate male and female seminal fluids, corpuscularly composed of course, are separated from the blood of the parents, meet in the womb, activate each other through mutual fermentation, and give rise to a fetus epigenetically through the setting up of currents and through subsequent enduration and a spinning-out of the fibers that constitute the solid organs. Whatever the subject, what Descartes has to offer are less explanations of fact (though he made some dissections) than "Cartesianized" (that is, corpuscular and nonpsychistic) versions of explanations borrowed, all too often, from earlier authors. His conclusions were, with exceptions, too derivative and too ingeniously deductive; but, more explicitly than those of any other author, they showed the significance of the revolution—a revolution ultimately micromechanical and nonpsychistic—in which physiology was incipiently involved.

Descartes taught that the material causes of life-as-action must be sought in the body's subvisible organization ranging downward from structures just below the level of vision to elementary material

[45] *Principia,* 4.1.

particles. He especially stressed the importance of features *just too small to be seen*, whereas most (though not all) ancient and pre-Cartesian renaissance students of the body had tended to leap deductively from the smallest visible structures all the way down to ones irreducibly small. To this overleaping of intermediate structure there had been some exceptions (allusions especially to subvisible pores or passages in the tissue) but they had not been widely accepted because they lacked detailed explanatory power and because they were overshadowed by the views of Aristotle and Galen who taught that the tissues were homogeneous blends of the four elements or the four dynamic qualities (hot, cold, wet, and dry). The details of the architecture imagined by Descartes were promptly challenged on theoretical grounds, and on empirical grounds as well (microscopes began to be used very shortly after his death). But as later thinkers rejected the details he described, they as promptly replaced them by others. Few later authors—not even the vitalists who were to appear in such variety and numbers—would doubt the existence and importance of an elaborate subvisible organization. The search to describe it correctly was unremittingly pursued, as later chapters of this book will show. Today, at the electron-microscopic level and beyond, that search continues.

Part 3

LIFE AND AIR

· 19 ·

Life, Motion, Heat, Air
Backgrounds of Later Seventeenth
Century Thought

The advance of biology occurs in part through the evolution and interaction of related streams of ideas—using "ideas" to mean here, specifically, solutions developed in successive eras to persistent, fundamental problems. Such a stream develops partly independently through new discoveries and insights and partly interdependently through contact with other simultaneously developing currents. The effect of collision may be a deflection or acceleration of the currents involved, or the annihilation of one or both, or, under suitable conditions, the creation of new ones. The result, in any case, is a complex dynamic flow of thought in channels swelled by tributary knowledge and altered by intellectual interaction.

The above points are strikingly exemplified in the recurrent interplay of ideas about the nature of *life,* of *motion,* of *heat,* and of *air.* Their interactions are complex and have largely occupied us in earlier chapters. We shall review them here, though only in outline, because they offer a perspective on late seventeenth-century theories of life and matter.

Life and Motion

Living things have always been identified as such partly by the movements they display, and biologists of every era have tried to understand these movements correctly. Comparing early with recent ideas, we may note a remarkable reversal in the scientific treatment of this subject. Greek science often gave psychistic, or biological, interpretations of what we think of today as nonbiological movements. Anaximander's *apeiron,* Empedocles' *aphrodite,* the *nous* of Anaxagoras, and the motive cosmic life-soul of Plato may be understood in part as survivals in early scientific thought of a prescientific disposition to attribute motion in general to life (*psyche*). Modern science proceeds in the opposite direction. It seeks nonbiological interpretations of biological movements. In chapters 11 and 15, we sketched some of the developments through which this reversal was brought about. The most decisive events were (*a*) the development of impetus theory which originated, effectively, in patristic thought and culminated in the era of Scholasticism; and (*b*), in later times, the Galilean-Newtonian revolution.[1]

The outcome of these developments was as important for biology as it was for astronomy and physics. In biology, the result was the mechanistic mode of analysis whose chief founders were Descartes and Harvey, the crucial step being Descartes' proposal for the elimination of *psyche* as the cause of plant and animal movements and for the virtual elimination of it, as a physiological factor, from man. The major consequence of the Cartesian position was the need it posed for a thorough reformulation of the problem of the cause of biological movements. Science had now to ask how (if not through soul) inertia is overcome and entropy is opposed in living systems. Historically, two answers, or two constellations of answers, to this question were offered.

1. *"Vitalistic" motive causes.* On the one hand, many theorists substituted for soul one or more "vital" principles, properties, or powers which were considered to be (*a*) natural rather than supernatural and (*b*) physical or biophysical rather than psychic. Though natural and nonpsychic, these principles, properties, and powers were considered to be essentially different in mode of action from any with which the physicist is familiar. The postulation of these extraphysicochemical agents was to reach a kind of culmination in the late eighteenth century at which time *vitalism* (as distinguished from animism) was introduced as a collective term for such theories.

2. *Corpuscular interpretations.* There developed in the meantime a second and scientifically more pregnant answer, or class of answers, to the problem of the causes of physiological movement. It viewed generalized or *mass movements,* of both living and nonliving systems, as summations of the *micromovements* of their corpuscular constituents. We have encountered one such solution in the essentially micromechanical physics and physiology of Descartes. But, in the later seventeenth century, as we are about to see, the corpuscular

[1] See chap. 15, n. 10.

theory was extended to embrace activities of a specifically chemical nature.

In the nonliving realm, the explosion of gunpowder received a corpuscular interpretation (see below). So did certain combustions, ebullitions, fermentations, and gas-forming activities, to mention reactions that chiefly interested the experimentalists of the second half of the seventeenth century. Our next six chapters will show how these reactions were seized upon by biologists of the period as bases for many different yet related theories of physiological motion, theories marked by an attempt to bring *biological, chemical,* and *corpuscular* interpretations together in a common synthesis. This point is significant because the slightly earlier evolution of chemistry out of alchemy had begun under the influence of a prevalently Aristotelian, hence anti-corpuscular, world outlook. When the atoms— or other sorts of corpuscles—were revived between 1600 and 1650, chemistry and corpuscularity had to be accommodated to each other. We shall see that commencing at about mid-century Robert Boyle, especially, made it a personal endeavor to try to reconcile these two ways of looking at matter. "That then which I chiefly aim at," he tells us in 1666, "is to make it probable to you by experiments (which I think hath not yet been done) that almost all sorts of qualities, most of which have been by the schools either left unexplicated, or generally referred to by I know not what incomprehensible substantial forms, may be produced mechanically; I mean by such corporeal agents, as do not appear . . . to work otherwise than by virtue of the motion, size, figure, and contrivance of their own parts . . ."[2] and he assures us that he means corpuscular parts.

Boyle's personal success in this endeavor was limited at best. His objective of reformulating chemistry in corpuscular terms was to be fully realized only a century and a half later through the classic quantitative studies of Gay-Lussac, John Dalton, and Avogadro. But his efforts, and parallel efforts of others, had an important influence in their immediate period, in that they became a central part of the general endeavor at this time to interpret vital phenomena in terms of corpuscular activity.

[2] Robert Boyle, *The Origin of Forms and Qualities, According to the Corpuscular Philosophy* . . . (first publ. Oxford, 1666), *The Works of the Honourable Robert Boyle* (London, 1772), 3:13; see also J. R. Partington "The Origins of the Atomic Theory," *Annals of Science,* 4 (1939):265–67.

It would not be correct to think of seventeenth-century physiology as suddenly attempting to reduce all vital activity to corpuscular motion. But particles of one sort or another were centrally present in the schemes of many of the major biomedical thinkers who were Boyle's contemporaries and successors. We shall hear different roles and different degrees of importance assigned to them by Hooke, Willis, Mayow, and Borelli in the seventeenth century, and by Boerhaave and Haller in the eighteenth. This point is especially well illustrated in the development of seventeenth-century ideas about the nature and significance of heat in living bodies.

Life and Heat

The second half of the seventeenth century saw an intensification of interest in connections between life and heat. At this time we hear life made variously dependent upon or analogous to ordinary fire or flame. These associations were, to be sure, not arrived at *de novo*. They were, rather, the new expression of an old tradition admirably illustrating how biology grows through the interaction of evolving streams of ideas. A cursory recapitulation of theories encountered earlier in our studies makes this point abundantly clear (though clarity was rarely a characteristic of the theories themselves).

The story of life and heat has been followed in an informative recent study by Everett Mendelsohn (1964).[3] Our own account has uncovered (among other links between life and heat or flame): the fiery *Logos* of Heraclitus, the animate fire atoms of Democritus, the glowing animal-gods of Plato's cosmology, and—of a more distinctly physiological character—the innate heat of later Greek thinkers from Aristotle to Galen. In the sixteenth century, we heard Paracelsus liken life to a fire, and we heard Fernel, reinvoking the Greek innate or vital heat, decide that this, unlike ordinary heat, arises in a special quintessence comparable to the ether in which the sun's heat arises.[4] We likewise heard both Harvey and especially Descartes accord to heat the key role in vital function, Harvey considering the central importance of circulation to be the distribution of life-giving

---

[3] Everett Mendelsohn, *Heat and Life* (Cambridge, 1964). Important parts of this subject are treated also by G. J. Goodfield, *The Growth of Scientific Physiology* (London, 1960).

[4] Charles S. Sherrington, *The Endeavor of Jean Fernel* (Cambridge, 1946), pp. 38–40; see also C. E. A. Winslow and R. R. Bellinger "Hippocratic and Galenic Concepts of Metabolism," *Bulletin of the History of Medicine,* 17 (1945):129.

heat,[5] whereas Descartes equated the life of animals with that fire-without-light that burns, according to his continuation of a very old tradition, in the heart.

The later sixteenth century was to demand, with Descartes, a more naturalistic view of animal heat and a more sophisticated theory of life-heat relations. But this required a new way of thinking about the subject, a willingness to consider animal heat as closely comparable in origin to heat of other sorts, and to view it as emergent from chemical or corpuscular action rather than as mysteriously and inexplicably immanent in living bodies. This viewpoint had been promoted—though not originated[6]—by van Helmont. In his system, we remember, fermentations of various sorts were made the bases of the body's activities, and fermentation was supposed to produce the heat that goes with them. We shall note, in the next several chapters, some of the signal responses to this important development. One of the most striking was a new effort—now actively experimental rather than merely rational—to find out to what extent, and in what ways, animal heat corresponds in its origin to heat of other sorts. A question of particular relevance was: why do both life and fire need air?

Life and Air

Recapitulation of classic ideas. In earlier chapters, we noted that near the beginning of the Greek scientific endeavor, Anaximenes made air (*aer, pneuma*) both the vehicle of the life-soul and a girding substance that holds together the living bodies of individual men and of the cosmos at large; and that other pre-Socratic thinkers— Anaximander and Heraclitus—also associated life with air, though it is impossible, on the evidence we have, to detail their different ideas on this subject. Other pre-Platonic theories we touched on were the Pythagorean concept that man (and the cosmos) live by inhaling a universal breath-soul and the Democritean idea that inhalation counters the potential outflow of soul atoms whose loss would lead inevitably to death.

In post-Platonic respiration theory, we witnessed a continued

[5] Walter Pagel, "William Harvey and the Purpose of Circulation," *Isis,* 42 (1951): 72.
[6] See Walter Pagel, "The Religious and Philosophical Aspects of Van Helmont's Science and Medicine," *Supplements to Bulletin of the History of Medicine* No. 2, (1944):4.

evolution and interaction of ideas about (*a*) air, (*b*) *pneuma*, and (*c*) heat or fire. Aristotle focused attention on a hypothetical innate heat which he considered indispensable to body functions and needing to be tempered by the cooling influence of air. In post-Aristotelian thought, air was related in various ways to the spirits or *pneumata* that were generally considered to be present in the body. Aristotle himself conceived of *pneuma* as innate rather than, as in later thought, a derivative of air. By contrast, the *pneuma* that was central to the Stoic world scheme was a variously proportioned fire and air mixture spread through the universe between and in material bodies composed of water and earth; in living bodies, *pneuma* was, for the Stoics, equivalent to life-as-soul.[7]

The most influential ancient pneumatologies were those of Erasistratus (third century B.C.) and Galen (ca. A.D. 175). The central development was the derivation of physiological "spirits" (*pneumata*) from atmospheric air. We have detailed these theories elsewhere (see chapter 10). It should not be forgotten that in addition to serving as the raw material of *pneuma*, air was viewed by Galen as preserving the vital heat through fanning, or ventilation, in the same way in which air makes a lamp continue to burn.

Life and air in early modern biology. Early post-medieval thinking about the biological importance of air had its roots in the ancient ideas just surveyed. Paracelsus and Fernel, each in his way, made air a special supporter of the vital fire. For Francis Bacon (chapter 16), air was a form of mercury that united with flame to become the pneumatical component of living tissue. Bacon's thought thus showed the influence of both Stoic and alchemical doctrine. He made a vague forward thrust in his intuition that the air-fire ingredient was a cause of motion, specifying that air affects both strength (e.g., as in wrestling and leaping) as well as speed (e.g., as in playing the lute).

Niter. The most pregnant early seventeenth-century idea on the life-air problem, however, was the suspicion that air contained a "nitrous" substance sustentative of both life and fire. We have already noted how Paracelsus infused Greek notions of a vital power of the air into the later alchemical ideas.[8] But a historical problem

[7] See, for the impact of Stoic pneumatology, Gerard Verbeke, *L'évolution de la doctrine du pneuma, du Stoicisme à S. Augustin* (Paris, 1945); and Samuel Sambursky, *Physics of the Stoics* (London, 1959).

[8] Walter Pagel, *Paracelsus: An Introduction to Philosophical Medicine* (New York and Basel, 1958), pp. 187–88.

remains as to how this power came to be associated with niter. To this question a historian of our own times, Guerlac (1953), has suggested a plausible answer.[9] Guerlac points out that de Vigenère and Sennert (both in 1618) explained lightning and thunder as analogous to an explosion of gunpowder (manufactured in Europe since the early fourteenth century from charcoal, sulphur, and niter). The thunder-gunpowder analogy was developed with increasing explicitness by Jan Comenius (1643), Thomas Brown (1650), and Pierre Gassendi (1658). (Guerlac also follows the "gunpowder theory" of thunder and lightning from science to *belles lettres* where it is represented in a series of poets from Milton to William Thomson; and we still hear of it as a scientific theory in the closing pages of the second edition of Newton's *Opticks*.)

The notion that a nitrous something supports *fire* and *life* (as distinguished from thunder and the explosion of gunpowder) has been traced by Guerlac from the alchemists Seton (1604) and Sendivogius (1609) to Kenelm Digby (1660–61), and from Digby—in all probability—to its chief expositors Hooke (1664) and Mayow (1668) (see below, chapters 21 and 23).[10] Another interim proponent, in addition to Digby, was George Ent (1641), known also for his espousal of Harvey's theory of the circulation; and it is possible that Ent formed an alternate link—in addition to Digby—between (*a*) the alchemists' interest in niter and (*b*) that of the serious physiologists (Hooke, Boyle, and Mayow) of the second half of the century. It may be relevant that Boyle himself, probably during the 1650s, performed experiments on niter, undertaking its disintegration and redintegration. He convinced himself that one of the decomposition products was a substance still nitrous yet now become volatile (hence, possibly present in air). Finally, it was known by this time that niter had a "frigorifick" effect on ice-water mixtures, and it was considered probable by some that aerial niter had a similar effect on the atmosphere. This notion agreed well with the ancient but still widespread belief that the use of breathing was to cool the heat that is generated in the heart.

[9] Henry Guerlac, "Studies in the Chemistry of John Mayow: II. The Poet's Nitre," *Isis*, 45 (1954):243–55. See also, Leonard G. Wilson, "The Transformation of Ancient Concepts of Respiration in the Seventeenth Century," *Isis*, 51 (1960):161–72.

[10] Henry Guerlac, "John Mayow and the Aerial Nitre," *Académie international d'histoire des sciences* (Jerusalem, 1953), pp. 332–349.

Life and Air in the Late Seventeenth Century

The above sketch of developing ideas about life, air, and fire shows, if nothing else, that the experimentalists of the later seventeenth century did not approach these topics as novelties. What was new about the group whose acquaintance we are about to make—Boyle, Hooke, Willis, Borelli, Mayow, and Lower—was, first, an imperfect but active use of experiment to support their conclusions and, second, their varied responses to revived corpuscular theory. We shall hear all of these thinkers suggest, each in his own terms, that in both organic life and combustion, an interaction of some sort occurs between a special substance from the air and a special substance in the living or burning body.

On the surface, this common conviction might seem to set the stage for at least a preliminary idea of oxidation. In fact, no significant oxidation theory was to materialize for more than a century, and this we may ascribe to many causes. One was the failure of chemistry to acquire a real basis in corpuscular theory. We shall note later that Boyle's "Attempt to Make Chymical Experiments Useful to Illustrate the Notions of the Corpuscular Philosophy"[11] had a very imperfect success. Another obstacle to oxidation theory was the inadequate state of analytical chemistry, which was to remain too primitive, for approximately another hundred years, to permit distinctions among the several sorts of "air." Still another countercurrent to the development of oxidation theory was that famous conceptual red herring, phlogiston (due mostly to Stahl, 1684, 1703). In the phlogiston theory, by no means naive for its period, fire involved not a *using up* of something by the burning object but a *separation* of something from it. This something was thought to be "fixed fire," or phlogiston. We may postpone our account of phlogiston and its downfall, since it belongs to a later period (the eighteenth century).

The general consequence of chemistry's slow development was that micromechanical (or iatrophysical) and iatrochemical physiology remained for some time at least partly independent. Such iatrophysicists as Pitcairne, Boerhaave, Hales, and others were to offer theories of vital activity that were corpuscular without being chemical; attrition, friction, and similar micromechanical processes seemed to these

11 Abbreviated title of a work of 1669.

authors adequate as producers of animal heat and of vital activity in general.[12]

Respiration and Combustion after 1650;
a Brief Chronology and Preview

During the 1650s, disagreement prevailed about the role of breathing and about the nature of the body's need for it. Does breathing permit a cooling, and perhaps even a recondensation, of pulmonary

TABLE 10
CHRONOLOGY OF OXFORD PHYSIOLOGISTS

	At Oxford	In London

Willis   b. 1621   1640   1666   d. 1675

Hooke   b. 1635   1653   1662   d. 1703

Boyle   b. 1627   1654   1668   d. 1691

Mayow   b. 1641   1658   1670 d. 1679

Lower   b. 1632   1658   1667   d. 1691

blood (this was Descartes' persuasion)? Does it "disburden" the blood of waste vapors (a Galenic idea still favored by many)? Does it refine the blood mechanically or in some way promote circulation? Finally, does it permit the body to acquire some needed part of the air?

Each of these views found defenders, and it was amidst this confusion of opinion that the significant developments in late seventeenth-century respiration theory occurred. A recurrent theme in its development, we shall see, was the scientific interest in niter. In succeeding chapters we shall follow this interest as it expressed itself among a group of investigators all of whom were at Oxford at the same time and all of whom settled later in London. We shall sketch here, by way of background, the chief ideas of the actors in this drama, and suggest their probable personal interactions.

1660 (Boyle). Boyle attacked the life-fire-air problem experimentally by showing that neither life nor flame could last long in a

---

[12] Mendelsohn, *Heat and Life.*

partial vacuum (see below, chapter 20). This fact opposed the notion of air's role as merely absorbent. If air were merely absorbent, then in a partial vacuum, Boyle argued, the increased space between the air particles would permit the absorption of increased amounts of waste matter; hence life and flame would last longer in a partial vacuum than under normal conditions. Boyle suspected that Paracelsus had been on the right track in positing, in the atmosphere, a "little vital quintessence" (which, in Boyle's experiments, the vacuum pump removed along with the air). This was as far as Boyle carried the analysis in 1660. Not until later was he to identify his "vital quintessence" explicitly with niter.

1664–1667 (Hooke). The attribution of a nitrous aspect to air (not a new thought, we remember) was taken up during the early 1660s by Boyle's collaborator, Hooke (who may have acquired the idea from Digby). Hooke and Boyle worked together from shortly after Boyle's arrival in Oxford (1654) until Hooke left for London (1662) to become the curator of the Royal Society. Thereafter, Boyle and Hooke kept close contact until—and after—Boyle settled in London (1668).

Hooke's surmise, in 1664, was that air and familiar niter contain a substance common to them both; in combination, this substance dissolves the sulphur of the burning body (see chapter 21). Thus the solution of sulphur by niter seemed to Hooke to be the central process in fire and flame. In the same year, Hooke found an artificial way, using bellows, of blowing a steady stream of air through the immobilized lungs of a dog and showed that the animal could live thus for more than an hour (if the stream was not interrupted). Breathing must be useful, Hooke concluded, because it brings air to the lungs and not, as some claim, because it promotes circulation. A few years later (1667), Hooke ventured the guess that air supplies "some nitrous quality" central to respiration as well as combustion.

1668 (Lower). Hooke's Oxford sojourn overlapped that of the talented experimental physiologist, Richard Lower, whom we remember otherwise for his successful work on transfusion (see chapter 24). Both Hooke and Lower interested themselves in the color of the blood and wondered whether venous blood was brightened, in its passage through the lung, by the uptake of something from the air. Lower was able, through experiments that we shall hear about later, to make it very likely that something of this sort in fact occurs.

When he asked himself how an aerial substance could enter the blood without the blood escaping, Lower speculated that the substance in question was a spirit—specifically, nitrous spirit—spirits being able to pass where liquids cannot. In summer, for example, nitrous spirit passes out from a glass of iced wine, according to Lower, and forms a frost on its external surface.

1670 (Willis, Mayow). During his residence in Oxford, 1658–1667, Lower associated himself as assistant with the iatrochemically-oriented physician and neuroanatomist Thomas Willis (chapter 22). During the late 1650s, Willis had viewed niter as containing sulphur imprisoned within it as a "nitrosulphureous ferment," and he thought of fire as a volatilization of this imprisoned sulphur. Later (1670), building partly on Hooke and Lower, he adopted the quite different idea of a combative interaction, in combustion and in respiration, between *internal* sulphur existing in the burning or living body and *external* niter introduced from the ambient air. Willis equated this combative interaction with life itself which he defined as a sort of fire or flame.

At about the same time (1668), however, another Oxford physician, John Mayow, entered the picture with views that have made him seem to some admirers a true predecessor of Lavoisier. We shall devote a later chapter (chapter 23) to our appraisal of Mayow and limit ourselves here to a simple statement of his central ideas. It was his view that air owes its expansive tendency, or spring, to the fact that individual air particles are linked to springy particles of a certain "nitroaerial" spirit. In combusion and respiration, sulphur particles escape from the burning or respiring body and dislodge the nitro-aerial particles from their association with particles of air. The air thus loses its spring, and the escaping nitroaerial particles constitute heat—and, under the proper conditions, light. We shall return later to the debate that has been conducted by historians about the extent to which Mayow's ideas anticipated eighteenth-century oxidation theory.

Recapitulation. Among the main events in the evolution of late seventeenth-century respiration theory were (1) Boyle's deduction from his experiments that air supplies a substance necessary to life; (2) Hooke's guess that this substance is present also in niter and that it acts specifically as a solvent of the body's surphur; (3) Lower's experimental demonstration that the brighter color of arterial blood is due to something blood gets from the air—something that he views as a "nitrous spirit"; (4) Willis' notion (not very different from Hooke's) that a nitrous substance from the air induces in living or burning bodies a volatilization of their sulphur; and (5) Mayow's rather different conception that volatilized sulphur dislodges nitro-aerial particles from the particles of air.

Beyond these developments we need only note that, in their later treatments of combustion and respiration, Boyle and Hooke, while not departing from their earlier conclusions, showed the influence, at least in their terminology, of the Oxford group as a whole. Boyle was ready by 1672 to suppose that the necessary substance supplied by the air was perhaps of a nitrous nature. Hooke, in his final formulation of the problem (1682), still saw air as a solvent; but he now ascribed its solvent action to an "aerial or volatile nitrous spirit," and this formulation owed something, as far as the choice of words was concerned, to Mayow.

The Oxford-London group's ideas about respiration and combustion interest us primarily because of their bearing on the concept of life and of life's dependence on special kinds and arrangements of matter. To see how the group viewed these and related questions, we shall briefly visit each member in turn.

· 20 ·

Life and Corpuscular Motion

ROBERT BOYLE [1627 · 1691]

We remember Robert Boyle primarily for his studies of partial vacua and the pressure relations of gases; but taken in their totality his interests were encyclopedic. He wrote about magnetism, electricity, heat and cold, respiration, combustion, fermentation, putrefaction, phosphorescence, fluorescence, bioluminescence, and the properties of many metals and many nonmetals. He made appreciable contributions to the understanding of the expansive force of freezing water, the propagation of sound, specific gravity, and refraction. To his scientific studies he added inquiries into philosophy, theology, logic, and medicine. His will founded a lecture series to argue Christianity's case against "atheists, theists, pagans, Jews, and Mohammedans." The diversity of his concerns is suggested by such treatise titles as "Cosmical Suspicions," "A Dissuasive from Cursing," "Relations about the Bottom of the Sea," "An Attempt made upon Gnats in our[1] Vacuum," "An Essay on Seraphick Love," a "Discourse upon the Mounting, Singing, and Lighting of Larks," "Reflection upon a Child that Cried for the Stars," and so on and on. The whole list of treatises numbers several hundred and fills, in the collected Works, five volumes.[2]

One properly thinks of Boyle as a pioneer experimentalist. As such, he was in the best tradition of the Royal Society, which he

---

[1] Boyle preferred to call the condition achieved in his receivers "our vacuum" or the "Boylean vacuum" so as to remove himself from the contemporary debate between defenders of the plenum (Descartes) and the absolute vacuum (Toricelli and others).

[2] *The Works of the Honourable Robert Boyle in Six Volumes to which is prefixed The Life of the Author: A New Edition* (London, 1772). (First publ. London, 1744 in folio and in five volumes only.)

For an engaging and informative personal appreciation of Boyle, see Marie Boas (Hall), *Robert Boyle and Seventeenth Century Chemistry* (Cambridge, 1958), esp. chap. 1. In this connection, see also Harold Fisch, "The Scientist as Priest: A Note on Robert Boyle's Natural Theology," *Isis,* 44 (1953):252–65.

helped to establish. Most of the scientific topics about which he wrote he knew directly through a remarkable range of experiments he made personally. ("No body ever made more experiments"—Boerhaave.)[3] However, not all of his conclusions were based on his experimental results. He occasionally depended on borrowed opinion. Part of what he borrowed was official doctrine, laid down by eminent predecessors; part of it was rumor and common hearsay; and part of it, according to Boerhaave, Boyle obtained through "a kind of commerce, or communication of secrets." Moreover, an important aspect of his endeavor was an extended critique of the opinions of other students of nature.

Matter

The most marked feature of Boyle's theory of matter was its foundation in corpuscularity.[4] His interest in material particles was influenced by Gassendi's revival (in 1647)[5] of Epicurus's theory of atoms. Thus, Boyle stood—through Epicurus and Gassendi—in line of descent from Leucippus and Democritus. Boyle and Gassendi were not the first modern physicists to espouse the corpuscular philosophy (though Boyle was perhaps the first to name it that). Variations of the view of physical bodies as composed of minute particles had been elaborated in a variety of ways by Bruno, Galileo, Sennert, Descartes, and Beeckman, among others. The modern historian of science, Thomas Kuhn, suggests that natural inquiry has tended to progress in alternate cycles of revolution (periods in which new major conceptual patterns, or paradigms, emerge) and "normal science" (periods in which existing paradigms are extended to larger and

[3] Hermann Boerhaave, *A New Method of Chemistry; including the History, Theory, and Practice of the Art: Translated from the Original Latin of Dr. Boerhaave's Elementa chemiae* . . . , trans. Peter Shaw (3rd ed.; London, 1773), p. 55. Two helpful studies of the revival of atomism are J. R. Partington, "The Origin of the Atomic Theory," *Annals of Science*, 4 (1939):245; and Marie Boas, "The Establishment of the Mechanical Philosophy," *Osiris*, 10 (1952):412–541. An earlier important treatment is that of K. Lasswitz, *Geschichte der Atomistik* (Hamburg and Leipzig, 1890). See also A. G. van Melsen, *From Atomos to Atom*, (New York, 1960).
[4] The corpuscular theory is developed by Boyle in several treatises, including *The Skeptical Chymist: or Chymico-Physical Doubts & Paradoxes, Touching the Spagyrist's Principles Commonly call'd Hypostatical* (London, 1661); and *The Origins of Formes and Qualities. (According to the Corpuscular Theory)* (London, 1667).
[5] Pierre Gassendi, *De vita et moribus Epicuri libri octo* (Lyons, 1647); however, Boyle seems to have been chiefly familiar with Gassendi's *Philosophiae Epicuri Syntagma* (London, 1660).

larger arrays of data). In one sense, Boyle's enterprise was a "normal-scientific" working out of the neocorpuscular paradigms set up by his immediate predecessors, and modified by him.

But Boyle's chief chemical work, *The Skeptical Chymist* (1661), was devoted largely to raising doubts about, and effectively if not totally repudiating, the chemical elements and alchemical *tria prima,* and was in this sense revolutionary rather than normal. Take, for instance, his ideas about water. In his laboratory, he raised plants hydroponically and then mistakenly ascribed their entire gain in weight to absorbed and transmuted water. He emphasized the variety of plant substances—for example, poisons so caustic as to raise blisters and combustible oils—that are producible through the alteration of water. Boyle believed the transmutability of water to be especially well illustrated by grafts; thus, a white-thorn tree can produce peaches, marvelously differentiated out of a "phlegmatic liquor [sap] that seems homogeneous enough, and but very slenderly provided with other manifest qualities than [those of] common water. . . ."[6] His conclusion from these observations was that water is transmutable, hence not strictly an element according to Boyle's definition thereof (see below).

As to Paracelsus' *tria prima,* various experimental and dialectical considerations convinced Boyle that they were not elementary but intertransmutable and corpuscularly complex. He thought that if the *tria prima* were to be regarded as primary, in any accepted sense, one would certainly have to add water (or phlegm) and earth to the Paracelsian trio of mercury, sulphur, and salt. Hence the proper number would not be three but five.[7] But Boyle doubted whether the five should be considered elementary, for he defined elements as "certain [putative] Primitive and Simple, or perfectly unmingled bodies; which not being made of any other bodies, or of one another, are the Ingredients of which all those call'd perfectly mixt Bodies are immediately compounded, and into which they are ultimately resolved."[8] This definition was not, and was not meant by Boyle to be,

[6] "Considerations and Experiments touching the Origin of Qualities and Forms," *Works,* 3:69–72.

[7] The belief in five elements (the *tria prima* plus two) is traced by Partington back through Willis to Sebastian Basso (1621); see J. R. Partington, *A History of Chemistry* (London, 1961), 2:305–6; see also Boas, *Robert Boyle,* p. 85.

[8] *The Skeptical Chymist,* p. 350; see also *The Skeptical Chymist in The Philosophical Works of the Honourable Robert Boyle, Esq.; Abridged, methodized, and disposed . . . by Peter Shaw, M.D.* (London, 1738), 3:261.

new. The relevant point to him was the nontransmutability of elements as so defined. Under this quite specific definition, the *tria prima*—and water and earth—failed to qualify.

Boyle's chemistry thus stressed as its central themes, first, that elements are by definition nontransmutable; second, that a virtually unlimited interconvertibility of all substance is ultimately possible; and, third, that the status of all supposed elements is therefore doubtful. As Kuhn emphasizes, Boyle's thesis was "that anything can be made of anything. . . ."[9]

These points seemed especially apparent to Boyle when he thought about them in terms of his corpuscular theory. He saw the smallest particles (*minima naturalia*) as differently cut subdivisions of universal matter. He considered the sensible properties of things to be ultimately reducible to the "bulk, figure, and either motion or rest" of the *minima naturalia*—an Aristotelian term, lengthily debated during the Middle Ages, here given by Boyle an Epicurean interpretation.[10] The smallest particles make up into different clusters (*prima mixta*), dissociable only with difficulty, Boyle thought, each cluster being specific to a familiar substance such as water or a metal.[11] The critical point for him was that seemingly pure substances are generally corpuscularly complex. "The chemist's salt, sulphur, and mercury themselves are not the first and most simple of bodies, but rather primary concretions of . . . particles more simple than they . . . by the different conventions or coalitions of which . . . are made those differing concretions that chemists name salt, sulphur, and mercury."[12] Neither is water an element. Boyle saw no reason "to conceive [that] the water mentioned by *Moses,* on which the spirit is said to have brooded . . . was simple and elementary water. . . ."[13] Boyle thus drew from his hydroponic experiment a conclusion opposite to that reached by van Helmont. To van Helmont—whom, incidentally, Boyle admired—the very fact of water's transformability made it appear to be a first principle.

Boyle's system seems, in perspective, estimable (in its efforts to

[9] For recent helpful analyses of Boyle's chemistry see Thomas S. Kuhn, "Robert Boyle and Structural Chemistry in the Seventeenth Century," *Isis,* 43 (1952):12–36; Boas, *Robert Boyle,* pp. 98–205; and Partington, *History of Chemistry,* 2:486–549.

[10] "An Excursion about the Relative Nature of Physical Qualities," *Works,* 3:29. For the history of the *minima naturalia,* see Dijksterhuis, *The Mechanization of the World Picture* (Oxford, 1961), p. 205.

[11] "An Excursion," *Works,* 3:35.

[12] "Imperfections of the Chemist's Doctrine of Qualities," *Works,* 4:281.

[13] Gen. 1:2; *Works,* 3:287.

replace occult with micromechanical properties) yet unsuccessful (in its failure to harmonize corpuscularity with chemistry). Why, historians have asked, did not Boyle's approach forthwith usher in the age of corpuscular chemistry? A partial answer may be that Boyle placed too much emphasis in his micromechanics on the motile properties of the particles and the restlessly labile condition of matter. Boyle's dynamic microworld, as Kuhn observes, was constructed without the stable foundations—namely elements—on which the new chemistry would be built more than a century later. Plato had provided the needed clue—a limited number of elements each represented by a corresponding particle. But Plato's proposal would only be effectively applied in the early nineteenth century. Meanwhile particles were often invoked, as we shall see—but only in a semieffective manner.

Living Matter

We have already noted that Boyle derived the full complexity of the plant body from the water that it absorbs. Boyle extended this complexity even to such properties as the plant's "scents, colours, tastes, solidity, medicinal virtues, and divers other qualities manifest and occult."[14] He "from thence . . . inferr[ed] that the same corpuscles which, convening together after one manner, compose that inodorous, colourless, and insipid body of water, being contexted after other manners, may constitute differing concretes, . . ." not excluding the bodies of animals.[15]

Such thinking raises, as Boyle recognized, a fundamental question. Does all this organization emerge from a "casual conflux of atoms," or does it require an extramaterial "guide"? We have encountered this question recurrently, commencing with our study of the pre-Socratics, several of whom posited a directive entity of one sort or another (see Anaxagoras, Empedocles). Boyle himself was of the opinion that men too eagerly leap at extramaterial explanations— "substantial forms," for example, or souls—when in fact material explanations would suffice. Boyle would have us pay closer attention to the minute parts as the principal actors in the drama, whether or not soul is pulling the strings.[16]

[14] "Considerations and Experiments touching the Origin of Qualities and Forms" (first publ. 1666), *Works,* 3:72.

[15] Ibid., *Works,* 3:70.

[16] "Free Considerations and Experiments touching The Origin of Qualities and Forms" (first publ. 1667), *Works,* 3:119.

Thus, "I very well foresee it may be objected that the chick with all its parts is not a mechanically contrived engine, but fashioned out of matter by the soul of the bird lodged chiefly in the cicatricula, which by its plastick power fashions the obsequious matter and becomes the architect of its own mansion." Nevertheless, "let the plastick principle be what it will, yet still, being a physical agent, it must act after a physical manner; and having no other matter to work upon but the white of an egg, it can work upon that matter but as physical agents, and consequently can but divide the matter into minute parts of several sizes and shapes, and by local motion variously context them according to the exigency of the animal to be produced. . . ."[17]

Boyle's intention was both to dispense with soul where it seemed unneeded and to give it, where needed, if not actually a physical character, a physical *modus operandi*. In 1667, he compared the organism to a wind- or watermill, with wind and water representing *soul,* soul acting here in its classic role of *causa motiva* but possessing, like wind and water, a material cast. In vegetables, soul is a fluid, Boyle said, and one of its chief functions is the replacement of displaced materials.[18]

If Boyle's world was a material world, divine forces were nevertheless acknowledged to have set it in motion originally, and to exercise interventive governance in such phenomena as epigenetic development. He focuses here on a question recurrent in science from Greek times until the present day, namely whether the Transcendental Guide merely set up the cosmos and then withdrew or, instead, intervenes on proper occasions. Thus Boyle did

> not at all believe that either these Cartesian laws of motion, or the Epicurean casual concourse of atoms, could bring mere matter into so orderly and well contrived a fabrick as this world; and therefore I think that the wise Author of nature did not only put matter in motion, but, when resolved to make the world, did so regulate and guide the motion of small parts of the universal matter, as to reduce the greater systems of them into the order they were to continue in. . . . But the world once framed, and the course of nature established, the naturalist . . . has recourse . . . in explicating phenomena . . . [to] only the size,

[17] "Considerations and Experiments," *Works,* 3:68.
[18] "Free Considerations," *Works,* 3:123–24.

shape, motion (or want of it), texture, and the resulting qualities and attributes of the small particles of matter.[19]

Boyle was specific in including, among things so created, plants and animals.

Emergentism

A central theme in biology, including twentieth-century biology, is the concept of emergence: namely, that the entire system has functional capacities that its separate components lack. This is true, from the modern viewpoint, whether one considers the organism as a whole or its constituent elements—organs, tissues, cells, or intracellular molecular systems. At each level the action of the whole differs from the sum of the actions of its isolated parts. This phenomenon is crucially important at the boundary between living and nonliving complexes since here the question arises whether life itself is anything *more than* an emergent activity of properly integrated components.

Certain scientists in each era, whom we now think of as vitalists, have answered this question affirmatively. They have insisted upon introducing a nonmaterial vital agent which materialists and mechanists have considered superfluous. For strict materialists and mechanists, activities that are specific and unique to the whole, though not duplicated in the separate parts, emerge automatically from the arrangement and interaction of those parts. Life itself is, for them, emergent.

At times Boyle held out, not as an indispensable but as a probable interpretation of the organization of living things, a distinctly emeggentist view. He saw organization—and the activity dependent on it—as emerging automatically from the immanent properties of properly configured particles. He offered this as a possible alternative to the Peripatetic "doctrine of subordinate forms." That doctrine taught that specific form (presiding over the whole object, living or nonliving) operates through subordinate forms (different for different parts or organs) to control the composition of the individual.

To Boyle, for whom form was "no more than a convention of accidents," formal causes seemed unnecessary. Besides, "for my part, that do not acknowledge in many bodies . . . any thing substantial

---

[19] "An Examen of the Origin and Doctrine of Substantial Forms" (first publ. 1667), *Works,* 3:48.

distinct from matter, I confess I do not readily conceive, which way this dominion attributed to the specifick form is exercised. . . ."

In a stab at an emergentist formulation, Boyle informs us that he

> should therefore rather conceive the matter thus; when divers
> bodies of differing natures or schematisms come to be associated,
> so as to compose a body of one denomination, though each
> of them be supposed to act according to its own peculiar nature,
> yet by reason of the coaptation of those parts, and the con-
> trivement of the compounded body, it will many times happen,
> that the action or effect produced will be of a fixed nature
> and, differing from that, which several of the parts, considered
> as distinct bodies, or agents, tended to, or would have performed.[20]

Thus it seemed possible to Boyle to view form not as something "presiding" but as the *most conspicuous* ("eminentest") aspect of things, and, in both nonliving and living systems, as "resulting from the coexistence of such [and such] corpuscles or parts after such [and such] a determinate manner." Form is mistakenly regarded, he says, as if it "were some distinct and operative substance that were put into the body as a boy into a pageant, and did really begin, and guide, and overrule the motions and actions of the *compositum.*" Boyle's reference in this passage is primarily to inanimate bodies, but he suggests that it may apply to living things as well.[21]

"I fear," Boyle tells us, that "we sometimes attribute to the specifick form or soul things that may be well enough performed without it by the more stable modification of the body, befriended by an easy concourse of natural agents." Indeed, "even in animals some things that are confidently presumed to be the proper effects of the animal's soul, may be really performed by the texture of the body, and the ordinary and regular concourse of external causes."[22]

Mechanicism and Biomechanicism

Mechanistic analysis was extended in the seventeenth century to phenomena at all levels from *atomos* to *cosmos,* and to both living and nonliving systems. Boyle was a mechanist exactly in this very general sense. We suggested earlier that his century produced mega-,

20 "Free Considerations," *Works,* 3:120–21.
21 Ibid.,135.
22 Ibid., 125–26.

and meso-, and micromechanical sciences simultaneously. Boyle himself was explicit on this point, asserting that "the mechanical affections of matter are to be found and the laws of motion take place not only in great masses, and in middle sized lumps, but in the smallest fragments. . . ."[23]

Mechanicism went hand in hand, for Boyle, with his rejection of imposed forms and immanent occult qualities. He was equally opposed to "nature" conceived as a directive intelligence, and to the idea of soul-as-form, in the Aristotelian sense.

Mechanicism and emergentism are linked in the world view that he developed. At the corpuscular level, the properties of *mixta* differ from, but depend upon, the mechanical movements of their constituent *minima*. At the cosmic and at all intervening levels the same rule is applicable. The machine analogy is used both implicitly and explicitly. Implicitly, in the sense that Boyle's cosmos, set in motion initially by God, now moves automatically—except where God interposes. Explicitly, in the sense that Boyle openly compares the world and its parts to engines, machines, and bits of technical or scientific apparatus.

He tells us that "according to our doctrine, the world we live in is not a moveless or indigested mass of matter, but an *Automaton,* or self-moving *engine,* wherein the greatest part of the common matter of all bodies is always . . . in motion. . . ."[24] And in a score of passages, Boyle applied this rule to inanimate organization at every level on the scale of magnitudes. As a biomechanicist he was equally explicit. He followed Descartes in considering "the body of a living man, not as a rude heap of limbs and liquors, but as an engine consisting of several parts so set together that there is a strange and conspiring communication betwixt them. . . ."

Boyle was impressed, for example, by the way in which small stimuli may "tricker" major motile responses. As in a gun, "a very weak and inconsiderable impression of adventitious matter upon one part may be able to put on some other distant part, or perhaps in the whole engine a change far exceeding what the same adventitious body could do upon a body not so contrived."

As examples of trigger action he cites especially the major effects

---

[23] "Of the Excellency and Grounds of the Mechanical Hypothesis" (first publ. 1674), *Works,* 4:71.
[24] "An Excursion," *Works,* 3:34.

following from intrinsically minor sensory or other stimuli, for example, laughing in response to having the feet tickled, wounds aching in foul weather, gross responses to small amounts of irritants (including some that we should class as allergens), and the general benefits derivable from specifics used in small doses.

Micromechanical Bases of Life-as-action

There were two fundamental ways in which seventeenth-century mechanistic biology expressed itself. First, in a critical way, it refuted such nonmechanistic entities as "forms," "qualities," "nature" (defined anthropomorphically), "faculties," and "souls" (considered as motive causes, plastic agents, or guides to organization). Second, in a creative way, it detailed the mechanisms—especially the micromechanisms—from whose action organized life was supposed to emerge. It must be confessed that Boyle was more involved in the first of the processes, that is, in the critical demolition of nonmechanistic hypotheses rather than in detailing mechanistic substitutes. He stipulated that soulless mechanics can explain locomotion, the concoction of food, the expulsion of excrements, and the production of milk and of semen. But he mostly failed to detail the machinery requisite to these activities.

Boyle did supply hypothetical micromechanisms, however, for several activities including nutrition. In the "Essay of the Porousness of Animal Bodies," we are told that "the corpuscles of the [alimental] juice [from the blood] insinuate themselves at those pores they find commensurate to their bigness and shape; and those, that are most congruous, being assimilated, add to the substance of the part, wherein they settle, and so make amends for the consumption of those, that were lost by that part before." Here, then is a highly mechanical and very unsoulful arrangement—but one that scarcely goes farther than Greek corpuscular ideas.

A roughly similar provision is made for plants,
in which, of the various corpuscles, that are to be found in
the liquors, that moisten the earth, and are agitated by the heat
of the sun and the air, those, that happen to be commensurate
to the pores of the root, are, by their intervention, impelled
into it, or imbibed by it, and thence conveyed to other parts of the
tree, in the form of sap, which passing through new strainers
(whereby its corpuscles are separated and prepared, or fitted

to be detained in several parts), receives the alterations
requisite to the[ir] being turned into wood, bark,
leaves, blossom, fruit, &c.[25]

Boyle extended his micromechanics beyond nutrition to encompass embryonic development (which he treated only sketchily) and diseases (treated at length). He specifically rejected soul, but occasionally acknowledged an unexplicated plastic power, as "architect of the body's mansion."

On medical subjects Boyle had much to say, and the rationale of his medical system was, again, primarily though not totally mechanistic. The Hippocratic doctrine of nature-as-healer (*natura est morborum medicatrix*) seemed acceptable to Boyle only if "nature" was used merely as a "compendious form of speech" (referring to body mechanics) but not if nature was to be made a "provident internal principle." Getting well, according to Boyle, was like the recovery of the needle of a sea compass after it has been jarred from its northerly alignment.[26]

It is the body's remarkable behavior at the crisis of the disease, Boyle realized, that seems to challenge mechanistic interpretation. "The universal opinion of physicians is, that [at the crisis] it is that intelligent principle they call nature, which . . . watches her opportunity to expell . . . [morbific matter] hastily out of the body. . . ." Boyle disagreed with physicians holding this view. He attributed crises, rather, "to the wisdom and ordinary province of God, exerting itself in the mechanism partly of that great machine the world, and partly of that smaller engine the human body. . . ." Specifically, he would explain everything in terms of (*a*) the "particular condition of the matter to be expelled," (*b*) the "peculiar disposition of the primitive fabric of some parts of the patient's body," and (*c*) "some unusual change made in the construction of these parts by the disease itself. . . ." Boyle acknowledged that specificity in therapeutics (a given drug is effective in a given disease) posed a special problem for corpuscularism, but he was quite sure that this very specificity was corpuscular in origin.[27]

Boyle fully admitted that God had initiated physical movements,

---

[25] "An Essay of the Porousness of Bodies" (first publ. 1684), *Works*, 4:761.

[26] "A Free Enquiry into the Vulgarly Received Notion of Nature Made in an Essay Addressed to a Friend" (first publ. 1686), *Works*, 5:230.

[27] Ibid., 211–15. See also, Robert Boyle, "On the Reconcileableness of Specific Medicines to the Corpuscular Philosophy," *Works* (1744), 4:301.

and had set up the laws that govern them. Moreover, he acknowledged that these laws may "sometimes be varied by some peculiar interposition of God." By and large, however, "many things . . . that are commonly ascribed to nature, I think, may be better ascribed to the mechanisms of the macrocosm and microcosm, I mean of the universe and the human body."[28]

Respiration

Boyle's studies of respiration consisted chiefly in subjecting a variety of animals to the discomfiture of the *vacuum Boyleanum.* The "lasting" of various species of animals was thus tested, among them larks, linnets, and green finches; "new kitten'd kitlings"; mice and shrews; "lusty frogs" and toads; numerous insects; vipers; snails, glow-worms (which would not glow in Boyle's vacuum); leeches; and a great many others. He tested adult animals, as well as eggs, larvae, and pupae. Plants, too, were subjected to decompression. The effect of evacuation was also tried on various chemical or biochemical reactions, especially combustions, gunpowder explosions, fermentation, and putrefaction. Some of the experiments were witnessed by wordly ladies and gentlemen and thus acquired the aspect of a "Roman holiday" (Thorndike's phrase).

Boyle's biological experiments were by no means all designed to verify specific or new hypotheses about respiration. They were not true *experimenta lucifera,* in Bacon's sense. Some merely confirmed experimentally what no thoughtful person had ever questioned, that both life and flame depend upon air. But some of them challenged the Galenic idea that breathing serves primarily to remove excrementitious vapours (also called steams, fuliginous recrements, smokes, superfluous serosities, etc.). The excrementitious particles had generally been supposed to be taken up into the interstices of the atmosphere. But were this the case, evacuation ought, by enlarging the interstices, not to extinguish life and flame—as it does in fact—but rather to promote them. Boyle concluded that air must serve life and flame in some other way, and he was led to revive an idea proposed earlier, he acknowledges, by Paracelsus, viz. that the air contains a special substance essential to, and utilized by, both ordinary flame and the *flamma vitalis.*

As initially announced (in 1660) Boyle's ideas on this subject were somewhat tentative and vague. We hear, first, that the air includes "a

[28] "A Free Enquiry," see n. 26 above, p. 230.

little vital quintessence." Boyle mentions in this connection the submarine made for King James by Cornelius Drebbel who, like Paracelsus, postulated a special substance in the air that fits it for respiration. Drebbel thought himself able to release this substance into the air by unstopping a vessel filled with a certain liquor. Concerning the precise manner in which the vital substance of the air functions, Boyle remained modestly agnostic. He did not deny that it condenses and cools blood passing through the lungs (Descartes' and Harvey's theory) or that it may purify the blood in the lungs. But he was "apt also to suspect that the air doth something else in respiration, which hath not been sufficiently explained; and, therefore, till I have examined the matter more deliberately, I shall not scruple to answer the questions that may be asked me, touching the genuine use of respiration, in the excellent words employed by the acute St. Austin, to one who asked him hard questions: Mallem quidem (says he) earum quae à me quaesivisti habere scientiam quam ignorantiam: sed quia id nondum potui, magis eligo cautam ignorantiam confiteri, quam falsam scientiam profiteri [Of the things you have inquired of me, I should indeed rather have knowledge than ignorance: but, since I cannot have it as yet, I choose to confess a careful ignorance than to profess false knowledge]."[29]

Only in 1672, twelve years later (and, incidentally, eight years after the publication of Hooke's *Micrographia*) did Boyle mature his concept of the relation of air to ordinary flame and to the flame of life (*flamma vitalis*).

"The difficulty we find of keeping flame and fire alive, though but for a little time, without air, makes me some times prone to suspect, that there may be dispersed through the rest of the atmosphere some odd substance, either of a solar, or astral, or some other exotic nature, on whose account the air is so necessary to the subsistence of flame; . . . . And indeed it seems to deserve our wonder, what that should be in the air, which enabling it to keep flame alive, does yet, by being consumed or depraved, so suddenly render the air unfit to make the flame subsist."

Expressing surprise that such exhausted air retains its springiness, Boyle went on to observe that "this undestroyed springiness of the air seems to make the necessity of fresh air to the life of animals . . .

---

[29] 'A Digression Containing some Doubts Touching Respiration,' "New Experiments Physico-Mechanical Touching the Spring of the Air, and its Effects, Made for the Most Part in a New Pneumatical Engine" (first publ. 1660), *Works,* 1:113.

suggest a great suspicion of some vital substance, if I may so call it, diffused through the air, whether it be volatile nitre or [rather] some yet anonymous substance, sydereal or subterraneal, but not improbably of kin to that, which I lately noted to be so necessary to the maintenance of other flames."[30]

Inconclusiveness, Skepticism and Superstition in Boyle's Biology

Boyle called for precision in the definition of life—which others had "ascribed not only to all sorts of animals and plants, . . . [but] . . . to stones and metals . . . to separate souls, angels, good and bad, and to God himself." People were not agreed as to whether life exists in unhatched eggs, in hibernating flies, in animals devoid of blood. "Nor are the boundaries and differences betwixt the life of a plant and that of an animal . . . settled and defined. . . ." Unfortunately, Boyle failed to answer his own call for clarity in all these matters. Our hopes thus raised for a definition of life, we find ourselves wandering in a verbal wilderness farther and farther from anything like a clear formulation.[31]

Boyle's intellectual inconclusiveness on this and other subjects is both understandable and historically significant. The earmark of his endeavor was doubt; he was ever entertaining "suspicions." In Boyle's era there was, in fact, an enormous task of question-asking, idol-smashing, and foundation-shaking to be carried out. The process of demolition was a relatively easier undertaking, in this period, than the process of reconstruction. It is not insignificant that Boyle designated himself, in the title of a major work, a "skeptical chymist." Such phrases in other titles as "experiments Touching . . . ," "Tentamina Quaedam . . . ," "Doubts and Paradoxes, . . ." "Suspicions About . . ." are indicative of Boyle's cast of thought. We would not suggest that Boyle's was an absolute or unreconstructed skepticism, however; he saw skepticism, rather, as an indispensable—and hopeful—prelude to the needed reconstruction of natural philosophy.[32]

Meanwhile, Boyle's own conclusions were not free of superstitious ingredients concerning which he evinced less skepticism than was his

[30] "Suspicions about some Hidden Qualities in the Air" (first publ. 1674), *Works,* 4:90–91.
[31] "Free Considerations about Subordinate Forms," *Works,* 3:127.
[32] For an appreciation of Boyle's scientific attitude, see Richard S. Westfall, "Unpublished Boyle Papers Relating to Scientific Method," *Annals of Science,* 12 (1956):63, 103.

general habit. Thus he envisoned "effluvia," which emanate from the earth, from the stars, and from various physical objects, as both causing and curing diseases. A turquoise may alter its color according to the health of the wearer, he thought. Certain persons petrify directly upon dying. The fingernails of a well-kept cadaver grow and need paring for above a quarter of a century.

Boyle's manner of mixing science and religion is curiously instanced in a treatise on "Some Physico-Theological Considerations about the Possibility of the Resurrection":

> Since . . . the same soul, . . . is said to constitute the same man, notwithstanding the vast differences of bigness, that there may be, at several times, between the portions of matter, whereto the human soul is united:
> Since a considerable part of the human body consists of bones, . . . not apt to be destroyed by the operation, either of earth or fire:
> Since, of the less stable, and especially the fluid parts of a human body, there is a far greater expence made by insensible transpiration, than even philosophers would imagine:
> Since the small particles of a resolved body may retain their own nature, under various alterations and disguises, of which it is possible they may afterwards be stripped:
> Since, without making a human body cease to be the same, it may be repaired and augmented by the adaptation of congruously disposed matter to that, which pre-existed in it:
> Since, I say, these things are so, why should it be impossible, that a most intelligent agent, whose omnipotency extends to all that is not truly contradictory to the nature of things, or to his own, should be able so to order and watch the particles of a human body, as, that partly of those, that remain in the bones, and partly of those, that copiously fly away by insensible transpiration, and partly of those, that are otherwise disposed of upon their resolution a competent number may be preserved or retrieved? so, that stripping them of their disguises, or extricating them from other parts of matter, to which they may happen to be conjoined, he may re-unite them betwixt themselves, and, if need be, with particles of matter fit to be contexted with them, and thereby restore or reproduce a body; which, being

united with the former soul, may, in a sense consonant to the expressions of scripture, recompose to the same man, whose soul and body were formerly disjoined by death.[33]

With a little help from on high, the individual living system is susceptible, it would seem, to posthumous reintegration!

SUMMARY

Boyle asserts, in effect, that living and nonliving things alike are arrangements of a single universal matter. Universal matter as represented in both living and nonliving things is corpuscular. Vital processes may be considered as separations and recombinations of material corpuscles. In seeking the causes of such separations and recombinations in living things, previous theorists placed too much emphasis on such extramaterial agencies as forms, qualities, nature, soul. Organization may equally well be considered as emerging from the "coaptation" of the particles. Soul is not totally abolished as an organizing agent but where admitted tends to be given a material aspect. In any case soul is known by the material (corpuscular) activities it governs. Finally, air may contain an "odd" and subtle something that supports the *flamma vitalis* much in the way that it supports the flame of a candle.

Boyle was less influential for detailed elucidations in physiology than for a historically important viewpoint which he expressed as follows.

And the indefinite divisibility of matter, the wonderful efficacy of motion, and the almost infinite variety of coalitions and structures, that may be made of minute and insensible corpuscles, being duly weighed, I see not, why a philosopher should think it impossible, to make out, by this help, the mechanical possibility of any corporeal agent, how subtle, or diffused, or active soever it be, that can be solidly proved to be really existent in nature, by what name soever it be called or disguised ... and, if an angel himself should work a real change with the nature of a body, it is scarce conceivable to us men, how he could do it without the assistance of local motion. . . .[34]

---

[33] "Some Physico-Theological Considerations about the Possibility of the Resurrection" (first publ. 1675), *Works,* 4:200–201.

[34] "Of the Excellency and Grounds of the Mechanical Hypothesis," *Works,* 4:72–73.

· 21 ·

The Visible Substructure of Life

ROBERT HOOKE [1635 · 1703]

To read Robert Hooke's *Micrographia* (1665)[1] is to become acquainted with an engaging and—from the scientific point of view—impressive mentality. The direction of Hooke's endeavor was influenced, but not circumscribed, by members of the Royal Society, especially Boyle, with respect to whom Hooke assigned himself a secondary importance.[2] Yet Hooke went beyond Boyle in many respects. Hooke wrote rather obtusely at times, and usually prolixly, but without formality or pretense and often with considerable charm. To a greater degree than any of his contemporaries in the Royal Society he combined technical brilliance with intellectual perception, keen observation with shrewd interpretation, and a healthy skepticism with a powerful conjectural imagination. Anthony Wood says Hooke was "a Person of a prodigious inventive Head, so of great Virtue and Goodness. . . ."

He is best known for his discovery of "cells" (so called by him at times; more often designated *pores*), for his famous law (that the strain is proportional to the stress), for innumerable inventions, for important studies on celestial mechanics, and for a shrewd, but only partly original, guess about the nature of combustion.

---

[1] Robert Hooke, *Micrographia; or, Some Physiological Descriptions of Minute Bodies Made by Magnifying Glasses with Observations and Inquiries Thereupon* (London, 1665); reprinted in facsimile (New York, 1961) with the indexes of the editors of 1745 and 1780.

[2] Preface, *Micrographia*, p. xiv. Hooke says of Boyle: ". . . *it becomes me to mention [him] with all honour not only as my particular Patron but as the Patron of Philosophy it self; which he every day* increases *by his* Labours *and* adorns *by his* Example." However, Waller's *Life* quotes Aubrey to the effect that Hooke "read his master (R. B. Esq.) Euclid's Elements, and taught him Des Cartes' Philosophy." We hear also that Boyle presently recommended "Mr. Hooke to be Curator of the Experiments of the Royall Society, wherein he did an admirable good worke to ye Comon-wealth of Learning, in recommending the fittest person in the world to them." Robert T. Gunther, *Early Science in Oxford* (Oxford, 1930), 6:5.

As a child Hooke was sickly, studious, and inventive. He was sent at thirteen to be apprenticed to the painter Peter Lely, but soon left Lely to become a pupil at Westminster School where he "in one Week's time made himselfe master of the first six books of *Euclid,* to the admiration of Mr. Busby his Master, in whose House he lodged and dieted. He also did there, of his own accord, learn to play 20 lessons on the Organ, and invented thirty several ways of Flying. . . ."[3] At eighteen he entered Christ Church, Oxford, at about the time when Boyle moved there. Within a few years, Hooke had invented the balance spring and anchor escapement which revolutionized watchmaking.[4] At the age of twenty-seven, he became Curator of Experiments for the Royal Society with the expectation that he would "both bring in every Day [of the meeting] three or four of his own experiments and take care of such others as should be recommended to him by the Society."[5]

Hooke's work in this capacity was prodigious. To take a sample entry in Birch's *History of the Royal Society,* Hooke was "put in mind of [his] several tasks . . . [viz.] of perfecting his new quadrant; of producing a new sort of watch more exact than a pendulum-watch; of observing the parallax of the earth's orb; of prosecuting the magnetical experiments, first for finding out (*a*) whether gravitation be magnetical; and then (*b*) whether the magnet will attract at the same distance in water, as in air; and also (*c*) whether the lines of a loadstone's direction are truly oval." On that same day the "experiments appointed for the next meeting were: 1. The circular pendulum to be prosecuted. 2. The new watch to be produced. 3. Some waternewts to be provided."[6]

To a remarkable degree, Hooke did what was expected of him. Maintaining the pace week after week, and year after year, he concerned himself with heat, light, sound, and magnetism; with astronomy and mechanics; with the velocity of projectiles; with (in a typical week) ear trumpets, diving helmets, tanning, the microstructure of muscle, and the design of the Royal Society's building. Hooke was, among other things, an important architect who collaborated with Wren and independently designed buildings for the Royal College of

[3] Anthony à Wood, *Athenae Oxonienses,* "new" ed. (London, 1820), 4: col. 628.
[4] Gunther, *Early Science,* 6:11–20.
[5] Ibid., p. 23.
[6] Ibid., p. 278.

Physicians, Bedlam, Montagu House, and the parish church at Willen, Buckinghamshire, among others. Of special interest to us in our present quest were Hooke's numerous studies on key biological questions.

Hooke's Endeavor

Hooke's endeavor should be viewed in relation to three major trends of the science of the times: first, the extension to biology of mechanical, especially micromechanical and corpuscular modes of explanation; second, and affected by the first, continued speculation about the subvisible arrangements that give rise to visible actions; third, an intensified interest in a group of phenomena which different thinkers were weaving together in a variety of hypothetical relations, namely respiration, combustion, explosions, fermentations, and putrefactions. With respect to subvisible structure, Hooke was at the heart of an important development in which hypothetical micro-arrangements were being reinvestigated optically at least to the limits of the then available microscopes. Hooke presented his science in a somewhat fragmentary way, but what he said contains the germs of a general system that can be pieced together from different parts of *Micrographia* and from other writings.

Hooke on Matter

Perhaps because Hooke was influenced by Boyle's skepticism—the two often worked together—he suspended judgment on many issues. He was noncommittal on the elements, enumerating air, earth, and water, but only as possibly deserving this designation. He built his cosmos of (*a*) aether ("an exceeding fluid body, very apt and ready to be mov'd and to communicate the motion of any one part to any other part, though never so far distant. . . .") and (*b*) "massie particles" constituting solids and tangible liquids, as well as air. He said that "the air [as compared with earth] . . . is a certain company of particles of quite another kind, that is, such as are very much smaller and more easily movable by the motion of this fluid *medium* [aether]; . . ." Whether the aether itself was corpuscular Hooke did not "much concern [himself] to determine."[7]

Encouraged by the successes of his own microscopic studies, Hooke thought the corpuscular problem itself might ultimately yield

[7] *Micrographia,* pp. 95–97.

to optical methods. In the admired "Preface" to *Micrographia*, he expressed the hope that improved instruments might one day reveal *"the* figures *of the compounding Particles of Matter, and the particular* Schematisms *and* Textures *of Bodies"*—as well, incidentally, as "living Creatures *in the Moon, or other Planets. . . ."*[8]

Hooke pictured a vast scale of being with least particles at one end, human imagination at the other. A fundamental "property of Congruity" inclines particles to form tiny globules, Hooke believed, especially when one fluid body is "encompast with a *Heterogeneous* fluid. . . ."[9] Studying crystals microscopically, he became convinced that their basic plans could be understood as "angular" arrangements of such globules. Using a "composition of bullets" as well as "one or two other bodies" he built models to illustrate the most important sorts of material organization. Starting here, Hooke posited a material hierarchy whose organizing principles, in order, were supposed to be: *"Fluidity, Orbiculation, Fixation, Angulization, or Crystallization*[,] *Germination or Ebullition, Vegetation, Plantanimation, Animation, Sensation, Imagination."*[10] Regrettably, Hooke failed to spell out completely clearly how these principles were supposed to operate or how he thought they were different from, or related to, one another.

Hooke's micromechanical bias appeared in the analogy he drew between the growth patterns of plants and of certain minerals and chemical substances that display arborescent or other vegetationlike patterns. Such an analogy holds for certain crystals, he thought, including the "silver tree" known to the chemists (see Bacon and Maupertuis, chapter 29) and a certain porous "Kettering-stone" from Northamptonshire.[11] Implicit in these comparisons was the assumption that the inanimate forces responsible for crystal-formation were the same as or similar to those responsible for organic growth. Hooke also studied fossils and recognized them as such rather than as bodies organized through some mysterious "Plastick virtue."[12]

He pointed to the mechanistic implications of the geometric similarity of crystal patterns in frozen urine to the growth patterns

8 Ibid., Preface, p.viii.
9 Ibid., p. 85.
10 Ibid., p. 127.
11 Ibid., pp. 95, 130.
12 Ibid., pp. 107–12.

of fern fronds. Both patterns show not only branches but lateral, collateral, subcollateral, and laterosubcollateral branches; whence "if the Figures of both be well consider'd, one would ghess that there were not much greater need of a *seminal principle* for the production of *Fearn,* then for the production of the branches of *Urine.* . . ." Ferns must be relatively simple plants, Hooke concluded, adding that he had been unable to discover any seeds in his microscopic examination of these plants (later he changed his ideas on this subject).[13]

Hooke's Biomechanical Outlook

Hooke made microscopic studies of a spectrum of plants including molds, mosses, ferns, and a variety of seed-producing species. Primitive plants seemed to him to raise some fundamental theoretical questions. He noted that molds, for example, tend to arise at foci of fermentation or putrefaction. He studied, and drew, a mold with round sporangia (probably *Rhizopus*)[14] and another found growing in the spots on speckled rose leaves.[15] Do molds arise as a byproduct of putrefaction, he asked, or rather by seeds (or something comparable)?

Conceivably, Hooke speculated, some germinal principle that ordinarily contributes to the life of the rose could survive the breakdown of that plant and give rise to a more limited type of vegetation.[16] Hooke tested this hypothesis by what we might call a thought experiment. Wondering whether mosses, like molds, could arise in putrefying matter, he asked himself whether a machine could be imagined (a watch, for example) in which a similar thing might happen. Conceivably, in a broken watch, some still intact part (for example, the striking mechanism) could, when separated from the other (and broken) parts, resume its partial function.[17] Perhaps "putrefactive generation" occurs in the same fashion.

---

[13] Ibid., p. 90. In an appendix to his 1677 work on lamps, Hooke described the spores of mosses ("and the number in a grain weight of them cannot be less than [1,380,000,000] . . ."); he also published a letter from "a very good friend of mine at *Bristol,* the Ingenious and Inquisitive Mr. *W. C.,*" describing fern spores and sporecases. See "An observation about the Seed of Moss," *Lampas: or, Descriptions of some Mechanical Improvements of Lamps and Waterpoises. Together with some other Physical and Mechanical Discoveries* (London, 1677); reproduced in facsimile in Gunther, *Early Science in Oxford* (London, 1931), 8:198–206.

[14] *Micrographia,* plate 12, opp. p. 125.

[15] Ibid., p. 125.

[16] Ibid., pp. 124–25, 134–35.

[17] Ibid., p. 133.

Such reasoning strikingly illustrates the impact that mechanical thinking may have on biological interpretation. Such thinking led, in Hooke's case, to the very general conclusion that "as far as I have been able to look into the Primary kind of life and Vegetation, I cannot find the least probable argument to persuade me there is any other concurrent cause then [= than] such [as] is purely Mechanical, and [to the conclusion] that the effects or productions are as necessary upon the concurrence of those causes as that a Ship, when the Sails are hoist up, and the Rudder is set to such a position, should, when the Wind blows, be mov'd in such a way or course to that or t'other place; . . . ."[18]

It should be added that with characteristic tentativeness Hooke admitted alternatives to this mechanical explanation of putrefactive generation. Perhaps decaying matter contains spontaneously compounded animal substance (but such substances may themselves "depend merely upon a convenient constitution of the matter out of which they are made, . . ."). Or, perhaps animals eject a seminal principle into the putrescent matter.[19] Indeed, many animalcules apparently produced spontaneously may turn out to be immature stages of flying creatures, as in the case of a water gnat Hooke studied.[20]

These and Hooke's other studies of insects caused him to generalize on biomechanics as applied, in particular, to the reproductive instincts of insects. He begins his report of this study in an appropriately reverential vein. He says that "If we consider the great care of the Creator in the dispensation of his providences for the propagation and increase of the race, not onely of all kind of Animals, but even of Vegetables, we cannot chuse but admire and adore him for his Excellencies, . . ." But note how Hooke wants us to admire the Creator and not the creatures. The latter are marvelous in that their "strange kind of acting in several . . . [species] seems to savour so much of Reason. . . ." In fact, however, the reason is the Creator's and not the creature's, "it seeming to me most manifest that . . . [these animals] are but acting according to their structures and [that they perform only] such operations as such bodies, so compos'd must

18 Ibid., p. 130–31.
19 Ibid., p. 123.
20 Ibid., pp. 190–91.

necessarily, when there are such and such circumstances concurring, perform. . . ."

And so, Hooke continues, "when we find Flies swarming, about any piece of flesh that does a little begin to ferment; butterflies about Colworts, and several other leaves, which will serve to hatch and nourish their young; Gnats, and several other Flies about the Waters, and marishy places, or any other creatures, seeking and placing their Seeds in convenient repositories, we may, if we attentively consider and examine it, find that there are circumstances sufficient, upon the supposals of the excellent contrivance of their machine, to excite and force them to act after such or such a manner . . . ," it being unnecessary to impute reason to them.

From the historical point of view, Hooke's mechanical analysis of behavior is significant as an elaboration of Descartes' insistence upon the automatism, the soullessness, of animals other than man. He gives some idea of the possible mechanisms involved by suggesting that ". . . those steams that rise from these several places may, perhaps, set several parts of these little animals at work," and he compares the response to that of "the contrivance of killing a Fox or Wolf with a Gun," in an automated trap in which "the moving of a string is the death of the Animal. . . ." We hear that in this contrivance, which almost seems to be able to reason, "the Beast, by moving the flesh that is laid to entrap him, pulls the string which moves the trigger, and that lets go the Cock which on the steel strikes certain sparks of fire which kindle powder in the pann, and presently flies into the barrel, where the powder catching fire rarifies and drives out the bullet which kills the Animal; in all which actions," and this is the crux of the argument, "there is nothing of intention or ratiocination to be ascrib'd either to the Animal or Engine, but all to the ingeniousness of the contriver."[21]

Hooke's line of thought, here, is one which we shall encounter again as our story develops, namely that the remarkable adaptedness of living systems betokens not a directive intelligence operating continuously within them but rather an Initiative Intelligence which created them in the beginning. The point is interesting because it shows how the mechanistic point of view can be made an argument for, rather than against, religious assumptions.

[21] Ibid.

Hooke and the Transformation of Species

In the late seventeenth and earlier eighteenth centuries, there was a growing interest in the question of the fixity of species. Analysis of this problem, mostly rather theoretical, was the prelude to the explicitly evolutionary ideas that emerged from about 1750 onwards (see chapters 28 and 32). Hooke was among the seventeenth-century forerunners of speculation on these topics. Thus, it seemed to him that organisms produced by putrefactive generation would theoretically be able to reproduce and thus to give rise to entirely new species.[22] Hooke was not an out-and-out evolutionist but he was quite prepared to accept the idea of possibly radical transformation of animals through environmental effects. Thus, he thought that certain "wandring mites" that he brought under his microscope might be the "vagabond parents of the Mites we find in Cheeses, Meal, Corn, Seeds, musty Barrels, musty Leather, &c." He thought that

> . . . these little Creatures [the hypothetical wild ancestors], wandring to and fro every whither, might perhaps, as they were invited hither and thither by the musty steams of several putrifying bodies, make their invasions upon those new and pleasing territories, and there spending the remainder of their life, which might be perhaps a day, or thereabouts, in very plentiful and riotous living, might leave their off-spring behind them which by the change of the soil and Country they now inhabite, might be quite alter'd from the hew of their *primogenitors,* and, like *Mores* [Moors] translated into Northern *European* Climates, after a little time, change both their skin and shape. And this seems more probable in these Insects, because that the soil or body they inhabit, seems to be almost half their parent. . . .

The environment might thus act in almost the way that heredity acts. Note how climate can imitate the effect of "*Negro* women [who] besmeer the of-spring of the *Spaniard,* bringing forth neither white-skinn'd nor black, but tawny-hided *Mulattos.*"[23]

[22] Ibid., p. 124.
[23] Ibid., pp. 206–7.

Hooke's Discovery of "Cells"

We shall trace elsewhere late seventeenth- and early eighteenth-century ideas about microscopic building blocks of living systems. In 1665, Hooke published a description of something that he referred to as "cells," but a century and three quarters had to pass before the cell would be accepted as a structural building block or physiological unit. In the interim, attention was concentrated primarily not on cells but on particles or granules and on the fibers that such granules presumably constitute. During this interim period between Hooke (1665) and Schwann (1839) cells were assigned a variety of roles. They were, in particular, regarded—by their discoverer among others—as avenues of communication, "channels, provided by the Great and Alwise Creator, for the conveyance of appropriated juyces to particular parts." This applies especially, in Hooke's account, to "the pores in Wood, and other vegetables, in bones, and other Animal substances. . . ."[24] He applied "cells," notoriously, to vacuities seen in a slice of cork. His interests were wider.

Hooke examined not only the dead air-filled cells of cork but the fluid-filled "pores" of "the pith of the Cany hollow stalks of several other Vegetables: as of Fennel, Carrets, Daucus, Bur-Docks, Teasels, Fearn, some kinds of Reeds, &c." which " have such a kind of Schematisme, as I have lately shewn that of Cork, save onely that here the pores are rang'd the long ways, or the same ways with the length of the Cane, whereas in Cork they are transverse."[25] He interpreted cork as "a kind of Fungus or Mushrome. . . ."

If these are channels for fluid conduction, he acknowledged, one must ask how fluids travel from one cell to an apparently separate neighboring cell. Through valvelike openings, Hooke surmised, which would "open and give passage to the contain'd fluid juices one way [only]." In Hooke's opinion, better microscopes than his would be needed in order to discern these valves.[26]

Hooke's Influence

Hooke's contributions as a microscopist have been both over-estimated and underestimated. To assert that he "discovered cells"

[24] Ibid., p. 95.
[25] Ibid., p. 115.
[26] Ibid., p. 116.

is erroneous if "cell" is taken in anything like its modern sense. He discovered what he preferred to call "pores"[27]—perforations filled, in most cases, with "juyces." Before this concept of the cell as a perforation became the modern concept, a series of further discoveries had to be made and interpreted; these we shall touch on in subsequent parts of our narrative.

What Hooke contributed to biology, then, was not the cell as we understand it today, but rather *a new general insight* that was to be immensely influential in the effort to associate life with a particular condition of matter: an insight, namely, into the significance of organization at the microscopically visible level. He made the point that fluid-filled spaces were a part of that organization; he considered these spaces important; and he made a guess regarding their function. It is to his credit, rather than otherwise, that his provocative guess about what he saw was to be repeatedly reconsidered and revised until in 1839 another microscopist, and one of Hooke's intellectual descendants, produced a "cell theory" and, in 1860, still another successor produced a new concept of the cell as a nucleated protoplasmic mass. Thus Hooke did not suddenly discover, or claim to discover, but he did help to initiate the search for the microscopically visible substructure of life. He did not assert that the secret of life would be revealed by a study of its microscopically visible substrate, nor was he the only important microscopist of his era. But along with the others (e.g. Malpighi, Leeuwenhoek, and Grew) he contributed to the life-matter problem the recognition that it must be approached, in part, by an analysis of the *visible* microorganization of the matter that manifests life. The ramifications of this new approach, namely, the optical analysis of living systems, have not been exhausted even today.

Combustion and Respiration

Hooke built airpumps for Boyle and collaborated with him on his famous evacuation experiments. Beyond this, there are two principal reasons for including Hooke among the important seventeenth-century students of respiration. First, from observations on charcoal he

[27] Ibid., pp. 112–21. These expressions all occur in the famous "Observation xviii." He also used the term *cell* as well as *box* and *bladder*. Feather seemed to be a "kind of solid or hardened froth, or a *congeries* of very small bubbles consolidated in that form, into a pretty stiff as well as tough concrete, . . . [of which] each Cavern, Bubble, or Cell, is distinctly separate. . . ." (pp. 115–16).

constructed a prescient, twelve-point theory of combustion. Second, he investigated breathing in a series of experiments, at least one of which was successful and significant.

Hooke's first development of his combustion theory was given, in brief, in Thomas Birch's *History of the Royal Society* in the entry for Jan. 4, 1664.

> Mr. Hooke made an experiment tending to show, as he conceived, that *air is the universal dissolvent of all sulphureous bodies,* and that *this dissolution is fire* [italics added]; adding, that this was done by a nitrous substance inherent and mixed with the air. The experiment was, that he took a live coal, and put it under a glass vessel; whereupon the coal after a very little time, went out; but then being taken out, and exposed to the free air, recovered its burning.
>
> It being objected, that it was the agitation of the air driving the igneous particles into the combustible body, which made it burn and consume; Mr. Hooke answered, that experiment would show, that a burning body, though agitated, would be extinguished, if it had not a free access of fresh air. He added, that a combustible substance, kept red-hot, even in a fire as hot as to melt copper, would not waste, but as soon as fresh air was admitted, was burnt away and consumed.
>
> An experiment was mentioned, to show, that a burning coal wanting fresh air would keep entire; but brought into new air would fall in pieces.[28]

An account of this sort must be read with attention both to what it does and what it does not say. It does say that something from air can enter a body and—by dissolving its sulphur—cause it to burn. It does not say that air particles *combine* with particles of the body substance. Hooke set forth the theory in detail in the following year, 1665, in his *Micrographia*.[29] We must not leap to the conclusion that Hooke (or any of the contemporary students of combustion) anticipated Lavoisier or set forth anything like a modern theory of oxidation. The nitrous element was, for Hooke, an agitant and a solvent but not in our modern sense a reactant. Seventeenth-century development *predisposed* the subject of combustion to the

[28] The relevant entry is reprinted in Gunther, *Early Science in Oxford,* 6:233.
[29] *Micrographia,* pp. 102–4; see also pp. 46–47.

eighteenth-century discoveries, but they did not forestall those discoveries. How different Hooke's idea was from Lavoisier's appears especially in the more lengthy treatment given combustion in the later *Micrographia* (1665).[30]

In the interim, Hooke had begun to study respiration with the techniques of the experimental physiologist. At the meeting of November 2, 1664, he "proposed an experiment to be made upon a dog by displaying his whole thorax, to see how long, by blowing into his lungs, life might be preserved, and whether anything could be discovered concerning the mixture of the air with the blood in the lungs. It was ordered, that the experiment be made between that and the next meeting."[31]

According to Sprat, Hooke reported the outcome as follows.

In prosecution of some inquiries into the nature of Respiration in several animals; a Dog was dissected, and by means of a pair of bellows, and a certain pipe thrust into the Windpipe of the Creature, the heart continued beating for a very long while after all the Thorax and Belly had been open'd, nay after the Diaphragme had been in great part cut away, and the Pericardium remov'd from the heart. And from several tryals made, it seem'd very probable, that this motion might have been continued, as long almost as there was any blood left within the vessels of the Dog: for the motion of the Heart seemed very little chang'd after above an hours time from the first displaying the Thorax; though we found, that upon removing the Bellows, the Lungs would presently grow flaccid, and the Heart begin to have convulsive motions; but upon removing[32] the motion of the Bellows, the Heart recovered its former motion, and the Convulsions ceased. Though I made a Ligature

[30] Ibid. The twelve points made by Hooke are, in brief: that air is a solvent (menstruum) of sulphureous bodies; that heat is required for its solvent action; that that action produces still greater heat (fire); that fire emits light; that the active substance is a component of air present also in niter; that the dissolution is accompanied by volatilization; that there is an insoluble residue extractable from soot; that another insoluble portion is, nevertheless, volatile; that still a third insoluble portion forms ash which contains alkali; that the menstruum is quickly exhausted and must be often replenished; that copious replenishment of it produces a violent solvent action; and, finally, that (*a*) fire is not an element, and (*b*) flame is a mixture of air and volatile sulphureous parts of dissoluble or combustible bodies.

[31] Thomas Birch, in Gunther, *Early Science in Oxford,* 6:214.

[32] Misprint for *restoring?*

upon all the great Vessels that went into the lower parts of its Body, I could not find any alteration in the pulse of the Heart; the circulation, it seems, being perform'd some other way. I cou'd not perceive anything distinctly, whether the Air did unite and mix with the Blood; nor did in the least perceive the Heart to swell upon the extension of the Lungs: nor did the Lungs seem to swell upon the contraction of the Heart.[33]

By this "noble" experiment, repeated several times, Hooke showed that the primary function of breathing is not the mere circulation of blood but its ventilation which he not wholly incorrectly interpreted as having the dual function of mixing it with air (or some part of air) and clearing it of noxious fumes.[34]

Hooke was often involved, in association with Boyle and independently, in studies tending to link combustion with respiration. We are not surprised therefore by Birch's entry for June 27, 1667, where Birch takes note of a

discourse what quality it was, that made the air fit for respiration. Some thought it became unfit by being clogged and entangled with gross vapours. Mr. HOOKE was of opinion, that there is a kind of nitrous quality in the air that makes the refreshment necessary to life, which being spent or entangled, the air becomes unfit.

He related an experiment long since made before the society with a chafing-dish of coals set in a close box, wherein was a pair of bellows so contrived, as to blow the coal with that air only, that was included in the box: the air so kept had this quality, that after one whole day's time fresh fire would not burn in it, till the grosser parts thereof were precipitated.[35]

The question whether in the lung air enters, mixes with and alters the blood thus raised by Hooke was to be tackled fruitfully slightly later by Richard Lower (see chapter 24). Hooke suggested two

[33] Thomas Sprat, *History of the Royal Society* (London, 1667), p. 232. The relevant passage is quoted in Gunther, *Early Science in Oxford*, 6:215–16. The experiment is described also as "An account of an experiment made by Mr. Hook of preserving Animals alive by blowing through their Lungs with Bellows," *Philosophical Transactions of the Royal Society*, 2 (October 4, 1667):539–40.

[34] Entry for October 3, 1667, Thomas Birch's *History*, quoted by Gunther, *Early Science in Oxford*, 6:315; see also 7:403.

[35] Ibid., 6:309.

LIFE AND AIR

ways of approaching that problem. One, which he tried without success, was to make a direct connection from the pulmonary vein to the aorta, bypassing the lungs. Hooke attempted this several times, and kept reporting to the Society that he had failed but had thought of another way of trying, and would report again later. Hooke's other approach was to utilize the well-known color difference between arterial and venous blood. Venous blood that has stood awhile, he noted, brightens at the surface where it is exposed to air. On April 30, 1668, Hooke suggested that "it might be worth the observing by experiment, whether the blood, when from the right ventricle of the heart it passes into the left, coming out of the lungs, it hath not the tincture of floridness, before it enters into the great artery; which if it should have, it would be an argument, that some mixture of air with the blood in the lungs might give that floridness."[36] It was this observation which Lower would later succeed in making (see chapter 24).

During the ensuing decades, Hooke returned episodically to combustion, respiration, and the life-air problem. In 1672, he told the Society he thought the use of breathing to be that "by the air something essential to life might be conveyed into the blood; and something that was noisome to it, discharged back into the air. . . ."[37] In 1678, May 2:

> It was discoursed, [in the meeting] that the specifical use of air
> for respiration was difficult to guess it; for some experiments
> had proved that there might be a circulation without the motion
> of the lungs; and that a man might be stifled, though he moved
> his lungs and breathed, if it were not fresh air [in other words,
> air, not mere motion, is the useful aspect of respiration]. This
> was thought a good argument to prove what Mr. Hooke had
> asserted, that air was the pabulum of the animal spirits, and . . .
> was the principle cause both of the heat and animal motion;
> for that the blood in the lungs was both impregnated with fresh
> air, and so received an enlivening florid arterial colour, and
> also discharged great quantity of steams and fuliginous matter,
> that was contained in it.[38]

[36] Ibid., 6:331.
[37] Ibid., 7:403.
[38] Ibid., 7:486.

· 308 ·

And on the following January 9 (1679), we read that whereas Croone said that it was steams from breath and body that cause death in an enclosure, Hooke thought that, were that true, animals should live longer in rarer air [Boyle's experiment]. Hooke thought, rather, that death "proceeds from the satiating of the dissolving part of the air, so making the remaining part effete and useless for maintaining the life of animals, which seemed to have much the same nature with flame and fire, since the same effects seemed to happen to it."[39]

In his later Cutler lectures, Hooke expatiated more broadly on the details of the process. In a lecture of 1680, *On Light,* he said, "Bodics . . . that emit a Light of their own . . . are all Sulphureous, Unctuous, Resinous, or Spirituous Bodies, which will being first heated, be burnt or Dissolved by the Air, as a Menstruum. . . ." He had thus not altered his idea of fire as produced by solvent action. He added, further, that just as " 'tis the fresh Air that is the Life of the Fire . . ." so animals need "fresh Air to breath [!] and, as it were blow the Fire of Life. . . ."[40] In another Cutler Lecture (1682), Hooke ascribed the solvent action of air, in language reminiscent of Mayow (see chapter 23), to a contained "aërial or volatile nitrous spirit." In some cases, Hooke believed, this spirit may be present in the body to be burned which can then ignite in the absence of the air; such is the case with gun powder which can be fired in a vacuum or under water. Hooke supposed that the nitrous part of the air "may most properly be called the Vital part thereof, which supplies the Menstruum to burning and flaming Bodies; and [is] that which continues the Life Heat and Motion of all Animals and Vegetables."[41] It is clear, then, that Hooke regarded an air-induced solution of the body's sulphur as indispensable and central to vital manifestations.

NOTES AND SUMMARY

Hooke's microscopical researches dramatize the importance of visible microorganization in the expression of life phenomena. He saw life as emergent at a certain level in the grand hierarchy of material

[39] Ibid., 7:507.

[40] Robert Hooke, "Lectures of Light, Explicating its Nature, Properties, and Effects, Sect. I, containing those read about the beginning of 1680," *The Posthumous Works of Robert Hooke* (London, posth., 1705), pp. 110–11.

[41] Robert Hooke, "A Discourse of the Nature of Comets. Read at the Meetings of the Royal Society, soon after Michaelmas 1682," *Posthumous Works,* p. 169.

organization. In its primary manifestations (for example, as in molds) life is explicable mechanically. Hooke was an enthusiastic and active but not an out-and-out mechanist. At higher levels of vegetation he acknowledged a directive influence. This Hooke called "an *anima* or *forma informans,* that does contrive all the Structures and *Mechanisms* of the constituting body, to make them subservient and usefull to the great Work or Function they are to perform, . . ."[42] This brief allusion was all that Hooke had to say about this animistic and tcleological agency; the tone of *Micrographia* is, elsewhere, mechanistic. Hooke's theory of combustion—as solution of a body's contained sulphur by nitrous spirits present in air—bore directly on the life-matter problem in that this solvent action was central to life as well as to fire. Note then, that life was, for Hooke, a molecularly—not a cellularly—conditioned phenomenon.

We may conclude our study of Hooke by citing briefly a final illustration of his mechanical viewpoint. A remarkable semimicromechanism that interested Hooke was the single hair of the wild oat's beard which curls back on itself when dry, but straightens out when it becomes wet. The oat beard, he said, has been used by "children and Juglers, and it has been call'd by some of these last named persons, the better to cover their cheat, the Legg of an Arabian spider, or the Legg of an inchanted Egyptian Fly, and has been used by them to make a small Index, Cross, or the like, to move round upon the wetting of it with a drop of Water, and muttering certain words.

"But the use that has been made of it, for the discovery of the various constitutions of the Air, as to driness and moistness, is incomparably beyond any other; . . ."

Hooke wished that he had time to enlarge further upon the subject of the oat beard, "for it seems to me to be the very first footstep of *Sensation,* and Animate motion, the most plain, simple, and obvious contrivance that Nature has made use of to produce a motion, next to that of Rarefaction and Condensation by heat and cold."[43] The oat beard represented, for Hooke, living machinery at its simplest. At this level the living world seemed removed by single small steps from the world of the nonliving. We seem to see emergent life, here, in the very moment of its emergence.

Hooke's contribution to the life-matter problem is, in sum, an

[42] *Micrographia,* p. 95.
[43] Ibid., pp. 149–52.

imperfectly integrated but not inconsistent skein of ideas with three principal strands: first, the organism is a machine and life a mechanical consequence of organization; second, by implication, important aspects of that organization are microscopically visible; but, third, at the corpuscular level, and here Hooke shows a surviving link with alchemical thought, life is fed like a fire or flame, by the solvent action of air on the sulphur the body contains.

22

Life as a Subjugated Flame

THOMAS WILLIS [1621 · 1675]

Thomas Willis was a teacher, neuroanatomist, and physician who arrived in Oxford well before most of the other virtuosi, taught some of them, and, according to several early biographers, learned from them more than he taught. Anthony Wood (1691) said that Willis was "the most famous physician of his time" and that when he settled in London (after the fire) was "so resorted to, for his practice, that never any physician . . . got more money yearly than he." Wood praised the "natural smoothness, pure elegancy [and] delightful, un-affected neatness of [Willis'] *Latin style.*" (Willis' Latin seems wordy at times but no more so than, for example, Boyle's and Hooke's English written at about the same time.) Wood says that "after a great deal of drudgery, that he did undergo in his faculty, (mostly for lucre sake) which did much shorten his life, he concluded his last day in his house in S. Martin's Lane before-mention'd, on the eleventh day of Nov. in sixteen hundred seventy and five; whereupon his body was conveyed to the Abbey Church of S. Peter in Westm. and there interr'd. . . ."[1]

The effect of the younger physiologists at Oxford and others on Willis is apparent when one compares the first edition of his treatise *On Fermentation* (1659)[2] with works written approximately ten years later. We shall try to suggest, as we proceed, how Willis adjusted to his younger colleagues' ideas and how his own conclusions

[1] Anthony à Wood, *Athenae Oxonienses* (London, 1691), 1:402–3. For an appreciation of Willis, see H. Isler, *Thomas Willis: ein Wegbereiter der modernen Medizin, 1671–1675* (Stuttgart, 1965).

[2] Thomas Willis, *Diatribae duae medico-philosophicae, quarum prior agit de fermentatione sive de motu intestino particularum in quovis corpore*, 2d ed. (London, 1659), trans. S. Pordage, "A Medico-Philosophical Discourse of Fermentation; or, Of the Intestine Motion of Particles in Every Body" (Treatise I in) *Dr. Willis' Practice of Physick, Being the Whole Works, . . .* (London, 1684).

seemed, to him at least, to relate to the problem of the preconditions of life.

Our study of the history of physiology thus far has taught us to expect, in any reasoned physiological system, certain more or less standard components. We expect, first, an explicit doctrine of matter, and second, and based on that doctrine, conceptual models of such overt activities as nutrition, generation, locomotion, volition, perception, and reason. We shall find that in these respects Willis followed the classic pattern, a pattern prevalent since the time of Empedocles and some of the earlier Hippocratics. More specifically we expect, in the seventeenth century, the number of chemical elements to be few and these to be engaged in living things in special sorts of interactions. In these ways, too, Willis followed traditional patterns.

Matter

As to matter, Willis thought it necessary to choose among: (*a*) the "fourfold Chariot of the Peripateticks," (*b*) the "Conflux of Atoms diversely figured," and (*c*) "Particles of Spirit, Sulphur, Salt, Water, and Earth." Of these three "Opinions of the Philosophers," Willis preferred the third.

As to the four elements, he thought them to be concepts so dark (*crassus*) and unrevealing as to render them about as useful to build things out of as "to say an house consists of Wood and Stone." Atoms he rejected as being too "remote from Sense" and because they did "not sufficiently Quadrate the Phaenomena of Nature." Willis thus cast his lot with the chemists "affirming all Bodies to consist of Spirit, Sulphur, Salt, Water, and Earth and . . . the diverse motion, and Proportion of these, in Mixt things." We hear the five Principles fairly definitely characterized by Willis, but their properties, as presented by him, seem rather ill-assorted and farfetched. Note that Willis' rejection of atomism (ultimate indivisibility) does not prevent him from regarding matter as particulate.

Spirits, for example, are "highly subtil, and Aethereal Particles of a more Divine Breathing . . . Instruments of Life and Soul, of Motion and Sense, of every thing. . . . From the motion of these proceed the animation of Bodies, the growth of Plants, and the ripening of Fruits, Liquors, and other Preparations; they determinate the Form and Figure of everything, ..." How they do this Willis promises to explain later, but he does not get around to doing so in any very satisfactory way.

Willis' view of sulphur looks back to the alchemical linkage of this element with fire, but reinterprets this idea in corpuscular terms. "Sulphur is . . . of a little thicker consistence than Spirit. . . . The Temperament of every thing, as to Heat, Consistency, and aimiable frame or contexture, depends chiefly on Sulphur; from hence also for the most part arise variety of Colours and Odors, and fairness and deformity of the Body, also diversity of tastes." On the other hand, ". . . being more impetuously moved or stirred up, they [sulphur particles] bring on dissolution of Bodies, yea a flame and Burning (*imo flamman & conflagrationem*)."[3] We hear, too, that "all the time that Fire continues in the Subject, Sulphureous particles fly away in heaps" (*particulae sulphureae confertim avolant*), and that "fire is no other thing, than the motion and eruption, of these kind of Particles [sulphur] impetuously stirred up."[4] Many of the properties ascribed by Willis to sulphur were later to be assigned to phlogiston by its chief promulgator Georg Stahl (see chapter 25).

"Salt is of a little more fixed nature . . . [not] so apt to fly away; but bestows a Compaction and Solidity on things, and also weight and duration, . . . Not only the duration of the individual, but also the propagation of the Species, depend very much on the Principle of Salt, because the fertility of the Earth, the growth of Plants, and especially the frequent foetation, and bringing forth of young, in living Creatures, takes their Original from the Saltish Seed. . . ."[5]

The three foregoing elements are *active* (salt less so than the others); the remaining two, water and earth, *passive*. (The classification of principles into active and passive is Aristotelian.) "Water is the chiefest Vehicle of Spirit and Sulphur, by whose intervention, they consociate with one another, and with Salt. . . . When water is wanting, the active Principles meet together too strictly. . . ."

"Earth in Solids, fills the empty little Spaces and Vacuities, left by

---

[3] Ibid., pp.2–4.

[4] Ibid., pp. 36–37.

[5] Ibid., p. 4, where he goes on to say that: "Hence it is that Venus is said to arise from the Sea, and best is called Salacity. For Salt, having obtained a flux, gathers together, and stirs up into motion, the idle, or too much disjoyned little Bodies of Spirit or Sulphur, and excellently keeps them together with it self, for the producing the first ground-work of things." Note the Latin origins of "salt" (*sal, salis*) and "salacious" (*salax* from *salire,* to leap).

other Principles, for these [water and earth] hinder the active Principles [the other three] from a too straight embrace . . . ; also by its thickness, it retains too Volatile things: besides, it inlarges the due Substance, and magnitude in Bodies. The more Earth that abounds in any thing, it is so much the less active, but of longer duration. . . ."[6]

To summarize: with spirit we associate volatility, life, and form; with sulphur, consistency and certain sensible properties, as well as fire; with salt, weight and lastingness and, in people, propagation. These three are the active principles. The passive pair, water and earth, serve variously as a vehicle and governing matrix for these active principles. It must be admitted that all this forms less than an elegantly ordered chemical system. It is, of course, not new (a five-element system had been suggested by Sebastian Basso, 1621) nor is it derivable in itself from Willis' various chemical and pharmacological investigations. In his physiology, we shall find, Willis uses the Aristotelian idea of active and inactive elements only in an approximate and not in a detailed or systematic fashion. On the other hand, he develops and extensively exploits ideas about fermentation and combustion, and his use of these in his medical and physiological scheme has caused him to be considered a dominant influence in the iatrochemical movement.

Willis' Doctrine of Fermentation

In 1659, when Willis first offered a physiological theory, fermentation furnished the key to his system. Appropriating and altering the ideas of van Helmont and apparently of de la Boë, he used fermentation to explain a wide range of organic phenomena. To grasp Willis' way of looking at living things, we thus need to know what he believed fermentation to be. He put the matter as follows:

Fermentation is an intestine motion of Particles, or the Principles of every Body, either tending to the Perfection of the same Body, or because of its change into another. For[,] the Elementary Particles being stirred up into motion, either of their own accord or Nature, or occasionally, do wonderfully more [move] themselves, and are moved; do lay hold of, and obvolve one another: the subtil and more active, unfold themselves on every side, and endeavour to fly away; which notwithstanding

[6] Ibid., pp. 5–6.

being intangled, by others more thick, are detained in their flying away. Again, the more thick themselves, are very much brought under by the endeavour and Expansion of the more Subtil, and are attenuated, until each of them being brought to their height and exaltations, they either frame the due perfection in the subject, or compleat the Alterations and Mutations designed by Nature.[7]

More simply stated, fermentation is particle interaction in which the volatile tendencies of finer particles are checked by coarser particles which are themselves thereby refined and appropriately rearranged.

Having developed this model of fermentation, Willis made it the basis of practically everything. In its varieties, it seemed to him to be the essence of all that happens in a living organism (or one that has died). In plants, he saw fermentation as central to germination, budding, flowering, ripening, and decay. In animals, it was the essential process in digestion, chylification, sanguification, rarefaction of the blood and consequent diastole of the heart (Descartes' idea), motion of the blood, formation of vital and animal spirits, embryonic differentiation, and development. "The first beginnings of Life proceed from the Spirit Fermenting in the Heart, as it were in a certain little punct" (Harvey's *"punctum saliens"*). Also in the male sex, "from the seminal Ferment, happen abundance of heat, great strength, a sounding Voice, and a manly eruption of beard and hair; by reason of the defect of this, men grow womanish, to wit, a small Voice, weak Heat, and want of beard is caused."[8]

As a physician Willis was naturally interested in the pathological applications of the doctrine of fermentation. "We are not only born and nourished by the means of Ferments; but we also Dye; Every Disease acts its Tragedies by the strength of some Ferment."[9] Again, "The Doctrine of Fermentation being explicated it remains that we handle the chief Instance or Example of it, to wit, *Feavers*" [the subject of the ensuing treatise].[10]

In sum, "Having thus far wandered in the spacious field of Nature, we have beheld all things full of Fermentation. . . ."[11]

[7] Ibid., p. 9.
[8] Ibid., p. 13.
[9] Ibid., p. 14.
[10] See n. 2. This is the other diatribe, viz., *De Febribus,* trans. "Of Feavers" (Treatise II in) *Practice of Physick,* p. 47.
[11] "Of Fermentation," p. 14.

Microstructure and Microfunction

Willis, in his earlier theorizing, saw vital activities as depending on fermentive interactions of two sorts of spirits occurring in the intimate tissue spaces. His model was based partly on materials with which we are already familiar. In line with Galenic pneumatology, for example, the two spirits were "animal" and "vital." A key part of Willis' scheme was his picture of the microstructure of the parts, which he considered to be so arranged as to permit a proper encounter between these two sorts of spirits. Willis noted that the use of lenses revealed the parts to be extensively penetrated by ramifying nerve endings and small blood vessels whose branchings reminded him of "Ivy." At a more minute level, the organs seemed to him to possess a fibrillar organization. The fibrils he viewed not as continuations of nerves exactly but as somehow dependent upon them, their function being to assure the presence of animal spirits in the tissues. Here, he supposed, the animal spirits encounter vital spirits brought in by the arteries. The vehicle of the vital spirit is "nutricious juice . . . separated out of the mass of the blood, for the nourishment of the solid parts, and cleaving to them (whereby it may be better assimilated) like Dew."[12]

We hear that the "heart and Brain, with the Arteries and Nerves hanging to them, are primigenious parts and highly original, but these . . . distribute a twofold humor, viz. one spirituous and endued with very active Particles which perpetually flow, though but in a very small quantity, through the passages of the Nerves from the Brain and Cerebel; and the other slow and softer, which being every where laid aside through the Arteries from the bloody mass, is rendred more plentifully." So much for the source of the twin humors. What of their interaction?

Willis saw the humors or the spirits they contain as interacting in a fermentative way—with animal spirits as the more active component, and vital spirits as the less active. But the animal spirits seemed, in addition, to have a plastic, almost a directive, role to play. Willis said that the "latter [arterial juice], being of it self dull and thicker by much, is actuated by the former [nerve juice], and being imbued by it, as by a certain Ferment, acquires strength and power of growth or vegetation. But indeed the Nervous Juyce, forasmuch

12 "Of Feavers," p. 52.

as it diffuses with it self the animal Spirits, imparts to every part, besides the faculties of Motion and Sense, the determinations also of form and figure." In conclusion, "these twofold or twin humours, coupling together in every sensitive part, constitute a liquor truly nutritious, to wit, which is both spirituous and nourishable." The encounter of the two juices accounts not only for vital motions (about which we shall hear more later) but also, interestingly, for assimilation. As to the nature of the plastic or form-building activity of the humors, we receive only the rather unsatisfactory information that "both these Juyces, viz. the nervous and arterious, being married together, are as it were the male and female seed, which being mingled in a fruitful womb, produces the plastick Humour, by whose virtue the living creature is formed and increases."[13]

Repeatedly in these studies we have discovered physiologists to be analytically occupied with certain stubborn problems, classic questions, which elicit new answers as new information becomes available or new paradigms invade and organize the biological endeavor as a whole. Willis' answers to these questions—questions of sensation, of nutritive assimilation, of self-instigative motion may be thought of as reductive or pseudo-reductive in the sense that life activities in general are brought down to fermentative interactions of spirits regarded as corpuscular entities operating in a suitably structured arena.

Life and Fire

In the treatise *Of Fermentation* (1659), Willis made fire a volatilization of sulphur (a substance "exceeding fierce and untamed"). Niter did not seem to him, at this time, to be indispensable to burning, but, being aware of its role in explosions, he developed a theory about it. He saw niter as a compound in which sulphur is imprisoned. If such a nitrosulphureous compound is mixed with other sulphureous bodies, the combined sulphur, according to Willis, violently erupts.[14] This concept of sulphur *in* niter appeared also in Willis' simultaneously published treatise *Of Feavers* in which he enumerated various possible sources of the normal heat of the blood. Does this heat result from the blood's own composition which makes the blood

---

[13] Thomas Willis, *Cerebri anatome, cui accessit nervorum descriptio et usus* (London, 1664), trans. S. Pordage, "The Anatomy of the Brain" (Treatise VI in) *Practice of Physick*, p. 105.

[14] "Of Fermentation," pp. 34–35.

heat like wine in the vat? Or from a ferment in the heart ("for this is the chief fire-place")? Or, from a flame or "nitrosulphureous ferment"? Willis thought there was no major difference among these alternatives, all being varieties of fermentation.[15]

The idea of a single substance containing both sulphur and niter recurred five years later in Willis' celebrated work *On the Anatomy of the Brain* (1664). Here he suggested that nitrosulphureous particles from blood, when fired by contact with animals spirits stored in the fiber of muscle, explode like gunpowder. The result is muscular motion.[16]

In the same year that saw Willis' *Anatomy* (1664), Hooke offered a theory of combustion which, we remember, depicted the sulphur in burning bodies as being dissolved by a quite separate nitrous menstruum "inherent in and mixed with air." There was, for Hooke, no sulphur *in* niter. Hooke presently extended this theory, as we noted earlier, to respiration and suggested a "nitrous quality" in the air as necessary to life, "which [quality] being . . . spent, the air becomes unfit (1667)."

A similar view was promptly adopted by Willis, and we shall now see that his early preoccupation with fermentation is succeeded by a preoccupation with combustion. The latter replaces the former as the basic process of life. Indeed, Willis goes further and actually equates life with a certain sort of flame. The resulting idea may be taken as his fully matured theory of life-matter relations. In a new treatise of 1670 *Of the Accension* ["infiring," "inkindling," ignition] *of the Blood,* he says that "three things are chiefly and principally Essentials, requisite for the perpetrating [of] Flame. First, that there be granted to it, as soon as it is inkindled, a free and continuous accession of the Air. Secondly, that it may enjoy a constant sulphureous food. Thirdly, that its recrements both sooty, as also the more thick, be always sent away. So then if I shall show these things to agree after the same manner with life, as flame, and to those only, without doubt, I think that life itself may be esteemed a certain kind of Flame."[17]

[15] "Of Feavers," p. 52.

[16] "Anatomy of the Brain," pp. 105–6, and esp. 110–11.

[17] Thomas Willis, *Affectionum quae dicituntur hystericae & hypocondriacae . . . cui accesserunt exercitationes medico-physicae duae: I De sanguinis accensione. II De motu musculorum* (London, 1670), trans. S. Pordage, "Two Physical and Medical Exercitations, viz. I Of the Accension of the Blood. II Of Muscular Motion" (Treatise IV in) *Practice of Physick,* p. 27.

As to the difference between the flame of a lamp and the flame of life, ". . . what should hinder, but that the act of life or of that corporeal Soul (consisting in the motion and agglomeration or heaping together of most subtil and agil Particles) may be called a certain Burning or perpetual Fire of the blood Mass? Wherein although the accidents and chief qualities of common fire are implanted, yet the form of the fire is obscured, as being subjugated to a more noble form viz. of the corporeal Soul. . . ." Life, then, is a sort of subjugated combustion.[18]

Willis clings, henceforth, to the concept of two substances— "within Sulphureous, and without Nitrous"—and he considers their interaction peculiar to ordinary fire and the fire of life that burns in the blood. If we ask why the flame of life is subjugated and does not flare out and consume the body, Willis gives a religious answer. Divine Providence has "destinated to everything its form by which the whole is able to imprint its type on the parts." His argument is, perhaps, that the *form of the animal* subjugates the flame in the blood and thus suits it to its needs.

Willis claimed no originality in his equation of life with flame, citing as other proponents of the same view Democritus, Epicurus, Laertius, Lucretius, Hippocrates, Plato, Pythagoras, Aristotle, Galen, Fernel, Heurnis, Cartesius, Hogelandes, and Honoratus Faber.[19] This list embraces a wide range of ideas about heat and life, in most instances markedly different from Willis' idea.

Flame Replaces Fermentation

In the work published in 1670, Willis turned explicitly against fermentation as source of the life-giving heat. The heat's source is neither friction nor fermentation, since he is now convinced that these processes produce heat only in solids.[20] To be sure, the blood engages in various fermentative acts but—in contradiction to his earlier doctrine—the blood's fermentation, since it occurs in a liquid, is heatless and must be eliminated as the *causa vitae*. Having thus ruled out friction and fermentation, Willis decides that the life-soul is produced by "inkindling." Borrowing primarily from Hooke, but

18 Ibid., p. 33.
19 Thomas Willis, *De anima brutorum . . . exercitationes duae* (London, 1672), trans. S. Pordage, "Two Discourses Concerning the Soul of Brutes" (Treatise XI in) *Practice of Physick*, pp. 6–7.
20 "Accension of the Blood," p. 22.

altering Hooke's theory of niter as a solvent of sulphur, he makes the blood's sulphur particles erupt and unite with nitrous particles of the air. They do so not violently, however, but in that subjugated way that permits life to be equated with flame, that is, with Descartes' *feu sans lumière*. Thus, where fermentation was earlier made the source of all body functions, flame rather than fermentation turns out in the end to be the essence of life.

Soul

It was also at about this time that Willis matured his theory of soul, becoming especially explicit about it in the essay *On the Soul of Brutes* (1672). He found the knowledge of soul to be "hard and abstruse," overshadowed with "dark Blackness, not less than the shades of Hell it self."[21] Nevertheless he had much to venture on the subject. Historically, *On the Soul of Brutes* is interesting as a pioneering study of physiological psychology, following (though radically modifying) the effort of Descartes in the same area. Willis' system is a loose synthesis of Greek thought, Cartesian theory, and contemporary chemistry with ideas derived from his own anatomical observations of the brain and nerves.

Background. Biologists of every period have noticed that the higher animals live simultaneously on several levels, ranging from mere physiological subsistence to, in man's case, abstract (and, some believe, divine and even immortal) reason. The hierarchy of vital functions expressed itself in Greece in the concept of several souls or parts or, in Aristotle's case, faculties of the single soul. Descartes' theory, if accepted, abolishes all souls in animals, and in man all but one (the divine and immortal rational soul). The title of Willis' work points to a difference between his view and Descartes' on this subject. For Willis, animals have souls.

In the case of man, the soul is given the following organization. First, man has an immaterial, immortal, rational soul that is exclusive with him (here, Willis follows Descartes). Second, man shares with animals a mortal and material soul divided into three further parts. Of these, one, originating in the heart, is carried around by the (therefore living) blood. The other, distilled from the blood into the cortex of the brain, resides in (and in fact consists of) animal spirits; these are conveyed to the body by way of a juice in the nerves.

---

[21] "Soul of Brutes," p. 1.

The third soul is an abstract of the other two. It passes from parent to offspring and forms the new individual.[22]

Willis emphasizes that the exclusively "humane soul" has a "Dignity, Order, and Immortality" beyond the lesser power of the soul we share with beasts and that ". . . if the whole [soul of beasts] be compared with the faculties of the humane Intellect, . . . it will hardly seem greater than the drop of the Bucket, to the Sea."[23] The mortality of the bestial soul is implied by the fact that there would simply not be room to hold the souls of everything that ever lived. Where, for example, are the souls of the "Fleas, Flyes, and other Various Kinds of innumerable Insects," that plagued Egypt in the past?[24]

The picture, so far, seems a backward step from Descartes in the direction of Galen. But it is saved from being merely that by a further elaboration that takes account of current chemical ideas as well as Willis' own anatomical studies. The mortal parts of the soul—both vital and animal—are made of subtle and agile particles; in other words the soul is explicitly material. Willis thinks that as the vital part is dissipated through its flamy activity, it must be reconstituted by an access of food, sulphureous from within, nitrous from without.[25] The animal part consists in the animal spirits distilled from the blood into the cortex (or "barky part") of the brain.

It is difficult to be sure how Willis viewed the precise composition of the two corporeal parts of the mortal part, since his numerous allusions to the subject do not come to rest on a single clear idea. We hear that the corporeal soul keeps "falling off" and having to be renewed; that it differs in different people according to the constitution of the blood "as it is more or less sulphureous, spirituous, saltish, or watery," and according to the nourishment supplied; that the soul as a whole "depends upon the temperament of the bloody mass, and the degree or manner of its accension or kindling," and the soul is the specter or "shadowy hag" of the visible body.[26]

Willis probably wishes us to consider the soul to be "many active, chiefly spirituous and sulphureous Particles, with some other saline being predisposed to Animality or Life. . . ." He says that when these come together "in a fit fire-place or furnace" they come alive either spontaneously or when inkindled by another soul.

[22] Ibid., pp. 6–7.
[23] Ibid., p. 39.
[24] Ibid., p. 4.
[25] Ibid., pp. 6–7.
[26] "Accension of the Blood," p. 26; "Soul of Brutes," pp. 6–7.

The animal part of the soul is, like the vital part, material. Its "constitutive parts" *are* animal spirits. "What they (animal spirits) are is hard to be unfolded; because we can hardly meet with any thing in Nature, to which they may be compared in all things." Rejecting substances such as vinous spirits, turpentine, and hartshorn (because they do not form the "Images of their Object"), Willis thinks that animal spirits may be like light rays (which *do* form images of objects). Animal spirits are "lucid and aerial." We hear more about them in connection with Willis' theory of sense perception to which we shall return after hearing from Willis on the third (genital) member of the corporeal team. We may let him speak for himself.[27]

But besides these two (flamy and lucid) members of the Soul, fitted to the individual Body, a Certain other portion of it taken from both, and as it were, the Epitomy of the whole Soul, is placed apart, for the Conservation of the Species: This is as it were an Appendix of the vital flame, growing up in the Blood, is for the most part Lucid or Light, and consists of Animal Spirits: to wit, which being Collected into a certain band, and having got an appropriate humour, viz. the genital, are hidden in the spermatick Bodies, to the end indeed, that when opportunity shall serve, that Band of spirits, as it were a little Brand not yet inkindled, may be able from thence to be drawn into fit fire, and to be inkindled into another Vital Flame, the formatrix of a new animated Body.[28]

In the history of biology, few subjects so long resisted physical analysis or were so evocative of vague and vitalistic interpretations as the continuity of life as expressed in reproduction. Willis' statement on this subject is not as vacuous and verbose as the translator's quaint English makes it appear. If one adopts, as Willis does, a pneumatically (hence for him materially) constituted soul analogous in two of its parts to fire and flame, it is consistent to invent a further, genital part which is (*a*) a genetic abstract of the others and (*b*) a smouldering brand that waits to be inkindled.

Willis' theory of sensation. Our studies of earlier authors point to the existence of a stable strategy of biological interpretation, a strategy used with modifications from Greek days forward. The procedure, already frequently noted, has been: to build conceptual models—usually micromodels—of living systems and to use them

27 "Soul of Brutes," pp. 23, 24.
28 Ibid., p. 22.

for the explication of such manifest activities as nutrition, generation, perception, and motion. In his highly eclectic and not very well-organized essay *On the Soul of Brutes*, Willis applies this strategy to the interpretation of sensation. As to sense in general, he says that although the blood is alive it is too much employed with the sustention of life to have leisure for "smaller Matters, or outward Accidents." But the animal or sensitive soul resident in brain and nerves is, through them, coextensive with the body and "perceives all Impressions either outwardly objected or raised up within."

Activity of the nervous system is a "certain Fluctuation or waving . . . in the Hypostasis [constitutive matter] of the whole Soul, or of the struck member; by which some Animal Spirits or subtil Particles . . . as a blast of Wind in a Machine, being struck run hither and thither, and so produce the Exercises of Soul and Motion. . . ." Willis' formulation does not convey a precise idea other than to state that the nerve has an active state represented by an agitated condition of its soul particles (animal spirits).[29] He does insist, however, that both motor and sensory transmission involve not a flow of particles along the nerve but a transmission of the activated condition along the line, since the soul is stretched through the whole and its particles perform their office like soldiers in array each acting in succession "without leaving his station."

Sensory discrimination is given a corpuscular interpretation not totally unlike that given by the Greek atomists. The pores of any sense organ and the disposition of animal spirits in it determine its sensitivity to different sorts of incoming particles. Light particles, being sulphureous, convey their image into the eye; air particles, being saline, to the ear and so on. There, they encounter different sorts of spirit particles, "naked" (for touch), pure "and as it were Chrystalline" (for sight), highly moveable (for hearing), "smered with Humor" (for taste and smell).

When the object impresses its character on the sense organs, "in the same instant, by a continued Series of the Animal Spirits, as it were an Irradiation, the Type of its Impression doth pass to the Head; and whils't the Spirits actuating the streaked Bodies [*corpora striata*] are in like manner affected by it, a perception of Sense, begun from the Organ, is formed." The *corpora striata* are, then, the common sensorium and seat of conscious awareness. However, the image is not painted on the *corpora striata* as on a table; there is rather a

[29] Ibid., p. 56.

differential wave motion of the animal spirits there which communicates itself to the spiritual particles of the hypostasis of the soul.[30]

SUMMARY

Fire, flame, and heat have been variously associated with life at different stages in the development of physiological thought. This is apparent even from our own very partial and selective study of the history of the life-matter problem. We saw that to Heraclitus and Democritus fire seemed, in different ways, immanently alive. And that, in later Greek biology, the "innate heat" functioned usually as an instrument of the life-soul. We heard both Lucretius and Paracelsus compare the living animal body to a burning log, and Francis Bacon speak of the flamelike ability of the body to repair itself while it is undergoing disrepair. Descartes told us that the life of animals is only a nonluminous fire in their hearts and fire in general is an agitation of particles of "subtle matter." Finally, the work of Willis' own contemporaries is marked by numerous suggestions, some merely speculative and others experimentally supported, of similarities between various vital activities on the one hand and combustion on the other (Willis is not the first of the Oxford group to speak of the *flamma vitalis*).

It cannot be our purpose even to sketch these theories at this point. We have mentioned them merely to remind ourselves of the complex tradition which Willis acknowledged and within which he introduced his identification of life with a "subjugated" flame. Willis sometimes gave resounding—but, in the cold light of reason, unconvincing— expositions of subtle and difficult matters such as the nature of life and of its physical basis. Yet the essence of his thought was not unimportant or ill-conceived. Even nineteenth-century biochemistry would carry forward the time-honored idea that life has its degradative as well as its creative aspect, and thinkers as diverse as Liebig, Bernard, Pflüger, and Verworn were to regard the degradative phase, or parts of it, as a "subjugated" analogue of combustion. These thinkers were also to acknowledge that degradation is in some sense a precondition of creation. Willis did not anticipate these later theories in a sophisticated sense, but his own theory was an early metaphor for the same intuited fact of life that certain nineteenth-century biochemists would express in a more rigorous but still metaphorical fashion.

[30] Ibid., pp. 57–58.

23

Life as a Combative Interaction
in the Tissues

JOHN MAYOW [1640 · 1679]

Many readers of Mayow's work believe that he came closer than any of his contemporaries to a Lavoisierian (or even post-Lavoisierian) model of combustion and respiration. Opinions about him vary from the approbation of Robert Gunther (1925) who makes Mayow "the discoverer of oxygen"[1] to the sharply negative assessment of T. S. Patterson (1931) who deplores the "extravagant praise" often accorded to Mayow's "useless speculations."[2] A moderate appraisal is that of Partington (1961)[3] who manages to restore to some extent the image of Mayow which Patterson's study tended to shatter.

Our present purposes may be served if we suggest (*a*) what Mayow's idea of combusion (and respiration) was, (*b*) how it compared with Lavoisier's (and later) theories, and (*c*) where it stood in the history of ideas about life and matter.

Mayow's Life

Mayow was a physiologist and physician who practised at Oxford, at Bath (in the summer), and eventually in London. Anthony Wood says:

> John Mayow, descended from a genteel family of his name
> living at Bree [Brae] in Cornwall, was born in the parish of
> St. Dunstan's in the West in Fleetstreet London, admitted scholar
> of Wadham coll. the 27th of September 1661, aged 16 years,
> chose probationer-fellow of All-s. Coll. soon after, upon the
> recommendations of Hen. Coventry esq; one of the secretaries

---

[1] R. T. Gunther, *Early Science in Oxford* (Oxford, 1925), 3:137.

[2] T. S. Patterson, "John Mayow in Contemporary Setting," *Isis,* 15 (1931):461–96, 504–543.

[3] J. R. Partington, *A History of Chemistry* (London, 1961), 2:576–636.

of state; where, tho' he had a legist's place and took the degrees in the civil law, yet he studied physic, and became noted for his practice therein, especially in the summer-time, in the city of Bath, but better known by these books, which show the pregnancy of his parts [Wood here lists Mayow's primary publications].

In 1669–70 Mayow moved to London where, after practising for some time, "He paid his last debt to nature in an apothecary's house, bearing the sign of the Ancher in Yorkstreet near Covent Garden, within the liberty of Westminster (having been married a little before, not altogether to his content) in the month of Sept. in sixteen hundred and seventy-nine, and was buried in the church of St. Paul in Covent Garden."[4]

Mayow's Chemistry

Mayow developed a five-element theory, the antecedents of which were alchemical and not dissimilar to those put forward by Willis. We noted earlier that the five-element theory was Stephano Basso's development of the Paracelsian trio. Mayow postulated two *active* elements (to Willis' three) namely sulphur and mercury to the latter of which he gave a new name (see below). Mayow's three *passive* elements were: salt (fixed or volatile), water, and inert *terra damnata*. Water he saw as a "suitable vehicle" which "with *terra damnata* contributes to building-up of the frame of things in due strength and consistency."[5] He thought that the active elements could combine with fixed or volatile salts to form more or less stable unions. Mayow considered the elements to be non-intertransmutable and hence, in disagreement with Boyle, not made of one universal substance.

Fermentation and Combustion

The key to all physiological, indeed all chemical, action was for Mayow, as for Willis, fermentation. But Mayow drew his own picture of this process. He postulated a subtle, agile, ethereal "nitro-aërial spirit" which he thought of as the counterpart of traditional

[4] Anthony à Wood, *Athenae Oxonienses* (first publ. London, 1721) (London, 1813–1820), 3:1199.
[5] John Mayow, *Tractatus quinque medico-physici* (Oxford, 1674), trans. A. Crum Brown and L. Dobbin, *Medico-Physical Works: A Translation of the Tractatus Quinque* (Alembic Club Reprint No. 7) (Edinburgh, 1907), pp. 34–36.

mercury. Fermentation involves, according to Mayow, a reciprocal agitation of the particles of the two active elements, the "nitro-aërial" and the "saline-sulphureous."

Fire, he said, is a sort of fermentation. Indeed, it "is nothing else than an exceedingly impetuous fermentation of nitro-aërial and sulphureous particles in mutual agitation. . . ." The nitroaerial particles may either be fixed with salt (as in niter—and this explains the explosion of gun powder in a vacuum) or they may come from the air (as in ordinary burning and breathing).

We hear in general that "nitro-aërial spirit and sulphur are engaged in perpetual hostilities with each other, and indeed from their mutual struggle, when they meet and from their diverse states when they sucumb by turns all changes of things seem to arise."[6] This was of course a very general formulation, and Mayow used it as an explicative scheme covering a wide array of phenomena including— in addition to fire—vinous fermentation, putrefaction, and calcination.[7] Mayow added that for fermentation to occur one of the two active elements must be united with salt. In violent effervescences (and gunpowder explosions) the nitroaerial particles are thus united, or fixed, and the sulphur is volatile, or free. In subdued fermentations, the reverse is true.

Combustion and Respiration

Mayow's most impressive scientific studies—impressive in that they were partly experimental rather than merely rational and conjectural—had to do with the supposed entrance of nitroaerial spirit into the blood during the latter's passage through the lungs. He said that "the lungs are placed in a recess so sacred and hidden that nature would seem to have specially withdrawn this part both from the eye and from the intellect. . . . Hence such an ignorance of Respiration, and a sort of holy wonder."[8]

Mayow thought that it might help to dispel this mystery to assume that in respiration the blood takes nitroaerial spirit out of the air. He supported this idea experimentally, using living animals in some experiments and burning objects in others. He placed the animal or burning object under an inverted cupping glass with its rim immersed in water. As the animal breathed (or the object burned),

6 Ibid., p. 35.
7 Ibid., pp. 20, 42–44; see also Partington, *History of Chemistry,* 2:590.
8 *Medico-Physical Works,* p. 183.

the water rose in the glass. Partington (1961) says that similar experiments on burning objects were described by Philo of Byzantium (ca. 125 B.C.) and Hero of Alexandria (ca. 100 B.C.); and in the seventeenth century by van Helmont, Fludd, Francis Bacon, von Guericke, and probably others.[9] Mayow's experiments were better designed than those of earlier workers, however, and were varied in a number of ways in an effort to rule out interpretations other than his own which was as follows:

Water rises in the glass, he believed, because the contained air, after use for combustion or respiration, is no longer able to resist the pressure of the outside air. This drop in resistance occurs not because there are fewer and fewer air particles in the glass as the experiment proceeds, but because the individual particles become progressively less springy, less elastic. Air particles are not simple, as often supposed, but branchy and, as it were, hooked together. Each is bent, to some extent, by the weight of the atmosphere pressing down on it, and it is the tendency of the bent particles to straighen out that gives the air its elasticity, or spring. This springy tendency may vary, and is less after the air has supported respiration or combustion. The central problem, then, is to discover how the air particles lose their spring in combustion and respiration.

The notion that the manifest expansive tendency of air is due to the forced bending or coiling of its particles derives from Boyle. Mayow gives it a partly chemical twist. His idea is that highly elastic air particles owe their springiness to the fact that they are tightly combined with particles of nitroaerial spirit. The nimble and solid nitroaerial particles, wedged or actually imbedded in the air particles, tend to make the air particles resistant to bending and incline them to straighten out and occupy more space.

This conception seemed to Mayow to permit the following understanding of fire. When sulphur particles are heated, they volatilize and, striking the air particles, break the bonds between the air particles and the attached particles of nitroaerial spirit. The latter break away in a highly agitated condition as heat or light. In another connection Mayow even suggested that the air particles might be breakable. If an object is actually flaming, the decreasingly springy air particles no longer resist the pressure of the ambient (still springy) air which therefore keeps moving in to feed the flame.[10]

[9] Partington, *History of Chemistry*, 2:595.
[10] Ibid., 67–88, 98.

Mayow supposed that a related but not identical process occurs in respiration. Elastic air with its associated nitroaerial spirit enters the blood as the blood courses through the lungs. Mayow at this point acknowledged experiments of Lower (see next chapter) showing that air actually enters the blood. Blood particles rub against these air particles, he thought, and cause them to give up their nitroaerial particles which then "mix most intimately" with the blood. In the blood, the nitroaerial particles interact with saline-sulphureous particles and "produce a very marked fermentation such as is requisite for animal life."[11]

Mayow, like Willis, thus follows Boyle's stipulation that chemistry and corpuscularity be brought together. He clings to the iatrochemical view of fermentation as fundamental to life. Fermentation—hence, life—is permitted, in turn, by respiration. Mayow's solution of the life-matter problem, like the solution of the other Oxford physiologists, falls within a frame of ideas that might be termed corpuscular chemiatrics. Mayow added further details to the picture, but the foregoing brief outline permits us now to compare his theory of combustion with that developed more than a century later by Lavoisier.

Mayow and Lavoisier

Mayow's model. In combustion, according to Mayow, volatilized sulphur particles of the burning matter cause a violent release of nitroaerial particles from the union of the latter with particles of air. The air particles thus lose their spring, and the agitated escaping nitroaerial particles constitute heat or light (they are like sparks struck from tempered iron which thus loses its temper).

Lavoisier's model. We shall see later that when Lavoisier learned from Priestley how to make a certain "eminently respirable" air, he considered this air to be a combination of (*a*) a certain weightless "matter of fire" with (*b*) an acid-producing, or "*oxygine*," base. Lavoisier supposed that in combustion, calcination, and respiration these two components of eminently respirable air become separated from each other. The base combines with and increases the weight of the burning or calcining object; the matter of fire (later, "caloric") diffuses away.

The two models are superficially similar in their affirmation that

11 Ibid., p. 102.

combustion and respiration set free something formerly united with air (nitroaerial particles in Mayow's model, and *matière du feu* in Lavoisier's). The two models differ in the crucial point that whereas Lavoisier stipulated a union of the residue (*oxygine* base) with the burning system, Mayow envisioned no such stable union.

The idea that air—or something in it—actually enters the body was not original with Mayow, having been suggested earlier by Boyle and Paracelsus. In antiquity, the entrance of air into the body was accepted by Galen, Erasistratus, the Stoics, Epicurus, Plato, and some of the pre-Socratics though they made no point of the body's election of a single component of atmospheric air.

Mayow's admirers point to his correct surmise that, in breathing and burning, it is not the air as a whole but rather a single constitutent among several that is removed and that Mayow thus correctly saw air as a mixture. But this was not quite new either, since Boyle and Hooke saw the used-up aerial nitrous substance in somewhat the same light, that is, as only a part of atmospheric air. Much is made of Mayow's experimental verification of the idea that breathing removes only a part of the air; and his experiments did importantly anticipate those of Priestley and Lavoisier in this respect. But Lavoisier's experiments were in a class apart from Mayow's. Lavoisier used the air's active component (oxygen) in a purified condition, combined it with mercury and then recovered it therefrom, and showed that it united with carbon to produce the substance ("fixed air") that is given off by animals in breathing. Lavoisier carried out all these and other relevant experiments on a careful, quantitative basis, and in a manner directly calculated to validate his hypotheses.

Perhaps the most conspicuous difference between Mayow and Lavoisier lay in the interpretations they placed on their respective experimental findings. In Mayow's time, there were several active traditions to which he endeavored to adapt his results. One of these was the alchemical—and iatrochemical—acceptance of sulphur and mercury as active elements. Another was the whole antecedent history of ideas about the interaction of sulphur and niter. And still another was the recently revived endeavor to explain physical transformations through corpuscular or microchemical models. Lavoisier did not accept but, rather, actively rejected existing ideas about elements, and was content to theorize without evoking a detailed picture of the particles involved. The result in Lavoisier's case was a certain parsimony of interpretation, and, at the same time, a certain freedom from complicating assumptions. Lavoisier's achievement, indeed, was the establishment of a new, rather than an adaptation to an existing, chemical tradition.

Fermentation and Life-as-action

Mayow used his model of mutual agitation of the particles of mercury and sulphur to explain, among other things, those phenomena whose ensemble is life-as-action.

Plant nutrition. He thought, for example, that nitroaerial particles, impelled by the sun's rays, enter the earth and by their vibrations lash their "most bitter enemy, terrestrial sulphur" which exists there "firmly united with fixed salt and nearly hidden and buried in its embrace; . . ." Next, "a rather notable effervescence is excited in the bosom of the earth . . ." and sulphur is separated from its "consort" (salt). Nitroaerial spirit now "succumbs . . . almost buried in her [the consort salt's] embrace."[12] That is to say:

nitroaerial spirit + terrestrial sulphureous salt → volatilized sulphur + nitrous salt (niter).

Plants need these two products (sulphur and niter), according to Mayow. If sulphur seems a surprising requirement, we need only remember the background of beliefs about sulphur and niter out of which Mayow crystallized his own conception. Mayow's theory owed something too to the sixteenth-century idea of sulphur as the oily, inflammable member of the elementary duo or trio. We can thus appreciate Mayow's attraction to the notion that "When, in this way, nitro-aërial spirit, effervescing obscurely with terrestrial matter, raises its sulphureous part to the requisite volatility and coalesces also with its saline part to form nitre, the elements of things are brought into a condition required for the formation of plants. For all plants seem to be composed of terrestrial sulphur in a sufficiently volatile and inflammable condition, and of nitro-aërial spirits held in the embrace of salt and subdued. . . ."[13]

Chemistry of animals. As for animals, "It is our opinion . . . that as in vegetables so also in animals, nitro-aërial particles are the principal instruments of life and motion."[14] The central act in digestion,

[12] Ibid., pp. 36–37.
[13] Ibid., p. 37.
[14] Ibid., p. 101.

in chylification, in various aspects of sanguification, in muscular contraction, in fact in all vital processes is fermentation, that is, a mutual agitation of nitroaerial and saline-sulphureal particles.[15]

When fermenting blood particles detach nitroaerial particles from air, the latter react with the blood particles in an effervescence similar to, but more active than, that which occurs in the earth. "For it is to be noted that blood consists of the same particles as the earth but in a more exalted state." The reaction is responsible for the blood's heat which is greater during exercise, incidentally, because more nitroaerial particles then enter the blood.[16]

Mayow developed a special micromodel for muscular contraction. A muscle comprises fleshy fibers whose axes are oblique to the axis of the muscle. Into these fleshy fibers are inserted smaller fibrils, parallel to the long axis of the muscle. The fleshy fibers are composed largely of capillaries (of which Mayow gives a reasonably accurate verbal description). Saline-sulphureous particles are filtered out of these fleshy fibers and stored in the neighboring fibrils.

The model, as outlined, suggested to Mayow a theory of the cause of contraction. "I think that the nitro-aërial particles springing forth from the brain into the motor parts effervesce there with the saline-sulphureous particles, and that muscular contraction is caused by their mutual agitation. . . ." The more we exercise the more of the two sorts of particles we use; hence, the harder we must breathe and the more we must eat to make up for the deficiency.[17] In muscle, saline-sulphureous particles proceed from the blood vessels, nitroaerial particles, "on the determination of the mind," from the nerves.

The actual contraction of the fibrils, "as far as I can make out from anatomical observation and from mental conjecture," is a matter of contortion or coiling. "And to these things we further add, that the motion of the nitro-aërial particles . . . is of a sort fitted for twisting the fibrils. . . ."[18] Evidence: a music string heated over a lamp tends to contract; when released, to coil. It does this because the nitroaerial particles have a "circumgyratory motion" (compare Borelli, chapter 24).

[15] Ibid., pp. 46, 246–278, esp. p. 253; see also p. 101.

[16] Ibid., pp. 102–111.

[17] Ibid., p. 235, see esp. p. 248.

[18] Ibid., p. 282. See, for a further account of Mayow on muscle action, E. B. M. M. Bastholm, "The History of Muscle Physiology," *Acta Historica Scientiarum Naturalium et Medicinalium* (Copenhagen), 7 (1950): 209–216, but this work should always be corroborated by comparison with the originals.

Identity of animal and nitroaerial spirits. Mayow would "further add that nitro-aërial particles, no less than the animal spirits themselves, are necessary for the existence of life. In fact it is difficult to conceive why animals should have such a necessity of breathing air, so that not for a moment can they live without it, unless the nitro-aerial spirits had a primary place in animal life and were the animal spirits themselves."

The brain is excellently fitted to collect and store nitroaerial particles because it contains no sulphur to agitate and waste them. When sulphur is introduced into the brain (as in alcohol and "the chemical spirits of vegetables"), the result may be drunkenness, or even madness and fatal convulsions.

Mayow acknowledged the objection that "it does not become the admirable artifice of the animal mechanism that it should be set in motion by an external principle." But to this Mayow replied that the cause of vital motion resides in the body's organization—and not in a unique or even uncommon motive cause. He said that "the artifice ... consists in this, that the parts of the body are formed with such perfect adjustment that quite stupendous effects are produced in it by common causes."[19] This idea, that the body is triggered into major responses by relatively minor stimulating causes, can be followed backward in time to Aristotle.

Life similar to but not identical with fire. Despite the fact that life and fire both seemed to Mayow to entail a separation of nitroaerial spirits from air, he poked fun at Willis' definition of life as a sort of "subjugated" fire. Thus,

> We do not need to have recourse to an imaginary Vital Flame that by its continual burning warms the mass of the blood, much less to affirm a degree of heat in the blood intense enough to produce light, from the rays of which, transmitted to the brain, the Sensitive Soul is supposed to be produced. I know not what the ancients dreamed about certain feral fires hidden in the urns of the dead, but now for the first time the vital flame, if such a thing can be, is kindled in the viscera of animals, so that we all burn like Ucalegon, and there is no reason why we should any longer wonder at a Salamander living in the midst of the flames. But really fire seems to be better adapted for

[19] *Medico-Physical Works*, pp. 250–259.

the dissolution and destruction of things than for the sustaining of animal life.[20]

Soul not the same as animal spirit. Nor did Mayow "wish to be so understood as if I thought nitroaerial spirit to be the sensitive soul itself: for we must suppose that the sensitive soul is something quite different from animal spirits, and that it consists of a special subtil and ethereal matter, but that the nitro-aërial particles, i.e. the animal spirits, are its chief instrument."

Mayow was not ready to go beyond Descartes to a materialistic-mechanistic description of soul. In a somewhat surprising reversion to Greek conceptions, he made the individual soul a portion of the soul of the cosmos. "For, indeed, as to the sensitive soul, I can form no other notion about it than that it is some more divine *aura,* endowed with sense from the first creation and coextensive with the whole world, and that a little portion of it, contained in a properly disposed subject, exercises functions of the kind which we observe and admire in the bodies of animals. . . ."

But only when the soul is incorporated in a body does it function as such. Mayow said on this subject that the "spiritual material, existing out of the bodies of living things, is not to be supposed either to perceive or to do anything but to lie quite dormant and inert, being much as is the case with the sensitive soul when the animal is buried in sleep." Soul is, then, imposed rather than immanent or emergent. Separated from the nitroaerial spirits in the body (where they act as animal spirits), soul is latent and inactive.[21]

SUMMARY

Mayow correctly saw air as supplying a life-sustaining component that enters the blood and is chemically active there and in the tissues. The result was a far cry from the account that Lavoisier was to offer a century later. Mayow's solution of the life-matter problem was a markedly seventeenth-century solution: the body seemed to him to be a physiological arena where nitroaerial and saline-sulphureous particles resolve their enmity. His picture of this activity—as a kind of fermentative combat—while not much like later oxidation theory, anticipated it in postulating a common energizing reaction, of sorts,

[20] Ibid., pp. 105–8.
[21] Ibid., p. 259.

for vital functions. Mayow supposed that for this reaction to occur effectively, so as to produce the appropriate vital motions, a proper microstructure was required, and he duly provided such a structure (though mostly for muscle). As to the reaction itself, he gave it a rather farfetched interpretation which was partly corpuscular and partly chemiatric.[22]

The conceptual materials used by Mayow were, in sum, borrowed, modified in varying degrees, and arranged in a pattern whose broad outlines were not new though the detailed piecing together was his.

[22] A helpful recent source of information concerning Mayow's chemical physiology is the series of papers by W. Böhm, "John Mayow und Descartes," *Sudhoffs Archiv für Geschichte der Medizin und der Naturwissenschaften,* 46 (1962):45–68; "John Mayow and his Contemporaries," *Ambix,* 11 (1963):105–120; "Die philosophischen Grundlagen der Chemie des 18. Jahrhunderts," *Archives Internationales d'Histoire des Sciences,* 17 (1964):3–32; "John Mayow und die Geschichte des Verbrennungsexperiments," *Centaurus,* 11 (1965):247–58.

· 24 ·

Other Late Seventeenth-Century
Contributions
to the Problem of Life and Air

R I C H A R D   L O W E R   [1631 · 1691]   AND
G I O V A N N I   B O R E L L I   [1608 · 1679]

RICHARD LOWER

Richard Lower's contributions to our story lie, first, in the successful outcome of an ingenious experiment he made and, second, in a separate speculation of his concerning the nature and locus of life. In addition to his successful transfusion of blood from an artery of one dog into a vein of another (1665), Lower is celebrated for anatomical researches on the nervous system published by Willis but by Willis properly credited to Lower. Following Willis to London, Lower outlived his senior associate by sixteen years and inherited his medical practice with the result that, according to biographer Anthony Wood (a personal acquaintance), Lower "was esteemed the most noted physician in Westminster and London, and no man's name was more cried up at court than his." He presently lost favor at court because of an ill-timed liaison with the Whigs.

Air Enters the Blood in the Lung

The experimental contribution that most interests us was Lower's demonstration (1669) that the difference in color between venous and arterial blood is due to a change occurring in the lung, a change resulting, presumably, from the entrance of air into the blood. Lower substantiated these ideas experimentally. For example, he exposed and immobilized a lung, insufflated it after Hooke's method, and collected blood from the pulmonary vein (blood that had not as yet reached the heart). He found that this blood had already acquired its vivid color.[1]

He concluded that the color was due to the uptake of "the nitrous spirit of the air" and not, as earlier supposed, to the heat of the heart or to admixture with vital spirits. Lower discredited the Cartesian—initially Greek—theory of a *feu-sans-lumière* in the heart. The heart did not seem especially warm when he touched it in vivesection. Nor did blood seem the sort of fluid to be volatilized and "go off like gunpowder." Even if blood were such a fluid, Lower doubted whether ebullition was the sort of controlled, orderly force needed to direct the blood to its varied destinations.[2] In any case, the heart beats even when removed from the body, or when blood is largely replaced with other fluids.[3]

Muscle Action

Lower equally doubted the occurrence of fermentative and ebullient activity in the tissues, for instance, muscle. Judging from his own excellent empirical studies of the fibrous constitution of the heart and other muscles, he concluded that muscles "seem to be designed so that they may accomplish their movements by the effort, and with the help, of fibres pulling from opposite ends" (a variant of Galen's formulation).

This observation seemed to Lower to argue against the idea of an ebullition in the minute muscle spaces (see below, Borelli). Such a contrivance seemed too violent and uncontrollable. Lower asks "what command has the [animal] spirit over us, if it is responsible only for the impulse to movement, serves only to fire the tinder, and arouses a tumult that it will be unable to quell at will?" In fact, experience teaches that we "are able to control our movements and keep them within such bounds as we please, . . ."[4] Note that Lower followed here the old tradition ascribing movement to an access of nerve-carried animal spirits, but he questioned the supposition that they operate through any sort of ebullition or explosion.

[1] Richard Lower, *Tractatus de corde item de motu & colore sanguinis et chyli in eum transitu* (London, 1669), pp. 165–71, trans. K. J. Franklin, "A facsimile edition of Tractatus de Corde, etc.," in R. T. Gunther, *Early Science in Oxford*, vol. 9 (London, 1932); quoted passages and references cited from Franklin's translation.

[2] *De corde*, pp. 61–66; Lower here reverses his own earlier belief that a fermentative activity occurs in the heart.

[3] He uses a mixture of beer and wine. Ibid.

[4] *De corde*, p. 78.

When Lower ligated the heart's nerve connections, the heart palpitated, quivered, and after a few days stopped beating.[5] This effect Lower ascribed to a stoppage of the flow of spirits from the "storeroom" (*promptuarium*) of the cerebellum. The intimate cause of muscle movement he considered unknown, perhaps unknowable. "I should here speak of the ultimate way in which the heart's movement is effected, but, as it is the privilege of God, alone, who comprehends the heart's secrets, to understand its movement also, I will not waste effort in examining it further."[6] The history of physiology in the late seventeenth and eighteenth centuries presents many examples of this sort of thinking—in which an investigator analyzes a phenomenon as far as possible and asserts that further analysis requires a transcendental intelligence.

Life and Heat

Although he deprived the heart of its special heat, Lower still thought of heat as essential to life, and in this connection came as close as he ever did to a theory of vital action. "The blood is, therefore, entirely responsible for the heat of the Heart itself and for the activity and life of our bodies, which its heat produces." As to the more precise role of this heat, Lower was at least sure that our body is "warmed by a fire that is more than fictitious or metaphorical, and so it would be worthwhile to explain at somewhat greater length how the blood becomes heated and in turn provides for the warmth of the whole body." The explanation here called for was not forthcoming from Lower, however, for he not only failed to assign air a role in the heating process but, in an access of modesty, explained that as "the learned Dr. Willis is giving the matter some thought in his book *On the Spirit, and also the Heating of the Blood,* I should not like to depart so far from professional courtesy as to forestall him in this matter."[7] Lower worked as an assistant to Willis, and by some historians is credited with doing much of the work on the basis of which Willis in part acquired his reputation.

Nutrition

In a series of ambitious experiments, Lower followed the flow of

[5] Ibid., p. 86.
[6] Ibid., p. 85.
[7] Ibid., pp. 73–74.

digested nutriment from intestine to lacteals to lymphatics to blood and so to the various parts of the body.[8] He occluded the common lymphatic duct in the esophageal region of the dog by exerting pressure on it and observed the backing up of chyle in the lymphatic system and its absence from the blood even of well-fed animals.[9] He also wounded these ducts and saw the chyle escape through the artificial opening. He sustained Pecquet (*contra* Aselli, and most other anatomists) to the effect that all chyle, hence all absorbed food, bypasses the liver to enter the blood entirely by way of the lymphatic system.[10]

In tracing the further fate of nutriment, Lower lapsed into the persistent pneumatology and unverified pseudochemistry of his period. Thus, the

> vital spirit and other active principles in the blood-fluid act
> on the constantly inflowing chyle, and break it up into very fine
> particles. When, for example, the chyle is very full of salt,
> sulphur, and spirit, these active spirits acquire freedom of move-
> ment as soon as its structure is loosened by fermentation,
> and they at once join with the blood constituents which are of
> similar or of related character. In the blood, indeed, (as in
> wine and other such fluids) the spirits, on acquiring control,
> dislodge and free their mass of all thicker and cruder particles
> which come into contact with them, and so render the
> remainder of the fluid clearer and purer.

> After the chyle is so perfected, it is completely fitted both
> for the restoration of the blood-fluid and for the nutrition of
> the body as a whole.[11]

There is a rather surprising contrast between such derivative theorizing and Lower's efforts, in connection with respiration, to proceed cautiously and on the basis of experimental evidence. This contrast

---

[8] Ibid., p. 199.

[9] Ibid., pp. 209–210.

[10] "Jam porro considerandum restat, quomodo necessaria ejus dispendia, atque (ut ita dicam) quotidianae expensae instaurantur; quod aliunde fieri non potest quam ex Chyli in illum influxu" (*De corde*, p. 193); and "Atque haec via sola atque unica est qua chylus e ventriculo & intestinis in ipsum sanguinem & cor infunditur: . . ." (p. 208); see also, p. 211 against blood formation by the liver. Pecquet announced his new ideas on this subject at the beginning of his *Experimenta nova anatomica* (Amsterdam, 1651), trans. *New Anatomical Experiments* (London, 1653).

[11] Ibid., pp. 218–19.

emphasizes the delay that takes place between the introduction of effective experimental physiology (in the seventeenth century) and of effective experimental biochemistry (a century later—and then at first only the chemistry of gaseous intake and output).

Blood as Locus of Life

In John Browne's *Myographia Nova* (1697) there appears "An Appendix of the Heart, with the Circulation of the Blood," offered as having been written by Lower. This posthumous appendix is, by contrast with the *De corde*, conspicuously speculative in tone. Here Lower says that when blood stands it separates into a watery matter which is the *succus nutritius* and a dark reddish matter which is blood itself. The *succus nutritius,* he says, is different from serum or urine. We hear also that "the other part of the *Mass* is the *Blood* its self, the Fountain and Original of Life, the *Primum Vivens and the Ultimum Moriens*" (the phrase is almost a quotation from Harvey). This blood, Lower continues, has "from its beginning Life in its self . . . without which, the *Artificer* can do nothing becoming the *Architect,* of his own House and Frame; every part being fitted for its own Reception and Habitation: . . ." We are told, further, that the blood has both a Local Motion providing for circulation and a "Vital one, by which it preserves its self."

The latter statement arouses our curiosity as to the nature of the vital motion in question, and this time Lower does not disappoint us. He says that "The Vital Motion is a constant Fermentation or Working of the *Blood,* by which motion all the most Minute parts of it are secretly divided, for the Reception of what is proper for it; and for the Expulsion, Amandation, and casting off of whatever is obnoxious or injurious to it; . . ." Further, the blood's "secret Agitation of its self and [its] Anatomical Division of the *Minimae Particulae* preserves it in its usual Vigour, . . ." The result is that "so long as its Fluidity continues, as the proper Effect of this its Vitality, it becometh brisk and lively, it causing the lively part of the blood to nourish and cherish the whole."

Lower is emphatic and explicit in making the blood the locus of life. The tissues die, he says, when they dry up and no longer let the blood send them its "dew of life." Then the blood, too, "gets into its self a kind of drynes, which makes it unfit and uncapable of receiving its own Nourishment; . . ." As a result, "for want of . . . [the] Vital

Fermentation it formerly enjoyed, it grows more dry and firm" and finally stops moving. Lower tacitly takes sides here on the question as to the relative importance of heart vs. blood, an issue raised earlier by William Harvey; Lower says, "The heart is made for the blood (the Seat of Life)."[12] This is not original but it is entirely unequivocal.

Summary

To sum up, Lower made no major new contribution to the life-matter problem acknowledged as such. He saw vitality as a secret agitation, a fermentative motion, secondarily of tissues, primarily of blood, involving a mechanical attrition of food particles entering with the chyle, the elimination of some end products and assimilation of others. It was his view that air enters the blood in the lungs and that the heat of the blood is indispensable to its life but how the heat is generated he left it to others to explain.

GIOVANNI BORELLI

It was because Giovanni Borelli seriously studied both mathematics and physiology, and was inspired to combine these two interests, that he became, after Descartes, the major founder of iatromechanics. Descartes and Borelli differed in their respective ways of thinking mathematically about living things; Descartes "constructed" the organism, after the manner of the geometrician, by proceeding from intuited axioms to rationalized details; Borelli often made use, though not invariably and not in the sector of his thought that concerns us here, of actual measurements. Born in Naples in 1607, Borelli studied at Pisa and later taught at Messina, Pisa, and Florence where, from its founding in 1657, he was especially active in the *Accademia del Cimento*. Returning to Sicily he became politically active against the Spanish occupation, and fled to Rome where, after a brief period under the protection of Queen Christina, he died in a religious house in 1679.

When, three quarters of a century later, animism was briefly revived in France and England, its exponents liked to point to Borelli, the founder of mechanical medicine, as an exponent of the motive power of the soul. This claim has a certain basis but taken out of

---

[12] John Browne, *Myographia Nova; or, a graphical description of all the muscles in the humane body, as they arise in dissection* (London, 1697).

context it could cause us to misinterpret Borelli's physiological out-look. Near the commencement of his great posthumous work *On Animal Movement* (1680) Borelli says: "That the soul is the prin-ciple and the efficient cause of the movements of animals is a fact of which no one, surely, is ignorant, since it is through the anima that the animate live (*animantia per animam vivant*) and through it, while life lasts, that motion perseveres; but once the animal has died, i.e. when its soul is no longer active, the animal machine is left en-tirely inert and immoble." Borelli continued in the general vein that, though appetite and election are concerned in animal movement, these act not directly but through instruments or virtues; it is through a locomotive faculty that soul exerts its motive effects.[13]

Our questions of Borelli will be to what extent, and in what sorts of motions, the motive faculty of soul is active. In the case of volun-tary motions, such psychic faculties are patently involved, he sup-posed, their role being to direct the flow of animal spirits to nerves in a manner appropriate to some desired response. But in the case of an involuntary motion, such as the beating of the heart, he found it difficult to be sure. He noted many examples of automatic rhythmic responses to continuous stimulation (such as the effect of a steady wind on water, on clouds, on fluttering banners, on quivering leaves). He also noted that certain manmade machines were contrived to re-spond in an oscillatory fashion to a nonoscillatory input. He thought the heart might be a natural example of such a machine. Perhaps the channels of the nerves have a spongy structure that causes them to introduce nerve juice into the heart tissue in drops and not in a steady flow. If so, the periodicity of the heartbeat could be con-sidered automatic.

But Borelli acknowledged the alternate possibility of a psychic faculty as the motive cause of the beating of the heart, or at least of its initiation. Perhaps in the preformed, wormlike creature recently found in the egg by Malpighi, the psychic faculty is present but asleep. Perhaps the heat of incubation wakes it. It responds to the in-commodious turgor of blood in its heart by inducing a contraction. What thus begins as a voluntary act presently becomes habitual and even unpreventable much as happens in the case of skeletal muscles whose action is initially conscious and difficultly learned but ulti-

---

[13] Giovanni Alfonso Borelli, *De Motu Animalium* (first published Rome, 1680–81); 2nd ed. (Leyden, 1685), part 1, chap. 1, pp. 1–4.

mately becomes reflexive. Such a psychic alternative to pure automatism in the case of the heart should not be laughed out of court.
Borelli was aware that hearts that have acquired the habit of in-
advertant pulsation can continue beating after removal from the
body. He attributed this to the persistent presence of the requisite
mechanism (described below) and presciently depicts these mechanisms as acting when irritated by a proper stimulus (*a stimulo
irritati*).[14]

The central point is that the immediately effective cause of the
continued beating of the heart is physical (it is, as we shall see, a
kind of explosion). What interests us in Borelli's viewpoint is—in
effect—the compatibility within it of an ultimately psychic with an
immediately and elaborately physical (chemical and especially
micromechanical) account of life-as-action.

It seemed to him "that automata have a certain shadowy sameness (*umbratilem similitudinem*) to animals in that both are organic
self-moving bodies which employ mechanical laws, and both are
moved by natural faculties." He continued by proposing, "Let us,
then, see whether we can trace the properties of *natural* [living]
objects by means of our knowledge of artificial ones."[15] In the
famous first part of the *De Motu Animalium* (1685), Borelli tested
this procedure on the mesoorganizational level; he applied it, with
evocative results, to muscle and bone. He recognized clearly, however, that gross muscular action is ultimately a problem of microorganization and microdynamics. And so, in the second part of his
De Motu (1680), he asks, among many other questions, what actually happens in muscle.[16]

In answering this question, he momentarily closely approached
the opposing iatrochemical view, since he saw the efficient cause of
contraction as a fermentative ebullition of the sort the chemists
envision (*fermentatio & ebullitio oriatur, similis eis, quae passim in
chimicis elaboratoriis observantur. . . .*)[17] This event was viewed by
Borelli as by Willis and Mayow, we noted, as an explosive but regu-
lated reaction between animal spirits and something brought by the
blood.[18]

[14] Ibid., 2.6 (cited by part and chapter), pp. 111–17.
[15] Ibid., 2.8, p. 164.
[16] Ibid., 2.1,3, pp. 1–55.
[17] Ibid., 2.3, pp. 45–55.
[18] Ibid., 2.3, p. 46.

But if the fermentative process seemed iatrochemical to Borelli the conditions of its effective operation appeared to be mechanical. The arrival of both blood and nervous juice is mechanically explained.[19] Nervous juice, the vehicle of animal spirit, does not flow through the nerve, but permeates its spongy structure and is extruded into the muscle as the result of a "convulsive concussion produced throughout the length of the nerve."[20] The response of the muscle is likewise mechanically portrayed. The muscle's texture is spongy, and, as a result of fermentation, in its spongy porosities, the muscle undergoes inflation (*inflatio*), hardening (*durities*), and contraction (*contractio*).[21] Indeed, wherever possible, Borelli formulates the kinetics of vital processes in micromechanical terms, even equating molecular phenomena to the actions of little machines. These corpuscular *"machinulae"* are developed in special detail in his theory of respiration.

Respiration

The functions of the breathing movements, for Borelli, were two: first, the inhalation and absorption of air; second, the comminution of the particles of the blood.[22] He supposed that in the breathing process the particles were rendered as miniscule as possible. It was in explaining the action of these ultimate air corpuscles themselves that Borelli carried his mechanicism farthest, and it was here, incidentally, that he developed something like a theory of life. We may allow Borelli to tell this story for the most part in his own words. He begins with a clock:[23]

"Facilitated by many, toothed, artificially interconnected wheels, driven according to certain laws by the motive force of a suspended weight, a clock is able to point out the courses of the sun and the moon and to effect other motions." The weight itself, however, is not enough. Acting freely, though it would cause the clock to complete its cycle, the clock's movement would "not be uniform nor equal to the movement of the sun and the moon. . . . To obviate which difficulty, it is customary to apply balanced weights on an oscillating

[19] Ibid., 2.4,5, pp. 64–109.
[20] Ibid., 2.3, p. 43.
[21] Ibid., 2.3, pp. 45–48.
[22] Ibid., 2.8, pp. 151–54.
[23] This and the ensuing excerpts in this section are, except where noted, a translation of most of *De motu,* 2.8.

pendulum which, forced by mechanical laws to move to and fro in equal time intervals, directs, regulates, and tempers both the violence of the motive cause and the movements of all the wheels. The result is operation conforming to the course of the sun and the moon."

Borelli next observes "that the life of animals, or their vital operations, consists in perpetual and uninterrupted movement. . . ." Life, for the machine-oriented Borelli, is the movement of the parts in conformance with mechanical law.

> . . . for the limbs are active as are all the solid, as well as fluid and spirituous parts, so long as the body moves and is carried in different directions; so long as it ingests, concocts and chilifies its food and turns it into blood, so long as it nourishes and replaces the parts that are lost; so long as it evinces sensible movements; in sum, while life lasts, nothing in the animal stays static.

It will not do, however, for these movements to be random or headlong; to be called vital, the movements must be of "set velocities, rhythms, and intervals." If "they are precipitately effected, they are no longer vital, and by the same token life cannot be maintained." The animal spirit, moreover, is so highly mobile that by itself "it would drive the organs with a raging and frantic motion (*furibundo & phanatico motu*) and would not accomplish the operations demanded by 'nature's purpose.' Whence as in a watch, so in the animal —the 'natural' automaton—a regulatory device must be added which by mechanical necessity will bridle the motive force lest it transgress the laws instituted by the Divine Architect."

What is needed is something "like the pendulum of a clock, which by its oscillatory force must regulate the motion of the blood and the spirits and not permit them to flow in every direction over a rash and raging course." In an attempt to specify the regulatory mechanism Borelli first considers, in Baconian fashion, all the things that seem indispensable to life: food? drink? sleep? warmth? blood? the heart? the *motion* of the heart, or of the blood, or of the lungs? None of these fills the need, since in the case of frogs and snakes, the body can live awhile without a heart (the frog even jumps), and the heart can live without body or blood.[24]

The sine qua non (as Boyle has shown, indeed, as everyone already

[24] Ibid., 2.8, p. 152.

knows) is: *air*. And air must be in some sense the regulator of every-thing else. "Whence we discern the great mystery of the need of animals for air; it is plain why the air particles must be continually insinuated into the blood while the animal lives: viz. because it is necessary that little aërial machines (*aeris machinulae*), mixed in with the blood, perform an oscillatory movement in the manner of the pendulum as described."[25]

Borelli sees the air corpuscles as tiny "spiral machines which can be compressed by an external force and then spring back sponta-neously, like a bow. . . ." Theirs is "an oscillatory motion of the nature of waves and pendulums."[26] In the biological as in the non-biological world, solubility, diffusion, freezing, capillarity, and other physical phenomena are thus accounted for in terms of corpuscular mechanics.[27]

"Blood particles contiguous to the thus vibrating aërial machines must necessarily be shaken with the same oscillatory motion; and by this primary motion all the parts of the animal are moved with a regular rhythm; they are both impelled and restrained no less than the wheels of the clock by the oscillation of its pendulum."[28] Only thus simultaneously driven and moderated can "the blood be carried in a perpetual flow in the fashion of a river to all parts of the body, there to become the cause and stimulus of movement, and thus the origin of life."

Borelli concludes his account of respiration with a note on the role of aerial machines in lower and higher animals and plants, all of which have at least a "certain semblance of respiration." In all of these there is an admixture of oscillatory air particles whose pen-dular motions are "the principal and most potent cause (*praecipua & potissima causa*) of life and vegetation. . . ."[29]

Summary

For Borelli, living bodies are machines. The life of the machine is the totality of movements exhibited by the moving parts and by

[25] Ibid., 2.8, p. 164.

[26] Ibid., 2.8, pp. 162–63. The same general ideas about air are developed in Borelli's *De motionibus naturalibus a gravitate pendentibus* (first published 1667?; 3d ed. Leyden, 1686), chap. 5, pp. 162–65.

[27] See James R. Partington, *History of Chemistry* (London, 1961), 2:443–44, the passage in which he alludes to Borelli's *De motionibus*.

[28] *De motu*, 2.8, is the reference for this and the following excerpts.

[29] Ibid., and chap. 13, pp. 273–74.

the machine as a whole. The whole machine is an assemblage of smaller component machines, and these of still smaller ones, and so on until we reach the corpuscular level. Even at this level we are likely to discover, as we do in the case of respiration, that the very particles are minute mechanical devices (*machinulae*).

All mechanistic theory becomes involved, sooner or later, with problems of impetus, or force. The parts of the organism, like the parts of a clock, require a cause of motion (*causa motiva*). The driving force of the living machine seems to Borelli to need restraint and regulation. The two major problems he poses are, thus, the problem of the cause of motion (the microcosmic equivalent of cosmic impetus) and the problem of regulation. Future physiologists will have much to say about these problems. Not until the nineteenth century will the *causa motiva*, the force that produces the impetus (*vis impetum faciens*), achieve a practically useful and theoretically fruitful formulation, with the statement of the energy laws. The problem of regulation which concerns Borelli so seriously will, almost as much as impetus, become a problem of the future. Borelli's screw-shaped aerial *machinulae* will, perhaps, not have much staying power; they, along with those of Descartes' subtle matter, are too gratuitously conceived. What matters historically, however, are less Borelli's solutions than the problems he sets out to solve. Like the cause-of-motion problem, the problem of regulation will stand close to the heart of the biology of a much later era. As to his general method, time will largely confirm his belief that, "just as the devine Plato called geometry and Arithmetic the two wings on which we fly to heaven. . . so we may affirm that geometry and mechanics are the steps by which we climb to a knowledge of the wonderful movements of animals."

Part 4

ANIMISM AND MECHANICISM
(1700–1750)

· 25 ·

Life and the Biomedical Soul

GEORG ERNST STAHL [1660 · 1734]

Among the elements of the new biology that took form in the seventeenth century—with its uneven strikings-out toward quantitativism, experimentalism, and corpuscular interpretation—no development was more pregnant than the effective elimination of the Greek physiological soul. The supposed lower-soul faculties governing nutrition and generation, still conspicuous in sixteenth-century physiology, had been dealt a blow (it would prove a death blow in the end) through the mechanical modes of thought heralded most eloquently by Descartes. Yet before their demise, these initially Greek ideas were to find, though in a new formulation, a final vocal defender.

Georg Stahl, German chemist and physician, wrote influentially on life and matter, his major contribution being in each of these subject areas an evocative scientific error (as judged, at least, by the directions ultimately taken by biology and chemistry). Stahl's two famous faux pas were: first, his theory of phlogiston, a substance supposedly set free by burning objects; and second, his notion of a "biomedical soul" which he saw as intervening in and determining every manifestation of life.

Admirers of Stahl praised the phlogiston theory, even after its downfall in 1775. It has often been pointed out that of two then plausible concepts of combustion—(a) that it adds something to, and (b) that it takes something away from the atmosphere—Stahl's choice was for its period as attractive, as explicatively effective and therefore as "right" in a sense, as the view (or various views) he rejected. The theory at least had the merit of provoking further experimentation leading to its own ultimate replacement by modern theories of oxidation.

Stahl's animistic ideas were to have a different history. When they were introduced, preliminarily in 1684 but definitively in 1703, iatrochemical and especially iatromechanical (and iatromathematical) theories very largely held the field. It is true that Descartes had not banished the physiological soul completely from later seventeenth-century science. In the interim, Swammerdam had made the soul a prime mover, and even mechanicists like Borelli and Perrault had given it a role in their otherwise predominantly physical views of organic function. Malebranche, who saw both matter and spirit as passive, would have had God, the only Active Substance, impart motion to matter, and intelligence and will to the spirit. But such ideas were less scientifically evocative than the flood of thought that increasingly interpreted living things along with lifeless ones in chemical and physical terms.

It was precisely this trend away from a presumptive animistic uniqueness in living systems that Stahl felt impelled to oppose. Medicine is the science of life, Stahl argues in essence, and physicians repudiate medicine when they refuse to ask what life is. Note, he admonishes, that many diseases are remedied by the spontaneous "autocracy" of some sort of vital guide or direction. We must study the mode of action of this guide, and base our therapeutic method on seconding its operations.[1] Stahl's interpreter Lemoine (1864) says that "in destroying the erroneous systems of his time [Stahl] brought the attention of physicians back again to life, and to its origins and its manifestations . . . though he erected on the ruins, in his turn, another ruinous system by attributing life to a reasonable Soul."[2]

The fact that Stahl's animism contained the seeds of its own "ruin" could blind us to the interim influence he exerted. Lemoine points to a (rather inconsequential) group of Stahl's students and immediate adherents, Carl, Gohl, Coschwitz, Juncker, Richter; and Gottlieb mentions Kundmann, Goelicke, Alberti, Madai and others whose names nobody now remembers. Many of these were degree candidates over whose examinations Stahl presided. In some of these

---

[1] George E. Stahl, "De scriptis suis ad hunc diem schediatismibus vindiciae quaedam, et indica" (first published Halle, 1707), bound with some printings of Stahl's *Theoria medica vera* (Halle, 1707, 1708), trans. and ed. T. Blondin, "Réclamations, défense, et indications," *Oeuvres médico-philosophiques et pratiques de G. E. Stahl* (Paris, 1859), 2:612, 667–68.

For copious appreciations of Stahl as a medical thinker, see the secondary materials included in the *Oeuvres*.

[2] [Jacques] Albert [Felix] Lemoine, *Le vitalisme et l'animisme de Stahl* (Paris, London, New York, 1864), p. 204.

examinations, the assignment was to develop or defend an idea or ideas that Stahl himself put forth. Lemoine likewise mentions a group of British and Scottish medical thinkers, among them Cheyne (to whom it seemed that man, as a perpetual motion machine, must possess an intelligent self-moving mover), "Bryan" (presumably the Daniel Bryan who wrote on the maladies of the mind), Frank Nicholls (whose *De Anima Medica* [1750] acknowledged Stahl as a source), William Porterfield, and Robert Whytt. Whytt's position (in his *Essay on Motion,* 1751) was to be that Stahl, by the extravagances of his theory, aroused opposition to the kernel of truth that there was in his scheme, namely that a sentient (though not, as Stahl insisted, a rational) principle informs our vital and involuntary motions (see below, chapter 33).

A pregnant aspect of Stahl's effect on later thinkers is suggested by the similarity between his soul-body theory and certain much later psychosomatic conceptions.[3] It is easy to show that Stahl was not only a precursor but a lineal ancestor—though only one of many ancestors—of nineteenth- and twentieth-century psychiatric theory. The most important effect of his thinking, however, was the base it provided indirectly for certain later developments to which the term "vitalism" would be attached (chapters 33–35, 43–45), and we shall see that some of the central events occurred at Montpellier. Stahl focused on the limitations of chemistry and physics and on the need for extraphysiochemical causal interpretations. His own alternative, the reinvocation of soul as imposed cause of life-as-action, was to have but little success. But many other interpretations—extraphysicochemical (though nonpsychic)—were forthcoming and were to be the subject, as we shall see, of nearly two centuries of debates. Of these debates Lemoine asserts that "it will always be Stahl's glory to have preceded and provoked them." Gottlieb (1943) says much the same thing of Stahl, emphasizing especially his evocative influence over the very different views of Haller.[4]

---

[3] For Stahl as a precursor of modern psychiatry, see G. Zilboorg and G. W. Henry, *A History of Medical Psychology* (New York, 1941), pp. 277–80. There are also notes on Stahl ("one of the four great fathers of psychiatry") in E. Harms, *Origins of Modern Psychiatry* (Springfield, 1967), pp. 41–43.

[4] For the influence of Stahl on eighteenth-century physiology, see B. J. Gottlieb, "Bedeutung und Auswirkungen des hallischen und kgl. preuss. Leibarztes G. E. Stahl auf den Vitalismus des XVIII Jahrhunderts, insbesondere auf die Schule von Montpellier," *Nova Acta Leopoldina,* n. s., (1943):425–502.

Matter

Matter was seen by Stahl in much the same way as by his predecessor and mentor, J. J. Becher, a man who seemed to Stahl "to have been designed for the real improvement of *Natural Philosophy*." Stahl agreed with Becher that the physical principles were four: water, and three sorts of earth. The three sorts of earth were thought to correspond in certain properties to the Paracelsian *tria prima:* salt (fusible earth), sulphur (flammable earth), and mercury (liquefiable earth). Though things were composed of these physical principles, they were not readily resoluble into them, Stahl said, at least "by the chemistry of our day."

Stahl conceived of each principle as being represented by a distinctive corpuscle, and of the corpuscles as arranged hierarchically in clusters, and clusters of clusters. "All *natural Bodies,*" he said, "are either *simple* or compounded" and he elaborated these distinctions as follows:

(1) "The *simple* are *Principles,* or the material causes of Mixts"

(2) The compound belong to three categories:

"[a] mix'd (if composed merely of Principles)

"[b] compound (if form'd of Mixts into any determinable thing); and

"[c] *aggregate* (when several such things form any other entire parcel of matter, whatsoever it be)."[5]

Stahl thus definitely acknowledged corpuscularity, but he did not regularly use it as a way of explaining chemical change. We shall see that it plays a certain role, however, in his solution of the life-matter problem. He spoke of corpuscularity as an "occult quality"[6] in a deviation from conventional terminology that is occult enough in itself.

Phlogiston. Stahl's phlogiston had an antecedent in the flammable earth or *terra pinguis* (*pinguis,* meaning fatty, oily, unctuous) of Becher.[7] With Stahl, phlogiston was important in combustion, but

---

[5] "The Structure of Matter," in Stahl's *Philosophical Principles of Universal Chemistry,* trans. Peter Shaw (London, 1730) ["principally taken" from Georg. Ernest. Stahlÿ, Consiliar. Aulic & Archiatri Regii, *Fundamenta chymiae, dogmaticae & experimentalis* (Nüremberg, 1723)], chap. 1, sec. 1, p. 3.

[6] G. E. Stahl, *Specimen beccherianum sistens fundamenta documenta, experimenta,* appended to Stahl's edition of Becher's *Physica subterranea,* edition of 1738 (first published 1703), p. 18.

[7] J. R. Partington, *A History of Chemistry* (London, 1961), 2:666.

also in other ways. First, he distinguished, though not as clearly as we might wish him to have done, between fire and the principle of fire. He said that *"fire—flamy, fervid, hot, and most potent—participates in the act of mixing as an instrument"* whereas "not fire itself but *the matter and principle of fire* participates *as an ingredient and material principle* in the very substance of the mixed." And Stahl added that "this [principle] I was the first to call phlogiston."[8]

The essential thing in combustion, he thought, is the liberation of phlogiston, which in escaping excites the movement that constitutes fire.[9] Flame may, but need not, be generated in the process. In the slow combustion ("calcination") of certain ignoble metals, for example, the heated metal may liberate phlogiston imperceptibly. Thus:

Metal → metallic ash (*calx*)[10] + *phlogiston*

This reaction was later to be rewritten by Lavoisier, as follows:

Metal + oxygen → metallic oxide + caloric

Stahl considered this process reversible. He thought it possible to reconvert such metallic ash (or as we should say oxide) into a "dense, solid, shining ductile, fusible metallic mass," but only if "the phlogistic substance is returned to them."[11] The phlogiston taken up by the metal in this process is supplied by fuel, liberating phlogiston as it burns. Thus:

Charcoal (a compound of phlogiston with wood ash)
→ wood ash + phlogiston

and the phlogiston thus freed is available for the reaction:

Metallic ash (*"calx"*) + phlogiston →
metal (a compound of phlogiston with metallic ash)

The phlogiston in this process is transferred from fuel to metal ash, and the metal ash is thus converted to shining metal:

---

[8] *Specimen beccherianum,* p. 19, trans. our own; for prior uses of phlogiston referring to the same and other phenomena, see Partington, *History of Chemistry,* 2:667–68.

[9] *Fundamenta chymiae dogmatico-rationalis & experimentalis, quae planam* (Nüremberg, 1732), p. 310; *Specimen beccherianum,* p. 160.

[10] With Stahl, *cinis* (ash); with later phlogistonists, *calx.* See Partington, *History of Chemistry,* 2:670.

[11] *Opusculum chymico-physico-medicum,* (Halle, 1715), trans. Partington, *History of Chemistry,* 2:670.

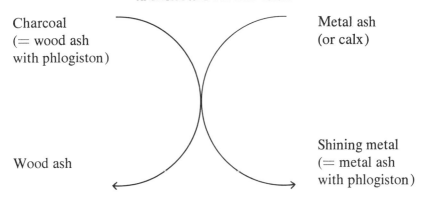

Charcoal
(= wood ash
with phlogiston)

Metal ash
(or calx)

Wood ash

Shining metal
(= metal ash
with phlogiston)

   Stahl used phlogiston in solving a wide variety of problems. For example, phlogiston seemed to him partly responsible for the solidity of certain bodies (whence their loss of solidity, often, when burnt) and for their color (which, too, disappears sometimes during burning).[12] Much of Stahl's interpretation of chemical change as involving a gain, loss, or transfer of phlogiston was, inevitably, misinterpretation. But the phlogiston theory led to so much new chemical work that a certain amount of useful knowledge was acquired incidentally. Patterson defends it as "the best theory possible in its era." According to Partington, phlogiston "smothered" the start that had been made by Boyle, Hooke, Lower, and Mayow toward a correct understanding of respiration, and the inquiry had in effect to start all over again a half-century or so later on.[13]

   Phlogiston and biology. Phlogiston influenced the history of biology by deflecting it temporarily from a forward-directed understanding of biological oxidation. Meanwhile, phlogiston was assigned a variety of roles in relation to life (see chapter 38). Today, we view the living system partly as a field for the complex interplay of energy transformations in some of which oxygen plays an enabling role as a hydrogen acceptor. But before that idea could be developed, oxygen had to be discovered (in the late eighteenth century), and the complexities of intermediary metabolism had to be unraveled (the effort to do this is still continued).

[12] Partington, *History of Chemistry,* 2:667.
[13] Ibid., 2:659. For a helpful account of the phlogiston theory, see also the series, J. R. Partington and D. McKie, "Historical Studies on the Phlogiston Theory," *Annals of Science,* vols. 2 (1937) to 4 (1939).

Stahl's Physiology

Phlogiston had no inhibitive effect on Stahl's own physiology since when he became a biologist he explicitly and almost violently repudiated his interest in chemistry. He rejected the "inane dreams of the ancients about the elements of the body . . ." and said that *Chymia tota, Theoriae Medicae, inutilis est,* . . .[14] "I feel," he wrote in a letter to Schroek, "the necessity of reconstructing medical theory and placing it on more solid foundations than concepts of mechanism and chemistry."[15] For Stahl, the gaps between vital and nonvital phenomena were so wide as to make chemistry an actual deterrent to the acquisition of physiological and medical knowledge. He urged attention to the phenomenon of movement in the body and to the causal basis of movement in an extraphysiochemical motive agent of some sort (soul).[16] Or so, at least, he protested (we shall see, in fact, that he did not personally reject chemistry as completely as these polemics against it suggest).

Living vs. Nonliving

Sensation and motion. Regeneration and reproduction. We shall see later that many leading eighteenth-century thinkers viewed life as manifested primarily through the two complementary phenomena of sensation and self-instigated motion. To these thinkers life was, or largely expressed itself through, feeling and movement. This dual way of looking at the organism was Greek in origins, but in its eighteenth-century manifestation it owed something to Stahl[17] who emphasized that living things—as opposed to nonliving ones—were "*locomotivae*" and "*sensitivae*." Stahl observed, also, that living things regenerate and reproduce, and that they do this not *fortuiter* as in minerals but in plants by forming seeds and in animals by com-

---

[14] *Paraenesis, ad aliena a medica doctrina arcendum* (Halle, 1706), usually found, variously paginated, with Stahl's *Theoria medica vera* (Halle, 1708). These two statements are found in the short "Argumentum" that begins the treatise. The *Paraenesis* also appears in translation as "La nécessité d'éloigner de la doctrine médicale tout ce qui lui est étranger," *Oeuvres,* 2:237.

[15] Quoted by Lemoine, *Le vitalisme,* pp. 33–34. See also, *"De scriptis suis,"* *Oeuvres,* 2:612, 667–68.

[16] "Réclamations," *Oeuvres,* 2:634, 652.

[17] "Pathologia," *Theoria medica vera,* sec. 1, memb. 1, p. 258.

parable processes governed by instinct, appetite, will, and desire. All of these and other differences seemed to distinguish living so deeply from nonliving things as to imply in living ones a wholly different, a more tenuous, fragile, flexible substrate which interests us especially because in it resides Stahl's conception of the kind of matter that can manifest life. It was this idea of a radical difference between living and nonliving bodies that seemed to Stahl to render chemistry, at least as ordinarily understood, "useless to medical theory." But note, in what follows, how difficult it was for him to free himself from his interest in physicochemical matters.[18]

Definition of life. To be motile and sensitive appeared to Stahl to require a combination of tenacity with softness and flexibility, a consistency at once mucous and fatty (*consistentia mucido-pinguis*) and

---

[18] Much of what we shall have to say about Stahl's idea of life and matter will be drawn from his *Theoria medica vera* (n. 1 above), and associated short medical and physiological treatises. Occasionally, Stahl discusses the organization of living things in his chemical works as well. What he has to say there, of which the following is typical, is certainly not easily understandable, but it shows an interest in the central problems.

Thus "we next proceed to *Vegetables;* which for their matter consist of *Minerals,* more or less alter'd, Thus they have their Life, as it were, from Water; and with this receive a certain *nitrous Salt,* containing a *bituminous Fat,* and a subtile *vitrifiable Earth.* But some, instead of a *Nitrous Salt,* attract a more acid one, carrying along with it a calcarious, earthy Substance; whence in a vegetable Compound the Mixts that compose it are more easily separated and demonstrated."

The above is followed by biochemical distinctions between trunks, leaves, flowers, fruit, and root. We also learn that:

"*Animals* appear to differ widely from the *Minerals;* all their *Atoms* being certain *Compounds* of various kinds of *Mixts,* variously alter'd and combined into a *specific Aggregate.* But the supposed *chemical Principles* are here found looser than in other bodies, from the necessary attenuation the *Elements* undergo in them; the *Preparations* they afford being very volatile; and reducible to the two most active *Principles, viz.* the *second* and *third,* under the form of *Oil* and *Volatile Salt:* the *terrestrial heavy part* subsiding to the bottom, and proving of a *calcarious,* rather than a *vitrescible nature;* from a certain saline substance that is intimately burnt into it. . . ."

We learn that some of these principles are found more copiously in some parts of animals than in others. "They are the most gross in the coarse and solid parts, if taken in that form; but thinner and more abundant in the finer; which then contain the greatest quantity of all, when they are resolved by putrefaction, and treated by a *combining operation.*

"The Grosser Parts are, (1) Bones, Horns, and Hoofs, (2) The next in order are Flesh, and the membranous Parts. (3) Then follow Fat and the Humours of the Body all which taken in substance, and treated with the same degree of Fire, yield more or less of a thin pure matter, as they are more or less gross and solid." "The Structure of Matter," *Philosophical Principles of Universal Chemistry* (see above, n. 5), chap. 1, sec. 1, pp. 18–20.

also moist. This simultaneously fat and moist consistency sufficed to explain the body's dissipative tendencies, since "water and fat, as is well enough known, serve to make no durable marriage, but are exceedingly prone to fermentative dissolution."[19] The system—of fat, subtle earth, and water—is stable at all only because the earth it contains is of such tenuous consistency as to be almost saline, so that it forms a kind of combination with water, a combination to which fat also adds itself (though only unstably). Everything chemical about the organism, in effect, militates precisely against its preservation. Everything inclines the body toward the corruption and disruption of each of its least particles (*minima singula corpuscula*). Life then is a *conservatio mixtionis corporis* against its tendency to decompose, the *mixtio* being the one needed for feeling and motion, reproduction and regeneration.[20]

Stahl was led, thus, to his own curiously intense formulation of the classic theme of displacement-and-replacement or destruction-and-reconstruction—the "remarkable paradox" which authors of all ages have made the central phenomenon of life. He said the sine qua non for vitality must be something that (*a*) maintains in matter its disposition to disorganization (its "imminent corruptibility") while (*b*) preventing that disposition from becoming a reality.[21] This something was for Stahl, notoriously, soul; not the vegetative or nutritive soul of the ancients but the immortal reasoning soul as we know it in man.[22] As to animals, Stahl took exception to the idea that they are *"nudas & absolutas . . . machinas. . . ."* They exist, as do men, in relation to purpose. His thought is difficult to follow, but he seems to extend from men to animals the unconscious physiological aspect

[19] "Medicinae dogmatico-systemicae partis theoreticae sectio I quam constituit physiologia," cited hereafter as "Physiologia," *Theoria medica vera,* sec. 1, memb. 2, pp. 268–69.

[20] Ibid., sec. 1, pp. 254–55, 268. Also "Paraenesis" (see above, n. 14), *Oeuvres,* 2:250. Also "De scriptis suis" (see above, n. 1), *Theoria medica vera,* p. 215 (*Oeuvres,* 2:665–68).

[21] G. E. Stahl, [*Dissertatio medica inauguralis*] *De Vita* (Jena, 1701), "Dissertation inaugurale sur la vie," *Oeuvres,* 6:461–74. Also: "Physiologia" (see above, n. 19), p. 254; and "De vera diversitate mixti et vive corporis . . . demonstratio, etc." (first publ. Halle, 1707), often bound, variously paginated, with *Theoria medica vera,* p. 25.

[22] "Physiologia," sec. 1, memb. 1 (passim, but see especially p. 260); also "De mixti et vivi," (see above, n. 21), pp. 53–55; and, for an early statement of the doctrine, *De sanguificatione in corpore semel formato* (Jena, 1684; Halle, 1704), "De la sanguification ou hématose dans le corps une fois formé," *Oeuvres,* 6:556–62.

of soul (*logos*) while withholding the consciously deductive and imaginative aspect (*logismos,* see below).[23]

A proper substrate for soul. A central question, in appreciating Stahl, is why so close a student of chemistry should have repudiated that chemistry in his study of life. His pretext was that the matter he studied in vitro behaved differently from the matter he studied in vivo. But this seems an only partly satisfactory answer, since the difference between living and nonliving matter, however great, would not justify a lack of interest in matter altogether. And Stahl himself acknowledges that the human soul cannot exercise its sentient or motive abilities except through the agency of the body.[24] But he believed that motion was the basis of life and that motion, though it needs to occur in a body of proper consistency, is due preeminently to soul. Moreover, Stahl's repudiation of chemistry was not nearly so complete as he seemed to think that it should be. In the details of his system we sometimes find him using his chemistry in spite of his antichemical protestations.

What Stahl rejected, or rather banned, from the living organism was the soulless chemistry of the laboratory. He often reasserted the necessity of distinguishing the merely mixed from the also living. Stahl saw an issue here and met it head on. Either materialism dispenses with soul entirely, or else it demands soul at the very center of every vital phenomenon. Stahl accepts the latter alternative. Since vital action is ultimately corpuscular movement, it is with corpuscular movement that soul is ultimately concerned. But our studies of other animistic thinkers from Greek times have made it clear that to postulate an immaterial agent in the living organism is not to dismiss the problem of a physical vehicle or target for that agent. Most animists have agreed that if the life-soul—however characterized in detail—is to express itself in a proper manner, a proper substrate is needed. Stahl shares this traditional viewpoint.

[23] "Disquisitio de mechanismi et organismi diversitate," (first published Halle, 1706) often bound, variously paginated, with *Theoria medica vera,* p. 31 (*Oeuvres,* 2:314–15); also *De la sanguification, Oeuvres,* 6:557–62. Among the many threads of psychistic theory that were spun out after and in spite of Descartes' attack on animism, we may mention Honoratus Faber who posits a single rational soul which *knows* directly but is *sentient* and *vegetates* indirectly. Faber thus partly anticipated —whether or not he influenced—Stahl. See *Tractatus duo de plantis et . . . de Homine* (Paris, 1666) 2d ed. (Nuremberg, 1677), p. 507.

[24] "Physiologia," *Theoria medica vera,* sec. 1, memb. 1, pp. 257–58.

In fact, he devotes a whole treatise to the difference between man (*a*) as a mixt and (*b*) as a living body. In bodies generally (whether living or not) the particles may either be mixts (if made up of simples, viz. water and the *tria prima*) or compounds (if made up of mixts). Structured aggregates may be homogeneous (presumably if made up of a single sort of particle) or heterogeneous (if made up of several). An aggregate, whether homogeneous or heterogeneous, may or may not be adapted to some use. If it *is* so adapted, it should be considered an instrumental or "organic aggregate." There does exist, Stahl acknowledges, a sort of order that owes its origins to the fortuitous concourse of certain matters (crystals exemplify this sort of order); but the instrumental order of a living body implies adaptation to purpose. Teleology, as L. J. Rather correctly observes (1961, see below, n. 26), is central to Stahl's biology.

Stahl points out some critical attributes of living bodies distinguishing them from mixts. Mixts may be aggregated homogeneously or heterogeneously. They are resistant to decomposition and have a tendency to longevity that is a function partly of their own constitution and partly of their surroundings. Living bodies are heterogeneous aggregates, tending to decomposition and shortness of existence. In living things, but not in nonliving aggregates or mere mixts, there is a constant process of decomposition and reconstitution. These distinctions, especially the flux of the body, imply a still further and crucial distinction, namely, the presence in living things of a formal immaterial causal agency (the biomedical soul).[25]

Life and cause. We have found the life-matter problem to be in part a search for the cause or causes of life. Causality is a subject on which Stahl had a theory of his own. Departing from Aristotelian theory (though not from a basically teleological bias), he distinguished occasional causes whose function is merely to do (*facere*) from efficient causes whose function is to do something (to accomplish) (*efficere*). In the latter, a strict reciprocity exists between cause and purpose. Herein the difference lies, according to Stahl, between mechanism and organism. Mere mechanism is disposed—by size, shape, location, or orientation, or by some mobile faculty or

---

[25] "De mixti et vivi," pp. 1–12; and *Opusculum chymico-physico-medicum,* pp. 226–36. For the strongly teleological aspect of Stahl's biology, see also "Physiologia," *Theoria medica vera,* sec. 1, memb. 1.

other—merely to act (*facere*); the idea of organism implies a mechanism existing as an instrument to accomplish (*efficere*) a particular end.[26]

Thus Lemoine paraphasing Stahl on some of these subjects:

1.) The body is a *mélange* of subtle earth, of fat, and of water, which cannot form a real union, but only a simple cohesion. Thus the *mélange* is essentially corruptible. The earth, almost as subtle as salt, promptly mixes with the water, producing a sticky consistency to which the fat can then adhere more easily.

2.) The reason for this *mélange* is that living bodies require for locomotion [he might have said for *all* vital motion] a consistency which is flexible, yet tenacious, without being fragile. The proportion of principles in the *mélange* is different in different parts of the body, bone, muscles, etc.[27]

The corruptibility of Stahl's "protoplasm" arises, as noted earlier, from the instability of the union between fat and water which Stahl undoubtedly had observed in his own laboratory. Subsequently to Stahl, incidentally, the instability of fat-water systems was to have a long history as a problem in the physicochemical interpretation of living things, and there are aspects of the problem still under inquiry today.

Temperament and texture. Stahl's occasional biomedical use of the chemistry and the microstructure both of which he so depreciates is well suggested in his treatment of temperament. Choleric persons seem to him to be predominantly sulphureous, their fluids (collectively, blood) subject to heating and to a violent, subtle sort of fermentation; phlegmatic types are watery and prone less to fermentation and inflammation than to salty and putrid decomposition; sanguineous, maintaining a mean among the others, are disposed to fluidity and vivacity; melancholics, being less sulphureous, to thickness and earthiness. To these chemical considerations, Stahl adds micromechanical ("textural") ones (sanguineous persons have spongy tissues, etc.), which are too detailed to permit our pursuing them here. He sees as his own contribution the identification of a

---

[26] "De mechanismi et organismi" (see above, n. 23), pp. 14–15 (*Oeuvres*, 2:-289–91). See also L. J. Rather, "G. E. Stahl's Physiological Psychology," *Bulletin of the History of Medicine*, 35 (1961):37–49.

[27] Lemoine, *Le vitalisme*, p. 37.

significant relation (*proportio organica*) between (*a*) the humoral temperament and (*b*) the microchannels (*meati*) of the tissue, a relation which determines how the soul can act "on, in, and through the body."

He erects a modified psychotypology based upon the way in which the humors, blood especially, move through the *meati* and upon the idea the soul obtains concerning that motion. Stahl explicitly acknowledges, here, a strict concomitance between "vital, animal, and moral" acts (*proportionis identitas inter actus tam vitales, quam animales, imo ipsos morales*) which, he says, it behooves us to understand. This phase of Stahl's thought seems far removed from his admonitions elsewhere that chemical and textural constitution are irrelevant to medicine in general.[28, 29]

Mode of action of soul. The completeness of soul's functional involvement, in Stahl's system, is summarized by Lemoine:

> The principle of life is the soul, not a special soul, but the rational soul, that which alone constitutes man, and is manifestly united to his body.
>
> The soul is not the life of the body; it cannot even be said to be alive, but only to give life; and it accomplishes this task of vivification, not by simple union with the body, but by real action. It is in this sense that we must understand the words of the Scripture: *Homo factus est anima vivens.*
>
> This life-giving act, the soul performs with complete intelligence in all details; it performs it by acting on all the organs, directing all their functions, using every appropriate means to arrive at its goal.
>
> The organs thus are not, as the name organ might suggest, merely simple instruments; it is the soul that makes the lungs breathe, the heart beat, the blood circulate, the stomach digest, the liver secrete; it is the soul that, while preserving the body, also makes it live and that, in order to preserve it, maintains corruptible matter in its [condition of] essential corruptibility yet keeps it from the act of corruption; and it is the soul, finally, that, to protect the body against actual corruption and to restore its losses, nourishes it and assimilates foreign substances to it,

[28] "Physiologia," sec. 1, membs. 4 and 5, pp. 292–311.
[29] "Paraenesis" (see above, n. 14), p. 54.

and makes repose follow movement and sleep follow waking.[30]

The historical significance of this view is clear enough from our study of Stahl's predecessors. The issue whether life-as-action does or does not entail an extraneous directive agent had already been posed by the pre-Socratics. During the seventeenth century, the pendulum had swung strongly and far—though not all the way—in the direction of an impersonal and antianimistic view of life. Stahl's "biomedical soul" expressed his own vivid reaction to this trend.

Stahl's *anima* is a threefold entity: an entity of action (*ens activum*), water being intrinsically passive; of motion (*ens movens*), motion being the basis of bodily function; and of intelligence (*ens intelligens*).[31] It exerts its physiological effects by causing and controlling movements. In the human economy, three movements are specially important: (*a*) circulation; (*b*) secretion (of both sound and decomposed humors); and (*c*) excretion (of the unsound ones).[32] The circulation has a conserving effect both on the blood and on the members it supplies.[33] Secretion (of lymph, milk, sperm, etc.) and excretion are assigned a special importance as "ultimate true instruments" of formal life. The washing-out of the tissues is indispensable to prevent the transmission of putrefaction from decomposed to healthy components.[34] The direction of flow of blood to and into the fleshy particles of the organs, and of *excreta* to the proper emunctories, is influenced by "tonic movements." The purpose of these movements, which are contractions and relaxations of the soft solids of the body, is to assure a proper localization of flow of blood and other juices so as to preserve the vigor and consistency of the parts. The tonic movements likewise encourage the elimination from these humors of abnormal solids they may happen to contain.[35]

Other vital phenomena are equally governed by soul. Nutrition, for example, is in essence a preparation of materials needed for the fundamental process of conservation. Thus hunger, salivation, deglutition, digestion, peristalsis, assimilation all are soul-governed ac-

---

[30] Lemoine, *Le vitalisme,* pp. 54–55.
[31] "De mechanismi et organismi," p. 34.
[32] "Physiologia," sec. 1, membs. 4 and 6; and Lemoine, *Le vitalisme,* pp. 44–45.
[33] "Physiologia," sec. 1, memb. 2.
[34] Ibid., sec. 1, memb. 2, p. 271, and memb. 6, pp. 319–82.
[35] G. E. Stahl, *Diss. epist. . . . de motu tonico vitali* (Jena, 1692), "Du mouvement tonique vital," *Oeuvres,* 6:518–24.

tions.[36] Presumably basic to all is corpuscular movement, and it is with this movement that soul must ultimately be concerned. It exerts its effects, moreover, without spirits or archei or any other intermediary agents, Stahl dismissing these concepts as "false," "inept," "sterile," and even "scandalous."[37] Soul intervenes at the corpuscular level, *directly*.

Reason and ratiocination. If we object that in ourselves the soul is not consciously concerned with the corpuscular (molecular) details of, say, digestion or assimilation, Stahl replies that neither is our soul consciously concerned with corpuscular details of gross locomotion. Nevertheless the soul is the cause of that locomotion. The difficulty can be removed by acknowledging two levels of knowing. One of these is a direct, intimate cognizance and judgment of such subtle elementary processes as tension, vibration, and fermentation (this is reason—*ratio*, or *logos*). At this level, the soul exerts, also, a motile influence corresponding to its intimate perceptions. The other is a rational drawing of conclusions from logical premises (*ratiocinatio* or *logismos*).[38] *Logismos* is supported by imagination and recollection. The key point is that "It is one and the same soul that acts both rationally and vitally" (*Eadem est anima quae et rationaliter et vitaliter agit*).[39]

SUMMARY AND CONCLUSION

Stahl has been generally characterized as personally harsh and inaccessible, intellectually recondite and contemptuous, spiritually pietistic, temperamentally "atribilious" (Partington), and intolerant. He divides men into two classes: the inept and the iniquitous.[40] To his credit are his defense of freedom of thought and of freedom to publish; he prefers that mediocrity and vapidity be tested in open controversy, rather than suppressed.[41] His writing, or that which bears his signature, whether in Latin or in a characteristic mixture of Latin and German, is congested, linguistically inelegant, syntac-

[36] Lemoine, *Le vitalisme*, p. 60.

[37] "Physiologia," sec. 1, memb. 1.

[38] "De mechanismi et organismi," pp. 25–26, 44–45. But see especially Stahl's *Propempticon inaugurale de differentia rationis et ratiocinationis* (Halle, 1701), "De la différence qui existe entre la raison et le raisonnement," *Oeuvres*, 6:445–52.

[39] See "Réclamations," *Oeuvres* 2:675; and "Différence . . . entre la raison et le raisonnement," *Oeuvres*, 6:447–52.

[40] "Réclamations," *Oeuvres*, 2:615.

[41] Ibid., 2:601–7.

tically disorganized, often too elliptic to be meaningful—in a word, *"illisible"* (Lemoine).

His two major contributions to science, phlogiston and the *anima physico-medica* or *biomedica,* are admirably illustrative of the irritant role in science of provocative "errors," ideas later generally rejected. Stahl's animism was so extreme as to have helped to sow the seeds of the eventual banishment from serious science not only of animism but of the vitalism which tended to replace it.

Stahl's physiology is thoroughly chemical, despite his protestations to the contrary. The kinetics of his chemistry of the organism are, however, distinguished by the intimate intervention of a conscious, volitional, rational soul. The behavior of the living system demands a special composition. Since vital processes are ultimately corpuscular movements, the *mélange corporelle* needs to be both soft and tenacious. This soft, tenacious material seems to Stahl to be a loose union of very subtle earth, water, and fat. But this *mélange* is inherently corruptible; whence soul is required to forestall corruption as well as to intervene in the most intimate way in everything the organism does.

There has been no definitive historical assessment of Stahl, and when one is undertaken it must come to grips with the influence of his spiritual pietism on his true theory of medicine. In our study of earlier (sixteenth-century) ideas of life and matter we noted the laborious process of disengagement by which science gradually freed itself from religion. Stahl stands as the last influential biologist for whom religion—as expressed in a soul which is at once physiological, rational, and immortal—acts partially but importantly, sometimes implicitly but nevertheless unmistakably, to define the author's scientific position.

· 26 ·

Micromechanical Models

HERMANN BOERHAAVE [1668 · 1738]

*Life and Matter in the late Ševenteenth and early
Eighteenth Centuries*

The seventeenth-century revolution in physiology was, in a central
sense, antianimistic: it entailed a repudiation of the Greek assumption of life-as-soul as cause of life-as-action and a working out of the
consequences of that repudiation. The situation that resulted from
this development was far from a uniform or settled mechanicism. In
the first place, we have already witnessed the vivid counterrevolution
launched by Stahl, a counterrevolution that failed in its immediate
goal but preceded and partly provoked a wave of later ones that were
to be called "vitalistic" (see below, chapters 33–36 and 43–45). In
the second place, the thinkers who abandoned animism moved off in
different directions so as to establish, especially, two major interpretive traditions, the chemical (with van Helmont as its dominant
earlier spokesman), and the mechanical (stemming from the quantitative and experimental mechanicism of Sanctorius, Harvey, and
Borelli, and the rational and deductive mechanism of Descartes).
Yet even this distinction held only in a limited way: it was never possible to classify all physiologists of the period (the late seventeenth
century) as either iatromechanists or iatrochemists. Thus Boyle's
objective, we noted, was not to *choose between* chemistry and mechanics but if possible to *bring them together*. In this he stood mid-
way between Descartes and Newton both of whom interpreted—
though not in detail—what we should term chemical change in terms
of particle theory. Indeed, many leading physiological thinkers of the
second half of the century—Hooke, Borelli, the Oxford virtuosi with
their interest in fermentation and combustion—mostly made use of
both chemical and mechanical concepts.

Yet despite these interactions and overlappings between iatrochemistry and iatrophysics and even occasionally between animism and mechanicism, the beginning of the eighteenth century was marked by the strongly contrasting viewpoints of two of its strongest minds. These viewpoints were: the animism of Georg Stahl (which we considered in the previous chapter) and the micromechanicism of the medical theorist Hermann Boerhaave (with whom we are about to become acquainted).[1] Both Stahl and Boerhaave combined their interests in medicine with interests in chemistry, but with results that were markedly different. Stahl disparaged the value of chemistry to medical interpretation, but used it all the same. When Boerhaave thought as a biologist and medical man, he left chemistry largely in the background, a fact explained in part by his having studied and written on biomedical matters before he turned to chemistry.

HERMANN BOERHAAVE

Hermann Boerhaave was an influential theorist and reformer who taught medicine and chemistry at Leyden during the first third of the eighteenth century. His contributions to both these sciences were matters less of discovery than of rational reinterpretation. His goal was to explain the patent manifestations of life in terms of a latent causality which he viewed as preeminently micromechanistic. Micromechanistic analysis was in itself not new; it had been a standard strategy with many earlier thinkers. But we shall see that in his use of this strategy Boerhaave set new standards of both thoroughness and ingenuity. He was explicit in adopting the classic procedure of combining direct observation (where the data permit) with inference (where the data are not directly observable). "There are two Methods which may be relied on as certain for the Attainment of our Profession . . . ," he said. Of these, "the First is an accurate *Observation* of all the Appearances offered to our senses in the human Body, whether in *Health, Disease, Dying,* or already Dead; . . . The *Second* is a strict Consideration and Discovery of the several latent Causes,

---

[1] For a recent appreciation of Boerhaave's contribution to biomedical methodology, see Lester S. King, *The Medical World of the Eighteenth Century* (Chicago, 1958), and *The Growth of Medical Thought* (Chicago, 1963); also Lester S. King's introduction to a facsimile reprinting of Haller's *First Lines of Physiology* (New York, London, 1966). For an appraisal of Boerhaave as a chemist, see J. R. Partington, *A History of Chemistry* (London, 1961), 2:740–59.

concealed from our *Senses* in human Bodies, by a just Reasoning; which is really necessary to prevent future ill Accidents, and Secure the good Events."[2]

Boerhaave summarized his physiological ideas in several celebrated tracts, especially *Institutes of Medicine,* which John Fulton (1938) said "quickly raised physiology from the nebulous sphere of the amateur to the dignity of an academic discipline."[3] Early editions of the *Institutes,* commencing in 1708, are mostly too sketchy and bareboned to convey Boerhaave's ideas distinctly, though they suggest his brilliance.[4] A more useful version is the richly annotated one brought out after Boerhaave's death by his celebrated student von Haller.[5]

Among Boerhaave's assignments at Leyden (after 1718) was the teaching of chemistry, a subject to which he ultimately contributed still another influential treatise, published only six years before his death. Since he developed sequentially his micromechanical and chemical approaches to the life-matter problem, we shall consider them in the same order here.[6]

Boerhaave's Mechanistic Biology

Boerhaave's biomedical system was pronouncedly, though not exclusively, mechanistic. Even in his later *Chemistry* he defined plants

[2] *Praelectiones academicae in proprias institutiones rei medicae. Edidit et notas addidit Albertus Haller* (Göttingen-Amsterdam, 1739–42; many subsequent editions, for which see Lindeboom, *Bibliographia Boerhaaviana* [Leyden, 1959], hereafter abbreviated *Praelectiones academicae;* trans. anon. *Dr. Boerhaave's Academical Lectures on the Theory of Physic, Being a Genuine Translation of his Institutes and Explanatory Comment, as they were Dictated to his Students at the University of Leyden* (London, 1742–47; many later editions), hereafter abbreviated *Academical Lectures.* Haller's edition is based on the Leyden edition, the last published within Boerhaave's lifetime, and reproduces Boerhaave's Latin text fairly faithfully. The anonymous English translation is best used in conjunction with the Latin original which it renders well enough but with occasionally prejudicial additions.

[3] Fulton's appraisal appears with other papers on Boerhaave in *Nederlandische Tijdschrift voör Geneeskunde,* 4 (Haarlem, 1938): 4807.

[4] *Institutiones medicae in usus annuae exercitationis domesticos digestae ab H. Boerhaave* (Leyden, 1708). For other editions and versions of this frequently republished digest, see C. H. Lindeboom, *Bibliographia Boerhaaviana.*

[5] *Praelectiones academicae.*

[6] *Elementa chemiae, quae anniversario labore docuit, in publicis, privatisque, scholis, Hermannus Boerhaave,* 2 vols. (Leyden, 1732); trans. Timothy Dallowe, *Elements of Chemistry being the Annual Lectures of H.B.* (London, 1735), hereafter abbreviated *Elements of Chemistry.* An earlier work, *Institutiones et experimenta chemiae* (1724 and after) is spurious (see Lindeboom, *Bibliographia,* p. 80).

and animals as "hygraulic machines," a view based indirectly on Harvey who "by the Discoveries which he demonstrated, overturned the whole Theory of the Ancients, and founded Physic upon a new and more certain Basis upon which it at present rests."[7] From his general program of mechanistic analysis, Boerhaave apparently wished to exclude, however, such conscious manifestations as "memory, Understanding, Reason, and the Knowledge of past and future Appearances. . . ."[8] He accepted, in this point, the dualism specified by Descartes: man = mind + body.[9] Mind and body act reciprocally, Boerhaave believed, but one cannot discover how they do so. As his commentator von Haller expressed it, "this Search after the Connexion between the Body and Mind not appertaining to a Physician, is to be rejected, among those which are useless to the Art."[10] Haller suggests that Boerhaave wants physic to set the mind-body problem aside and consider man "not as a metaphysical Entity, nor as a Mind, but as a living and animated Machine."[11]

It was Boerhaave's view that some parts of the machine are solid, others fluid. The solids, in turn, are either fibers, membranes, or vessels. Out of these components, especially vessels, various instruments (organs) are formed, each capable of a certain action when set in motion. Among these instruments ". . . we find some of them resembling *Pillars, Props, Crossbeams, Fences, Coverings;* some like *Axes, Wedges, Leavers,* and *Pullies;* others like *Cords, Presses,* or *Bellowes;* others again like *Sieves, Strainers, Pipes, Conduits,* and *Receivers;* and the Faculty of performing various Motions by these Instruments, is called their Functions; which are all performed by Mechanical Laws, and by them only are intelligible."[12] Fluid parts similarly operate in a mechanical fashion, agreeably to the "Laws of *Hygrostatics, Hygraulics,* and *Mechanics.*"[13]

This enumeration illustrates Boerhaave's interest in what we termed earlier mesomechanics, the point-of-view that compares the body to familiar, visible machinery. Actually Boerhaave, like Des-

---

[7] *Academical Lectures,* 1:41.
[8] Ibid., See Haller's n. 1, 1:64–65.
[9] Ibid., 1:65.
[10] Ibid., 1:70.
[11] Ibid., 1:53.
[12] Ibid., 1:80–81.
[13] Ibid., 1:85.

cartes, was primarily concerned with micro- rather than mesome-chanics. Between Harvey and Boerhaave in time there had stood the gigantic figure of Isaac Newton. Newton's emphasis on law—mean-ing mathematically orderly correlations between phenomena—was praised by Boerhaave but not much applied in his own approach to the organism. Rather, having drawn the foregoing rough comparison of gross structure to various technical artifacts and properly praised Newton's general canons, Boerhaave concentrated in a nonmathe-matical way on his own primary interest in microstructure and micro-dynamics.

It is not easy to be completely sure how Boerhaave pictured to himself the microstructure of the body. His presentation is elliptic, and he confesses to uncertainties in his own mind. The essentials, however, are as follows. The body is ultimately corpuscular. Cor-puscles are linked end to end as fibrils and fibrils are linked side by side as membranes. Membranes are ordinarily cylindrically arranged, that is, as the walls of tiny vessels out of which the tissues are fabri-cated.[14] Not seldom the smallest vessels are interwoven to form the sidewalls of vessels of a higher order (as in visible blood-vessels) some of which remain open whereas others are pressed to-gether to form higher order, solid fibers (for instance, ligaments). These larger fibers give the body strength, thickness, hardness, and rigidity.

The heart is the source of the arteries and their arteriolar ramifica-tions. But in addition to anastomosing with veins, arterioles give rise to small nonsanguiferous branchlets (and these to still others, etc.) the smallest of which, with invisible but still open channels, are the constitutive vesicular fibers of the tissues. These fibers carry extremely fine-particled natural spirits.

There is a special problem about nervous microstructure. In the cortex of the central nervous system, arterioles give rise to tiny (non-sanguiferous) tubules that are, in fact, nerve fibers and carry animal spirits, or nerve juice, to all the parts of the body. The question in Boerhaave's mind is whether the nerve endings may not actually constitute the solid substance of the organs—whether, in other words,

---

[14] This summary is based partly on Haller's notes, but the picture is undoubtedly about the one that Boerhaave had in mind. Ibid., 3:391–94.

the organs are ultimately actually nervous, or only nerve-like (*nervis aut vasculis his similibus*).[15]

This difficulty is mitigated to some extent by the fact that nerve fibers themselves are merely very fine nonsanguiferous branchlets of arterioles. And this brings us to what is the cardinal point about the whole scheme that Boerhaave develops. Not only blood, but a hierarchy of subtler fluids—serum, lymph, nerve juice (and perhaps natural spirits)—all circulate. All leave the heart together in the arterial system; blood returns through the veins; the subtler humors primarily through the lymphatics. Since the vessels involved are the prime constituents of the body, and since the heart supplies the impetus, we can well see what Boerhaave means when he defines the whole body as an hydraulic engine.

Pre-Boerhaavean ideas about fibers. Fibers as structural units are by no means Boerhaave's invention. We touched earlier on Praxagoras' belief that arteries terminate as "neural" microelements of tissues, as well as on Erasistratus' trivascular (vein-artery-nerve) hypothesis and Galen's picture of muscle as a complex of neural, ligamentous, and tendonous components. In the sixteenth century, fibers had been assigned a variety of constitutive roles by anatomists deviating in this or that way from the ideas of Galen; the group included Vesalius, Fernel, Fallopius, and Fabricius. We considered in some detail Descartes' theory of fiber formation and renewal. Descartes in his later physiological treatise stood effectively at the end of the premicroscopic era.

The microscope seemed to some but not all of its users to confirm the idea of fibrous organization. The best early case in point is that of Nehemiah Grew (1672) who was primarily interested in plants, and thought his microscopic studies revealed a fibrous structure there. At about the same time, another Englishman, Francis Glisson, invoked animal fiber as the seat of irritability. The body solids are mostly fibrous, according to Glisson. Irritability is the common capacity of fibers to respond to stimuli (irritants) by certain motions the summations of which are vital actions (see chapter 27).

---

[15] Compare *Academical Lectures,* 2:305, where "the nerves are expanded into either a sort of very fragile membrane or a soft pulp" (*expanduntur vel in speciem tenuissimae membranulae, vel in mollem pulpam*) with p. 345 where "we may believe that almost all the solid parts ot the body are interwoven ot nervous fibers and consist thereof" (*credemus fere, omnes partes solidas corporis contextas esse fibris nervosis, atque iis constare.*)

Debt to Baglivi. Boerhaave was influenced by but departed importantly from the pronounced iatromechanicism of Georgio Baglivi (1669-1707). The human body, according to Baglivi, was composed of solid parts, fluids playing only a secondary role, the form and function of these parts being basically those of their constitutive fibers. Baglivi likened the body to a little mechanical doll, whose hands, feet, and head move "admirably," without the impulsion of any fluids, simply because of the special assembly of their solid parts. Baglivi studied under the microscope, and recorded the details of, fibrous structure in both animals and men.[16] He concluded that the fibers must be grouped in two systems: (*a*) muscular (*fibrae musculares s. motrices*) originating in the heart; and (*b*) membranous (*fibrae membranaceae*) originating in the brain, or rather in the *pia mater* and *dura mater.*

The two motor centers, heart and *dura mater,* have much in common. Both pulsate; both derive their force from their own organization. The heart directs the flow of blood into all muscular fibers which constitute flesh, tendons, muscles and bones. The *dura mater* by its contraction and oscillating movements operates the nervous fluid in all the membranous fibers which compose the viscera, glands, nerves and membranous parts of the body. Muscular fibers range from soft (in the muscle) to hard as marble (in the bones). Membranous fibers are bloodless because they lie close together.

Baglivi had developed a complex theory of muscular contraction, ascribing it to the pressure exerted on the muscular fibers by blood corpuscles in the spaces between them. As to animal spirits, perhaps they function by altering the shape of the corpuscles, Baglivi thought, but he was not sure where animal spirits arise, or how the mind affects them, or whether they supply the power for contraction at all. Baglivi inclined rather to a purely mechanical explanation.

Boerhaave, like Baglivi, built a mechanical system in which the structural units were fibers, but the prominent feature of Boer-

[16] Georgio Baglivi, *Specimen quatuor librorum de fibra motrice et morbosa* (London, 1703). The same work (Basel, 1703) is also found bound with Baglivi's *De praxi medica* (Leyden, 1699) and appears in his *Opera Omnia* (Leyden, 1704). A part of it was published earlier as a letter to Alexandrum Pascoli, (Rome, 1700) and is bound with Pascoli's *Il corpore umano* (Perugia, 1700). See also E. Bastholm, *History of Muscle Physiology* (Copenhagen, 1950), pp. 178–89; Alexander Berg "Die Lehre von der Faser als Form-und-Funktionselement des Organismus," *Virchow's Archiv für pathologische Anatomie und Physiologie,* 309 (1942):394; and Charles Daremberg, *Histoire des sciences médicales* (Paris, 1870), 2:783.

haave's fibers was their hollowness and resultant conductile capacity; he saw them as tiny vessels (except in the case of those smallest and ultimate fibrils that compose the walls of the smallest vessels).

Corpuscular organization of fluids. Having made the solids micro-structurally fibrous, Boerhaave gave the liquids a globular or spher-ular microstructure. Viewed microscopically, blood "consists of red Globules swimming in a thinner and almost pellucid Serum. . . ." Boerhaave followed Leeuwenhoek in supposing that a red blood globule is divisible, under certain circumstances, into six yellowish serum globules and each of these into six still smaller pellucid glob-ules of lymph. Further divisions give us the constituents of other "limpid Humours, not concreating by Heat with the Milk, Urine, and all others gradually decreasing in the size of their Particles, but not yet accurately classed or enumerated."

In an important footnote, Haller details Boerhaave's picture of the lung as a separative and attenuative device. Large red globules go through the lung to the veins; smaller yellow serous globules turn off into serous vessels; and still smaller pellucid ones into even smaller lymphatic vessels which ultimately return them to the veins.[17] It was Boerhaave's idea that individual globules are mechanically sub-divided as they move in the bloodstream—especially in the lung—and are turned aside into smaller branch-vessels that they are minuscule enough to enter. This brings us to Boerhaave's theory of the significance of breathing.

Respiration. Boerhaave viewed the action of the lungs as pre-eminently mechanical. Other ideas—that air cools and condenses the blood (Harvey, Descartes, Sylvius), supplies the blood with a subtle niter (Boyle, Lower) or with ponderable elastic oscillating particles (Borelli)—he rejected. He did not dispute that air is necessary, how-ever, and as a mechanist Boerhaave suspected that some clue to its use might be afforded by the loss of elasticity air experiences when rebreathed in a closed container (see Boyle). The elasticity would serve the function of swelling and rounding out the "cells" (alveoli) of the lung. Boerhaave wondered, however, why persons breathing under water are "quickly *suffocated,* notwithstanding . . . [continued] alternate Contraction and Dilation." They ought not to drown if the function of breathing is exclusively micromechanical. Boerhaave concludes with the observation that "this is no easy Question."

17 *Academical Lectures,* 2:172 including Haller's n. 1.

In any case, the major effect of breathing is, for him, the mechanical attrition and blending of mixed chyle and blood particles. The fact that the lungs receive the totality of the blood, other organs only a part,

> discovers the Lungs to be the principal elaboratory of Sanguification, and for converting Chyle, which is all brought thither, into *nutritious Juices;* in this Organ, the whole Mass of Juices receive that Degree of *Fluidity* and Attenuation which fits them to circulate freely through all the smallest vessels which they cannot receive in any other Part; and it is therefore here chiefly that Blood and its Juices are prepared and fitted to make the several Secretions, and to perform all the Action of Life and Health which depend on the Efficacy of the Fluids.[18]

We hear, in summary:

> The Chyle therefore which has been prepared in the Mouth, digested in the Stomach, elaborated in the Intestines, secerned in the Lacteals, attenuated at the mesenteric Glands, and farther diluted and mix'd in the thoracic Duct, then blended with the venal Blood, afterwards more intimately mixed, dissolved, and *digested* or attenuated by the right Auricle and Ventricle of the Heart, is lastly impelled into the conical Branches of the pulmonary Artery, by whose Sides being resisted, and still urged on behind, it receives a violent Pressure, whereby its Particles take a *Configuration* suitable to compose all the solid and fluid Parts of the Body.[19]

Breathing, then, is primarily subservient to nutrition.

Nutrition. Boerhaave sees replacement as the most fundamental of biological actions. "It is therefore *necessary* for the due continuance of Life, and the Maintenance of the several parts of the Body in their sound or natural State, that as much or *as many of the same Kind* of solid and fluid Particles should be perpetually restored, as are continually wasted by the several Actions of Life, and this Restitution is called Nutrition."[20]

The wearing away of both solids and fluids is mechanical (it occurs

[18] Ibid., 2:111–39.
[19] Ibid., 2:91–92.
[20] See above, chap. 7, n. 23.

by abrasion in solids and attenuation in liquids). Health implies an equilibrium of decrement and recrement ("He that can perform this one Point, is acquainted with the art of preserving the Body immutable to extreme old Age. . . ." As to the importance of this phenomenon it seems "credible, that Nutrition is one of the most ultimate and most perfect Actions of Nature)."[21]

In the Latin editions of Boerhaave's *Institutes* the section concerned with this topic is entitled *Nutritio, Incrementum, Decrementum.* Decrement and increment have antecedents in Bacon's *Depradation* and *Reparation,* in Plato's *Anachoresis* and *Plerosis,* in the Hippocratic idea of consumption and repletion, perhaps even in Heraclitus' idea of the body as involved in a two-way transformation ("the way up" and "the way down"). On these points, therefore, Boerhaave enlists in one of the hardiest of physiological traditions— that which associates life with renovation. We have heard often before and will hear often again that the body keeps wearing away and replacing itself or is involved in some sort of two-way transformation.

How, more specifically, does recrement occur? The central consideration is that "the Parts and Organs of a human Body in Health have . . . the *Faculty* of converting the various Aliments into a Matter similar to their own, and fit for *augmenting or restoring* such Parts of the Body as are decay'd or consumed."[22] The same point had been made in somewhat similar terms by Galen, though Boerhaave was not a Galenist in the sense of postulating a series of attractive, alterative, and assimilative faculties as such. He pictured food as first physically comminuted by the cutting and grinding action of the teeth, and gave copious detail about the effectuation of mastication through the muscles of the jaws, tongue, lips, and cheeks (the many muscles are all identified by name).[23] Saliva serves to "assimilate" food; mix its oily and aqueous parts; dissolve its salts; and excite in it a fermentation along with "an intestine motion" of the food particles caused by their admixture with air.[24] The acknowledgment of fermentation seems rather iatrochemical for Boerhaave, but Haller's comment points out that when one mixes food with air and heats it in an enclosed place, fermentation inevitably follows. Belching is one evidence of its occurrence.

---

[21] *Academical Lectures,* 3:340, 388–89.
[22] Ibid., 1:101.
[23] Ibid., 1:119.
[24] Ibid., 1:145.

In the stomach, according to Boerhaave, the swallowed aliment "does there quickly begin of its own accord to ferment, or putrify; according to the different Nature of the Aliment or Disposition of the Stomach; and is either way wonderfully changed into an ascescent, alcalescent, rancid or glutinous Mass." This mass is "Chyle, fit to supply [compensate for] (*a*) the Abrasion of the Solid, and (*b*) Consumption of the fluid Parts of the Body. . . ." Chylification is promoted by gastric juice, by muscular movements of the stomach (much detail on muscle action), by heat, by the mechanical impact of the nearby aortic pulsations, by the solvent action of nervous juice, and by the compressive effect of the peritoneum and of the diaphragmatic and intestinal muscles (all named).[25]

In the intestine, chyle is mixed with three fluids, *cystic bile* (dissolves oil and mixes it with water, "cleanse[s] off viscidities," opens lacteal passages, stimulates the appetite, acts as an assimilative ferment, etc.); *hepatic bile* (same effects in less degree); and *pancreatic juice* (not a ferment, mixes and dilutes the chyle, obtunds its acrimony, abates its viscidity and bitterness).

Equipped with this understanding of digestion, Boerhaave dismisses such "obscure and dubious *Hypotheses* or *Postulata*" as "a vital, innate, or digestive Heat" and volatilizing "acrid Ferment"; "the operating *Archaeus*" (invented by van Helmont and termed by Boerhaave a "spiritual Cook"); and the "false and imaginary . . . peripatetic Qualities and *galenic* Faculties; . . . and innumerable other false and *pernicious* Hypotheses misleading from the truth."[26]

Boerhaave admits that an easily imbibed moiety of the chyle is sent to the liver, with the primary effect of supplying the raw material for the production of bile. But most of the chyle enters the lacteals and bypasses the liver on the way to the heart. En route, it is further acted upon by the assimilative juices absorbed with it from the intestines.[27]

Within the body two agencies, the movements of the viscera and vessels and the heat they produce, introduce "divers Changes continually in the healthy circulating Serum of the Blood, till at length some Part of it becomes so subtil a Humor as is required for the Business of Nutrition, in order to repair immediately that which is wasted; and it is this last most subtil Humour, prepared from the

[25] Ibid., 1:184, 198, 201, 211.
[26] Ibid., 1:307.
[27] Ibid., 1:336.

Serum of the Blood, which at length becomes the true and immediate Matter of Nutrition."[28] We have already noted that Boerhaave gives even breathing a primarily nutritive function.

Boerhaave goes on to picture the intimate process of replacement. The attenuated nutritious fluid finally finds itself moving through those minute vessels that are the microconstituents of the body. These vessels are so tenuous as to differ little in stability from fluids; of all body parts, they are the "least compact" and "nearest to a State of Dissolution."[29] Their texture is, thus, readily abraded, an effect due in part to the activity and heat of neighboring organs and in part to the motion of the nutritious fluid itself. Particles are swept away to leave the body as excretions or exhalations. Resulting vacuities in the vesicular membranes are filled in by properly corresponding small particles obtained from the nutritious fluid.[30] The same juice that wastes the vesicles supplies them in the proper proportion. This homeostatic arrangement makes clear the "wonderful Wisdom of the *Architect*." It is, as Haller avers, an "artifice . . . worthy to be admired by all Mortals, and is not to be matched in any other Machine whatsoever. . . ."[31]

Micromechanics: Further Applications

Perhaps the primary outcome of our whole study thus far is its demonstration that physiologists have proceeded in every era according to a well-accepted strategy of interpretation. They have studied the visible differentia of life—nutrition, generation, perception, volition, motion—and have tried to understand them in terms of an invisible but inferable causality. True to this tradition, Boerhaave sought the causes of macrophenomena in the microphenomena that lie beyond the reach of vision. He and his commentator point out more than once that certain initially invisible structures were first discovered deductively but became visible later with the introduction of proper aids to vision (for example, Harvey deduced, Malpighi saw, the channels connecting the arteries with the veins). We have already suggested how Boerhaave applies his microdynamics to the phenomena of nutrition and respiration. We shall consider next, briefly, how he applies it to muscles, nerves, and glands.

[28] Ibid., 3:372–73.
[29] Ibid., 3:383–84.
[30] Ibid., 3:360–88.
[31] Ibid., 3:388.

Muscles and nerves. The central questions about muscle and nerve, respectively, are: what is contraction? and what is conduction? We shall expect Boerhaave to give microdynamic or even micro-hydrodynamic answers to these questions. In fact, he conjectures that the contractile elements of muscles are expanded nerve fibers containing nerve juice pumped there from the heart by way of the brain.[32] These nerve elements are interwoven with larger artereolar vesicles and branches of them.[33]

Reviewing certain experimental results mostly obtained originally by Galen by such means as neuro- and myosection, ablation, and ligature,[34] Boerhaave concludes, unoriginally, that when a limb is immobile its muscle antagonists are pulling against and tending to elongate each other without effect because the pulls are equal. Injury to a nerve or to the spinal cord above the level at which that nerve emerges causes paralysis. Contraction is due to "even the least excess of Force in either" of the two antagonists.[35]

From "duly considering the several *Phaenomena* aforementioned," the reader must agree that the cause of the contractile rounding-out of the fiber must be "an influxive thin, fluid, motile body," whence the "*nervous* Liquor . . . is to be acknowledged for the true *Cause;* nor is it difficult to understand its Manner of Action."[36]

The contraction of a given muscle results from increased speed of delivery to it of nervous juice (along with a concomitant decrease to the antagonist). Haller elaborates this picture with the suggestion that voluntary responses occur very rapidly "not because a certain quantity of the nervous Juice is impell'd at that time all the way from the Brain into the Muscles, but because all the nervous Tubuli being full, an Impulse communicated to the Liquor at one End of the Tube will thrust out its Globules at the other End in the very same Instant of Time; as we know by placing a Row of ivory Balls close to each other upon a Table, and then by striking upon the outermost Ball at one End, the furthermost at the other End will instantly recede or run off with the Velocity first communicated, without any sensible Succession through the intermediate Balls; and

---

[32] Ibid., 3:175–76.
[33] Ibid., 3:184–85.
[34] Ibid., 3:191.
[35] Ibid., 3:194.
[36] Ibid., 3:216–22.

if a Tube be full of Liquor you no sooner urge more in at one End but it instantly runs out at the other."[37] (For an earlier mechanical account of transmission, see Borelli, chapter 24.)

The motor nerve impulse, as here described, is a shock wave transmitted downstream through the nervous juice (or animal spirit). What of the sensory impulse? We must depend, here again, partly on Haller's elaboration. We should agree that it is perfectly possible for a shock wave to move against the current, and this is what Haller, and probably Boerhaave, see as happening.

As for the stimuli, they vary with the sense modality. According to Boerhaave, taste presumably, and smell certainly, depend on spirituous emanations; touch and sound stimuli are mechanical;[38] visual stimuli are focused light rays (Boerhaave is generally lucid in his account of physiological optics). These stimuli act on the surface of the nerve ending. "But this change in the (surface of the) Nerve is not sufficient to produce Sensation; but it is farther required, that the same Change be *propagated* through a free Nerve to some Part in the Medulla of the Brain, and from every *single* Nerve to a corresponding particular Part in the Medulla of the Brain; and this we learn from Ligatures, Wounds, and Corruptions of the *Nerves* and Brain." Propagation is presumably mechanical (a shockwave passes upstream). Haller terms the phenomenon "a Repulse of Spirits against their Origin." Sensory discrimination depends on the varieties of objects, the motions they apply to sense organs, the fabric of those organs, the "faculties" peculiar to each nerve, and the parts of the Medulla where different nerves arise.[39] The picture represents a marked improvement over that advanced by Descartes.

Secretion. This activity may serve as another example of Boerhaave's microdynamic mode of interpretation. Boerhaave rules out mere size of pore as the ruling factor in secretion as well as the Helmontian and Paracelsian ferments his attitude toward which is suggested by his comment that "nobody could ever assign the Cause,

---

[37] Ibid., 2:317.

[38] Ibid., 4:13-54. Concerning touch we learn that "a certain Motion is impressed upon . . . (the sensory) Papillae, the Effect of which is conveyed to the common Sensory (*sensorium commune*) (4:6–7)"; concerning sound, see 4:157–208. The external and middle ear mechanisms are on the whole admirably described: "From all which is evident that the sonorous Rays in the trembling of this (cochlear) Membrane are conveyed, by means of these agitated Nerves to the Common Sensory, and there excite the *Idea* of Sound."

[39] Ibid., 4:226–32.

Origin, Matter, Place, Mixture, Efficacy, Proportion, Continuance and Effects or Uses of such Ferments."[40] He describes the microstructure of the glands and the agitation, rotation, compression, and collision of arterial blood arriving there. Secretion is a separation under pressure of thin pellucid blood particles less in diameter than the orifices through which they leave the blood. Since glands are supplied with nerves, there is a possible admixture of secerned juice with nervous juice. The filtered humor passes not through the sidewalls but into tiny nonsanguiferous branchlets which either unite like veins and ultimately become the lymphatics or debouch into a common cavity with its own emissary canal.

The brain as a gland. The cerebral cortex is seen (following Malpighi) as a typical gland. Arteries, subdividing, give rise both to sanguiferous veins and to smaller serous arteries. Serous arteries subdivide to form both serous veins and smaller lymphatic arteries. Lymphatic arteries subdivide to form both lymphatic veins and smaller vessels which enter the medulla as nervous tubuli.[41] These tubuli are continuous with the filaments composing the peripheral nerves.[42]

As to the nerve juice, its particles are the most "simple, *dense* or *firm,* subtle and *moveable* of any Juice throughout the whole Body."[43] How small the particles of a liquid may be is suggested by the fact that albumin particles can enter the tiny blood vessels of the chick embryo at the beginning of its development and can even circulate within the presumed tubules of a seminal animalcule![44] The cerebellum is the seat of control of internal functions which are the indispensable concomitants of life itself. Perhaps the brain (*encephalon*) secretes animal spirits, and the cerebellum vital spirits (and the heart, via the arterioles, natural spirits). On these points, however, Boerhaave is somewhat tentative.

Generation

Students of generation have been interested at every period in two particular questions: What do male and female contribute, respectively, to the new individual, (*quid pater, quid mater, generendae*

[40] Ibid., 2:207–44.
[41] Ibid., 2:259–71.
[42] Ibid., 2:273.
[43] Ibid., 2:290.
[44] Ibid., 2:292.

proli facit?)?[45] and What kind and degree of organization does the individual possess at the very beginning of its development?

Boerhaave's answers to the first question mark him as a spermatist or animalculist and to the second as a preformationist (with a certain qualification to be noted). He fully recognizes the reality of the ovum, but he thinks its role is to receive and nourish a seminal animalcule.[46] For Boerhaave, the crucial phenomena are the initial constitution of this animalcule, and its subsequent development.

A careful reading of Boerhaave and earlier and later students of embryos shows that the issue of epigenesis vs. preformation is less simple than might be supposed. The difficulty is partly definitional. Strict epigenesis implies a strictly amorphous homogeneity at the beginning of development. Harvey's crystalline colliquament meets this criterion. Harvey and a century later C. F. Wolff are frequently cited as founders of modern epigeneticism. It is worth noting that neither of them contributes significantly to the causal interpretation of development. Harvey falls back on an unabashedly Aristotelian soul-stuff in the semen, while Wolff will evoke a vitalistic *vis essentialis*. The contribution of Harvey and his eighteenth-century successors is their identification and preliminary illumination of an ontological fact, the fact of differentiation, to which it is left to later generations to supply a causal interpretation.

Strict preformation, defined as the preexistence of a fully-formed miniature individual in the germ, is well represented by some of the spermatists (Dalenpatius and, later, Gautier who "saw" small horses, men, or birds in the animalcula of those species). But most seventeenth- and eighteenth-century investigators of embryology occupied intermediate positions.

Even Malpighi "did not," as Needham observes (1959), "maintain a perfectly equal swelling up of all the parts existing at the start, but rather an unequal unfolding, a distribution of rate of growth at different times in different parts of the body."[47] His contemporary Swammerdam was enough of a preformationist to believe in encapsulation.[48] Yet Swammerdam gave a far from naively preforma-

---

[45] Ibid., 5:230.

[46] Ibid., 5:68, 131, 139.

[47] Joseph Needham, *A History of Embryology* (New York, 1959), p. 168.

[48] He says, anticipating Bonnet, that "there is no generation but only propagation, the growth of parts. Thus original sin is explained, for all men were contained in the organs of Adam and Eve. When their stock of eggs is finished, the human race will cease to be." Ibid., p. 170.

tionist account of the early development of the frog. We shall discover that eighteenth-century "preformationists" likewise, among them von Haller and Bonnet, leave room in their accounts for a certain amount of progressive elaboration. Their approximate contemporaries Buffon and Maupertuis occupy a peculiar position in believing that specialized particles of the male and female seminal fluids unite rather suddenly to form the new individual. Buffon calls this first formed embryo a "mold" or "sketch" (*ébauche*) of what will develop later. It contains only an outline of future structure.[49]

It is an assemblage of rudiments that Boerhaave, too, sees in the seminal animalcule.[50] Influenced by Malpighi's studies of the chick embryo, he supposes—not entirely incorrectly—that when development begins, the pronounced divisions of the central nervous system are the first to unfold. (He differs from Malpighi, however, in assigning the critical role to the male; Malpighi was an ovist.) The heart follows shortly. Boerhaave specifies that the viscera as such are not there at the beginning (except perhaps as still folded rudiments). This view of development, basically preformationist, still leaves room for a step-by-step increase in complexity. Haller makes much of the progressive character of developmental events; all animals pass early through a vermicular phase; the higher animals live first a fishlike, then an amphibious, then an aërial existence.[51]

Microdynamic aspects of development.   Boerhaave has already indicated (in describing the nerve juice) that the animalcule has a microvesicular structure that enables it to take in nutrient seminal fluid.[52] After it penetrates the ovum its food is the colliquament which the egg encloses. The egg is generally permeable to nutrient matters— in mammals to uterine fluids, in the chick to exhalations from the incubating hen. When the placenta develops, the fetus at first absorbs what would otherwise be menstrual blood.[53] When the digestive system is formed, the fetus takes in amniotic fluid by mouth, and this is digested by the usual juices and thus transformed into chyle.[54] The umbilical vessels permit an exchange between the fetus and the mother, useful partly in connection with the need for mechanical subtilization of the embryo's blood (in the mother's lungs, since

[49] Needham has organized these varying opinions in tabular form. Ibid., p. 184.
[50] *Academical Lectures*, 5:68.
[51] See chapter 27, n. 35.
[52] *Academical Lectures*, 5:75.
[53] Ibid., 5:175.
[54] Ibid., 5:175.

the embryo's lungs are inactive). All these matters are developed in impressive detail, more than is relevant to the focus of our present study.

We may finally mention briefly Boerhaave's scheme for the more intimate (corpuscular) aspect of the developmental process. Morphogenesis is, not exclusively but importantly, a hardening process in which fluid parts become solid, and soft solid parts become harder. Solidification results when particles are properly apposed so that, as a result, their cohesive properties become effective.[55] Boerhaave makes a special and fairly extensive study of albumin, demonstrating its properties to his students at the lecture table; albumin has all the qualifications to serve as a morphogenetic precursor substance.[56] The process of stiffening continues through life. Old age is a matter of progressive induration and fibrosis.[57]

Boerhaave's Chemistry

Boerhaave provided no effective synthesis of the foregoing micromechanical or bio*physical* theories of life and matter and his subsequent bio*chemical* ideas on the same subject. His rule, in his chemical system, was to borrow as sparingly as possible from the theoretical speculations of earlier chemists. He proposed to eliminate as irrelevant to medicine "many useless and fallacious hypotheses," among which his editor and annotator Haller assumed him to mean: Pythagorean numbers; monads; chemical elements; alchemical principles and *archei;* universal matter (of the Democritean-Epicurean variety); and the subtle matter of the Cartesians. Equally useless in Boerhaave's view, were such metaphysical considerations as first and final causes, occult qualities, and substantial and seminal forms.[58] Instead of dealing with such matters, chemistry has to do, he said, with the art of effecting desired changes in a predictable manner (an object which, as Partington observes, applies equally well to chemistry and to cookery). In sum, Boerhaave subjected chemistry and the chemical aspects of physiology to a thorough conceptual cleansing—in the course of which, we shall see, some scientific babies were thrown out with the bath water.

[55] Ibid., 3:394, 407.
[56] *Elements of Chemistry*, 2:235.
[57] *Academical Lectures*, 3:355–59, 363, 399.
[58] *Academical Lectures*, 1:52, 71-73.

Classification of kinds of matter. Boerhaave did not deny the necessity of classifying the objects of his chemistry, and a glance at his scheme reveals his debt to tradition. Bodies are either fossils (also termed minerals), vegetables, or animals. The fossil or mineral class subdivides five ways into (*a*) metals, of which he acknowledges six (or seven if mercury be included); (*b*) salts, "sometimes called concreted Juices" (there are seven of these); (*c*) sulphurs, "true" and "common" (along with such allied substances as arsenic, petroleum, naptha, jet, and amber); (*d*) semimetals; and (*e*) stones.[59]

As for plants and animals, Boerhaave objected to the pretensions of chemists claiming specific knowledge of the blend of elements in this or that organized body. He did accept the challenge of distinguishing between the three kingdoms, but not in strictly chemical terms. We shall note in Boerhaave's response to this challenge that some of the principles he attempted to repudiate he subsequently found to be indispensable. We are impressed, as so often before, with the slow and labored progress made by chemistry in its effort to disembarrass itself of its alchemical inheritance. Boerhaave was ready to accept certain parts of alchemy which were, in his opinion, useful or at least well-intended.

The vegetable kingdom. Boerhaave thought that plants differ from fossils (minerals) in utilizing the "water, spirits, oils, salts, and whatever lies hid in the bowels of the earth. . . ." These are "applied to" the roots of plants by the motive effect of "subterraneous, artificial, and celestial Fire." Circulating through the manifold vessels of the plant, they are concocted and perfected and so become the "proper juices of the plant." The leaves, especially, perfect the juices and return them so altered to the parts of the plant, an idea expressed earlier in somewhat the same way by Malpighi.[60]

A still further perfection occurs in flowers where nectars, scents, and honey are produced. Within the seeds is formed a "fine, pure, volatile spirit, which is the ultimate production of the plant: This the alchemists have stiled, the Spiritus Rector, the Inhabitant of Sulphur, the *Archeus,* and the servant of nature." Boerhaave thinks that this "spirit being quickened by an [otherwise undescribed] active power may possibly breathe a vital principle into the Juices that

[59] *Elements of Chemistry,* 1 (the two volumes are bound together): pp. 19–36.
[60] Ibid., 1:36.

nourish the embryo, and stamp upon it the character that distinguishes the family; after which everything is changed into the proper nature of that particular plant."[61] Boerhaave thus seems to use his version of the *spiritus rector* to account for the idiosyncratic distinctiveness of the individual, for its fidelity to its species, and within the species for the similarity of offspring to parent, that is, for inheritance. Boerhaave's use of the term *vital principle* is somewhat troubling in a framework of ideas otherwise largely physical or physicochemical.

Boerhaave also calls his directive spirit a "particular and peculiar juice" which "possesses the true nature and vertues of the plant. This juice," we are told, "can scarcely be reduced to any class of things yet known, and therefore must be looked upon as something perfectly singular." It owes much, conceptually, to the partly Stoic and partly alchemical idea that all things have both a body and a quintessential spirit and something also, perhaps, to the idea of the radical humour. Rejecting the chemists' claims to specific knowledge concerning constitution or *krasis*, Boerhaave insists that all that the chemists have really discovered in plants is: (*a*) the *spiritus rector;* (*b*) a fine oil that is the seat of this spirit; (*c*) three salts (one acid, one neutral, one alkaline); (*d*) a soap and accompanying saponaceous juice; (*e*) another, earth-associated oil; and (*f*) a "pure and simple Earth."[62]

The animal kingdom. As plants appropriate and alter matter taken from the earth, animals appropriate and alter matters taken, directly or otherwise, from plants. Animals contain: (*a*) water; (*b*) a distinctive salt similar but not identical to *sal ammoniac* (one of the acknowledged seven); (*c*) various oils (as in marrow and fat); and (*d*) an earth "which serves as a *Basis* to the whole body, connects all the particles together, and retains the fluids within their proper bounds." Animals, like plants, contain a *spiritus Rector,* "a kind of Aura, or Vapour, that is proper only to that particular body; and that is of so subtil a nature that it discovers itself only by its scent, taste, or peculiar effects. This Spirit expresses the true genius of the Body in which it resides; and it is this chiefly that accurately distinguishes it from all others." Somehow this spirit volatilizes, is mixed with air, descends with dew, snow, rain, or hail, and so "by this revolu-

[61] Ibid., 1:38.
[62] Ibid., 1:40.

tion returns into new bodies, in order to govern them and render them active."[63]

The *spiritus rector* is invisible, volatile, entangled in oil, and present in such minute amounts to be detected only by taste or smell (as when "your Hounds . . . will single [it] out, and pursue [it] over a great deal of ground, and through a vast confusion of tracks. . . ." Without quite espousing the idea personally, Boerhaave says others assure us that the spirit "if you cherish it with a pregnant warmth, and support it with proper nourishment . . . will still increase in activity, and in a wonderful manner continually acquire new strength for the production of an Offspring like itself." Whether there is a *spiritus rector* in metals—confined there in fixed sulphur and "exceedingly efficacious in the cure of diseases"—are questions to which Boerhaave protests he is unequal. In any case, Boerhaave's chemistry has not fully renounced its alliance with alchemy.[64]

Chemical change. Chemical change is, for Boerhaave, simply motion. "The whole business of Chemistry therefore is to unite, or separate; nor is there a third thing that is capable of performing; and hence all its operations may be reduc'd hither, without exception." In his view, to consider such a proposition as too reductive is to misunderstand the variety that combination can produce. Boerhaave touches here on the thorny problem of the appearance in compound systems of properties that do not appear in their separate components (the phenomenon of emergence). And he raises the related question whether components persist as components once compounded. He observes that "from the application of one Body to another, there often appears a new power [e.g. magnetism, solubility, etc], which before lay intirely concealed." Similarly, when a compound is resolved, the end products exhibit properties they never showed in combination. These products—water, spirits, salt, oil, and earth—have been called elements but are in fact further resoluble (Boyle, we remember, held a similar view). We have no evidence that even these ever existed as such in, say, an animal body. Boerhaave makes the further point—of great consequence in the later history of chemistry—that once an organized body is resolved it cannot be reconstituted from the products.[65]

[63] Ibid., 1:47.
[64] Ibid., 1:48–49.
[65] Ibid., 1:45–47.

Boerhaave on the Intellectual Position of Chemistry

Boerhaave discusses the implications of chemistry for medicine, the mechanical arts, natural philosophy, and natural magic. On the latter score, he shows that what would otherwise appear to be magical occurrences (such as raising a massive tower and then dashing it to the ground, or causing letters to glow in the dark) may have chemical explanations (in these two cases the use of gunpowder and of English phosphorus respectively). He eschews as "old wives' fables" and "the reveries of idle persons" such occurrences as "the foretelling of things future; . . . the creating, removing, or assuaging distempers by numbers, words, signs, figures, inarticulate murmurs, charms, little images, a look, or laying on of hands; . . . riding through the air at one's pleasure; . . ," etc.; his list is quite exhaustive. But he is too cautious to dismiss as impossible the insights ascribed to the *Magi*. He is "disposed to believe, that there lies hidden in the bosom of futurity an infinite number of things, that will at some time or other be clearly revealed to mankind, of which at present there is not the least shadow of appearance. Who can deny, that there may be beings, who have a faculty of looking more intimately into nature than the most sagacious human mind was ever capable of? Who can demonstrate but that these Spirits, without corporeal assistance, may be able to get acquainted with Bodies, understand their powers, perceive the chain or order of causes, see things present, foresee things future, and know things that are past? Nor is there any absurdity in supposing, that these Daemons may insinuate their thoughts into human minds. . . ."[66]

Boerhaave praises the saner alchemists for their wish to learn to imitate the laws of nature, especially the law that like is begotten of like. He benignly forgives the alchemists for having died without having achieved the capacity to distinguish experimentally between their right and wrong hypotheses! He acknowledges that a soberer science has not confirmed what the alchemists claimed for the Philosopher's Stone, but he says that as there "are inconveniences in being too credulous . . . so there are in being too skeptical. It is the business of a wise man to try every thing, and abide by that which he finds to be true, nor ever to prescribe limits to the power of the omni-

---

[66] Ibid., 1:63.

potent Governour of the universe, or of the natural beings which he has created."[67]

Conclusion

Our concern in this series of studies has been with answers given, in successive generations, to the questions *what is matter? what is life?* and *with what condition of matter is life associated?* In a definition of life given by Boerhaave in his *Institutes,* life and condition of matter are equated. He says that ". . . by *Life* I would be here understood to mean, in the common Sense of the Word, that Condition of the several fluid and solid Parts of the Body, which is absolutely necessary to maintain the Commerce between that and the Mind to a *certain Degree,* so as to be not perfectly removed beyond the Power of being restored again."[68] In other words, when the "commerce" between mind and body subsides below a critical threshold, its ebb proves irreversible and the organism is ipso facto dead.

It is a little difficult to be sure how precisely Boerhaave means this definition, which he offers with a certain diffidence and tentativeness. Mind-body interaction is in any case a too limited formulation in that it excludes things that have life without mind. Boerhaave's cryptic allusion to the irreversibility of death is pregnant enough, but Haller raises some practical objections to it.[69]

Two things are clear enough about Boerhaave's view of the nature of life. We hear nothing about a physiological life-soul, and we hear a great deal about bodily activity. Boerhaave would probably not have disagreed with his famous student's footnoted suggestion that life is the "Sum or Aggregate of all the Actions resulting from the Structure of the several Parts in the human Body; when all those Actions are performed with Ease and Perfection, it is called Health." With Boerhaave the interest in life-as-soul has been replaced virtually completely by an interest in life-as-action. Where this interest leads him as a theory-builder depends upon whether he responds as biophysicist and micromechanist or as a biochemist. We might agree that in the latter role he places too much confidence in the

[67] Ibid., 1:72–78.

[68] Ibid., 1:90.

[69] Haller's comment on it raises the difficulty that in certain cases apparently dead persons have revived (e.g., a "drowned" sailor when rolled over a barrel and a seemingly dead woman in America brought back by the saliva of a native who had chewed the bark of a certain tree).

archaically conceived *spiritus rector* evoked in explication of subtle and central biological problems which it in fact does not explain at all. Chemistry was not yet able to provide physiological insights nor was Boerhaave the man to make it more effective.

As a mechanist, Boerhaave gives us a hierarchy of corpuscles, fibers, membranes, and vesicles so "contexted" as to form the visible solids; these solids include higher-order vessels both open and compacted. Through the open vessels and the ranks of smaller vessels that branch off from them there circulate a whole hierarchy of humors whose fluidity increases with the minuteness of their globular corpuscles. The fluids are driven by the heart; the whole is thus a hydraulic machine.

We have referred often to the "grand strategy" that physiologists have followed from Greek days forward, their recurrent re-examination of classic questions in the light of new facts and ideas that permit new causal interpretations. That Boerhaave's causal interpretation is micromechanistic is, in itself, not new. What we have learned about Boerhaave in this chapter suggests, however, that he uses micromechanics more exhaustively and more consistently than any earlier biologist. In time, his still speculative scheme will be replaced by other "better" schemes. Early eighteenth-century physiology awaits three releasing developments that still are decades in the future: first, the birth of a significant quantitatively based chemical science (1760–1810); second, systematic experimental procedures of its own (episodic until the early nineteenth-century); third, effective adaptation of the microscope to the microanalysis of gross and complex structure (incipiently from ca. 1820).

Micromechanical Models, continued

ALBRECHT VON HALLER [1708 · 1777]

"It pleases nature to hide," said Heraclitus, in about 500 B.C., and he added that "Eyes and ears speak falsely to a mind that fails to understand them."[1]

We have noted many times in these studies that from its inception science has concerned itself with two universes of inquiry. There is an outer, eyes-and-ear world that can be seen, heard, felt, smelled, and tasted. But there is also a "hidden" world whose organization is so small, whose events are so subtle or slight, or so far away in time or space, that they are beyond the reach of sensation; this world is known through inference. It has been the general assumption of science that when the difference between these two worlds was one of size, the subvisible was in some way the more real—or more like the real—than the visible. On this assumption, the business of science becomes first to see clearly the visible world and then to reformulate ("interpret," "explain") it in terms of the more real world that lies hidden behind it. We have suggested earlier the use of the terms phenomenal and cryptomenal to designate these two universes of scientific inquiry.

The rise of microscopy in the seventeenth century had an important effect on this interpretive activity since it rendered visible some parts, at least, of the world that had previously been hidden. Descartes merely postulated a microorganization; three decades later, Robert Hooke actually saw one. Hooke expected, we remember, that many of the "occult qualities" of things would turn out under microscopic analysis to be "small machines." And to a certain degree time has borne out his expectation. This did not change matters fundamentally, however, since a hidden (molecular) world was still there, behind the new world that the microscope rendered visible. The task

[1] Fragments 107, 123.

of biology now became the dual one of interpreting the directly visible in terms of the microscopic and the microscopic in terms of the submicroscopic.

Haller's Microsystem

It was that dual task that challenged the Swiss physiologist Albrecht von Haller. Haller built both microscopic and molecular models of the organism and then tested these models by applying them to important overt manifestations of life.

Fibers. The living body, as Haller described it, was partly fluid and partly solid; the solid part, essentially fibrous. In many body parts, he saw the fibers as composing, in a manner to be noted in a moment, something termed *cellular substance.* He viewed muscle, for example, as fibrous not only at the gross visible level but microscopically, even its microscopic fibers being composed of still smaller ones—and so on, presumably, beyond the reach of the microscope. The invisible ultimate fiber was, for Haller, a linear series (or in some cases a platelike arrangement) of earthy particles stuck together, as in Boerhaave's formulation, with a kind of gluten or jelly. The earthy particles contain air, chalk, and iron, he thought; the gluten being made of air, oil, water, and volatile salts.[2,3] The body solids thus contain the elements earth, iron, oil, water, and air. Among the fibrous parts he included—in addition to muscle—bone, cartilage, membranes, vessels of all sorts, the cellular tissue, the visceral parenchyma or packing tissue, hair, and nails. Plants, too, were viewed as fibrous, and fibers were present in many metals.

The fiber had a special status for Haller as a basic building block for all body solids. *"Fibra enim physiologo id est quod linea geometrae, ex qua nempe omnes oriuntur."* Again, *"Fibra ... communis toti humano corpori materies est, ..."* Haller disclaimed originality in

[2] Albrecht von Haller, *Primae lineae physiologiae in usum praelectionum academicarum,* first published Göttingen, 1747; translation of the third Latin edition (Göttingen and Leyden, 1765) *First Lines of Physiology,* Edinburgh, 1801, paragraphs 1–24. (Note: an earlier edition of this translation [Edinburgh, 1786] has been reprinted with helpful bibliographic, historical, and biographical introduction by Lester S. King [New York, 1966]; all our references, however, are to the Edinburgh ed. of 1801.)

See also Albrecht von Haller, *Elementa physiologiae corporis humani* [hereafter abbreviated *Elementa*], 8 vols. (Lausanne, 1757–1766), vol. 1, bk. 1, sec. 1.

[3] *Elementa,* 1.1.2 (cited by volume, book, section and, in some instances, paragraph).

this connection and cited especially J. F. Schreiber and Bernard Connor, as well as Hermann and Kaaw Boerhaave, Gaubius, and Gorter.[4]

Our study of Hermann Boerhaave suggests Haller's primary obligation to him in the matter of fiber theory. We noted there, too, antecedents of eighteenth-century fiber theory, going back to antiquity. In Haller's own century, certain French theorists were to view the fiber as endowed with immanent vital properties (see Diderot and Bordeu). These ideas foreshadowed, though not yet completely explicitly, the search for common microcomponents of living systems, that is, for living "units." Other theorists (Buffon and Maupertuis) ascribed life or psychic properties of some sort to still smaller entities, certain particles which they saw as constituting all living bodies. There presently developed, and we shall return to it later, an increasingly articulate debate on the following issue: Is "living" only properly used in speaking of the organism as a whole?—or is it legitimate, also, to speak of the parts as "living"? In the latter case, What is the smallest part that may be properly thought to have life? The most convincing answer, coming less than a century later, was that of Theodor Schwann with whom the *cell* became both the universal formative unit and the smallest independently viable subdivision. Even Schwann's solution, though cogent, was not destined to solve the *Lebenseinheit* problem, however, since after Schwann other theorists asked whether something smaller than a cell—a cell fragment, a bit of extruded protoplasm, a "biogen," a virus particle—did not merit the epithet "living" (see chapters 47–50).

Cellular organization. Cellular substance (usually *tela cellulosa*) was understood altogether differently by Haller than by modern microscopists. He compared it (in a modification of the ideas of the seventeenth century plant microscopist Nehemiah Grew) to a sponge, a fleece, a meshwork of elementary fibers whose open spaces, or *areolae,* intercommunicate permitting fluid-flow from one to another. The more tenuous sorts of areolar (cellular) matter were built of linear fibers; the coarser sorts of a platelike or laminar type.[5]

What Haller saw and described as *tela cellulosa* was in part what later microanatomists were to call areolar connective tissue; he also probably included what we know as adipose tissue. But he extended cellularity as a principle of organization to the body in a much more

---

[4] Ibid., 1.1.1.
[5] Ibid., 1.1.2; *First Lines,* 7–12 (cited by paragraph).

general way. He viewed the cellular substance as existing in either a loosely textured condition or in various degrees of compaction—fat deposits exemplifying the open texture, membranes the compact type.[6] By a variety of techniques, especially the injection of air and various liquids, "cellulosity" was established for muscle membranes, the arachnoids, the pericardium, skin, and the enveloping sheaths of the viscera including the glands. The solid substance of organs comprised, for Haller, variously folded vessels (originally cellular) and fibers (including nerves) imbedded in the cellular *tela*. In sum, the cellular substance "certainly constitutes by far the greatest part of the body if, indeed, not the whole."[7]

As to its use, this substance, where open enough to retain its cellular appearance, was supposed to support, contain, and separate the body parts while permitting them adequate freedom of movement. The areolae or open spaces were supposed to intercommunicate continuously throughout the whole body. Haller described the containing solids as being moistened in places by a "watery vapour, gelatinous and somewhat oily, exhaled out of the arteries and received again into the veins." Elsewhere, the areolar spaces receive an arterial efflux that congeals into fat.[8, 9] Others before him, Haller assures us, noted and called by other names the arrangement which he and other recent authors term cellular.[10]

Applications of Theory

Having developed his microsystem, Haller made it a partial basis for an orderly analysis of the body parts, their visible structure, and their functions. His treatments of the circulation, of excretion and secretion, of respiration, of nervous action and muscular motion, of nutrition, of reproduction were, with an occasional unevenness, exemplary for their period. These subjects had all been treated before, but never as objectively and reasonably. Haller typically supported every general statement with specific experimental evidence. In the exhaustive and copiously annotated eight-volume *Elementa* (1759–1766), references are cited for all theories and experiments reported. The *Primae Lineae* (1751), prepared earlier for student use, covers

[6] *First Lines,* 9.
[7] Ibid., 22.
[8] Ibid., 21, 22.
[9] Ibid., 9, 16, 18; *Elementa,* 1.1.2.
[10] *Elementa,* 1.1.2.

many of the same subjects, but covers them more briefly, treats fewer topics, and has very few footnotes.

It would be quite impossible—however tempting—to explore in as brief a treatment as this the full reach of Haller's genius. But we can learn much about his system by noting how he applied it to four problems that had been recognized as central ever since the beginning of biological inquiry, the problems of nutrition, muscular motion, psychic action, and generation.

Nutrition. Nutrition, classically conceived, has as its central process the wearing away of the body and its continual replacement. Haller's analysis followed this classic conception. He saw fluids as "squandered" by way of the breath, by tears and other secretions, and by sweat, urine, and feces. The erosion of solids was mechanical rather than chemical, he supposed, being caused by pulse beats (small but numerous), by the friction of the body fluids, by abrasion at the edge of loose membranes, and by the wear-and-tear of muscular movement. Displacement (of solids) is a crumbling that occurs when the parts come unglued.[11]

"The fluid parts are restored by the aliments, and that pretty quickly. . . ." The solid parts are "almost repaired by the same means which we have described in the history of the fetus. A gelatinous juice is conveyed from the aliments through the arteries to all parts of the body and exudes into the cellular texture everywhere. . . ." The gluten repairs most of the organic parts, tendons, and membranes—being formed into new cellular substance in the adult as in the fetus.[12]

Haller's theory of old age was Boerhaavian. The process of induration starts at the beginning of life, he believed. In saying that this idea is Boerhaavian, we do not mean to overlook the fact that repeatedly since pre-Socratic times aging had been associated in a variety of ways with increased dryness and concomitant hardness of the body. In life's very climax, Haller said, the trend begins to reveal itself for what it ultimately is—the prelude to old age and death. A detailed and physicianly account of life's dénouement is given by Haller. It concludes with the statement, "I call that death, when the heart has become totally deprived of irritability."[13]

[11] *First Lines,* 957–58.
[12] Ibid., 959–60.
[13] Ibid., 959–61.

Haller had much more to say about mastication, salivation, deglutition, peristaltic action, gastric and intestinal digestion, absorption, chyle formation, every aspect of the circulation, and the discharge of nutriment from the arteries and its ultimate assimilation. No earlier author had presented these topics as sensibly and systematically as Haller. He used his cell-fiber-particle theory consistently in explaining these varied functions, but was never tyrannized by it. He presented physiological knowledge on whatever level—macro- or microscopic—he could, limiting himself to subjects on which evidence was obtainable.

Motion and Irritability

Haller viewed muscle as sharing with other fibrous parts a general contractile tendency which he called *vis mortua* or dead power, but this was not, in his opinion, the cause of active muscular contraction. The latter was due to an immanent power (*vis insita*) of the muscle endowing it with *irritability*.[14] Irritability had for Haller a restricted meaning rather different from that which it held for both earlier and later biologists. For Francis Glisson, who had fostered the use of the term three quarters of a century before, irritability had been a quite general property permitting all sorts of vital responses.

Glisson, a seventeenth century English physician and physiologist, had followed tradition in portraying the body solids as fibrous. In an early *Treatise on the Energetic Nature of Substance* (1672), he had argued in a hylozoic vein that even apparently inert matter manifests a certain kind of life—a *vita naturae*—of its own, since "all . . . self-subsisting substances are endowed with [a] perceptive, [b] desiderative, and [c] motor faculties of a sort." In his more influential later *Treatise on the Stomach and Intestines* (1677), Glisson offered something close to a general analysis of life-as-action. When matter with its three psychic properties is arranged as animal fiber, he said, that fiber thereby acquires irritability (the capacity, when irritated, to react in such a way that the irritant is no longer effective). Such reactions require (*a*) that the fiber perceive the irritant, (*b*) that it desire to reduce the irritation, and (*c*) that it move in such a way that its wish is fulfilled. Such reactions are fundamental to physiological responses in general. The applications of irritability were thus

14 Ibid., 391, 400.

much more numerous in Glisson's than in Haller's usage, since Haller limited the term, as we are about to see, to the ability of a single sort of body part, muscle, to contract in response to proper stimulation.[15]

In a now famous paper of 1755, Haller reported experiments testing hundreds of animals to find out what parts of them were (*a*) irritable (contractile) and (*b*) sensible—experiments he described as "a species of cruelty for which I felt such a reluctance, as could only be overcome by the desire to contribute to the benefit of mankind."[16] Any part of the body is, by Haller's definition *sensible* if, when it is stimulated, the organism is ipso facto aware of the fact (in the case of an animal, if its response suggests it feels pain). It is irritable if, when stimulated, it contracts.

Haller's experiments convinced him that there is only one sort of irritable body part (muscle) and only one sensible sort (nerve). Other organs that seem sensible contain nerves or nerve endings which are the organs' only really sensible components.[17]

Haller acknowledged the need for grounding irritability and sensibility in something more basic (in corpuscular composition). But he was reluctant to attempt this. Thus "the theory why some parts of the body are endowed with these properties, while others are not, I shall not meddle with. For I am persuaded that the source of both lies concealed beyond the reach of the knife and microscope, beyond which I do not chuse to hazard any conjectures, as I have no desire of teaching what I am ignorant of myself." He suspects that irritability "lies hidden in the intimate fabric" of the irritable parts in ways

[15] See Francis Glisson, *Opera omnia* (London, 1691), p. 160. See also the discussion of Glisson's theory in C. Daremberg, *Histoire des sciences médicales* (Paris, 1870), 2:650–72 (contains translation of excerpts from Glisson). See also E. B. M. M. Bastholm, "The History of Muscle Physiology," *Acta Historica Scientiarum Naturalium et Medicinalium* (Copenhagen), 7 (1950):219–25.

[16] Albertus de Haller, "De partibus corporis humani sensibilibus et irritabilibus," *Commentarii societatis regiae scientiarum Gottingensis,* 2 (1753) (read April 22, 1752): 114–58; trans. anon. *On the Sensible and Irritable Parts of Animals* (London, 1755), p. 2. (Note: this translation was based not on Haller's original but on a French translation by M. Tissot. This English version was republished with an introduction by O. Temkin [Baltimore, 1936]. Our references, however, are to the edition of 1755.)

[17] Ibid., pp. 31–32. Compare Garrison's misleading statement that Haller's "greatest single contribution to the subject [physiology] is his laboratory demonstration of Glisson's hypothesis that irritability (e.g., in an excised muscle) is the scientific property of all living or organized tissues," in Fielding H. Garrison, *An Introduction to the History of Medicine* (London, 1914), pp. 246–47.

that escape the power of available experimental procedures. He vouchsafes here, and develops the point at length, in his slightly later *Elements,* that as between the earthy particles of the fiber and the gluten that holds them together irritability probably resides in the gluten.[18]

A Digression in Defense of Haller on Irritability

Haller's attempt to restrict the concept of irritability was only temporarily successful, since other authors continued the idea of irritability as a universal property of everything living. After Haller, a variety of opinions competed for acceptance as to whether the proper understanding of life-as-action requires the assumption of only a single vital property or of several. Where two were recognized, usually sensibility and irritability, it was debated which of the two was fundamental.

The evolution of this debate—and especially of the irritability idea—would make a seminal subject for historical investigation, since from it we could learn a great deal about the rise in power and the final decline of a potent organizing concept.[19] Such a study is outside our present objectives, but we may at least allude to three later treatments of irritability, spaced widely in time, because by comparing them with quite recent (mid-twentieth-century) developments, we may obtain a certain perspective on Haller. The dramatis personae are pathologist Rudolph Virchow (1858), and physiologists Claude Bernard (1878) and Max Verworn (1913), with all of whom we shall become better acquainted in subsequent chapters.

Virchow accommodated the irritability idea to his personal view of cellular action. He inferred the existence not of one irritability but three, namely "nutritive," "functional," and "formative," corresponding to, and making possible, the nutritive, functional, and formative functions of cells. Bernard stipulated as the essential condition of plant and animal life their ability to react in a suitable way to the action of external agents. This capacity exists in its simplest form in their protoplasm, he supposed, protoplasm constituting the physical basis of life in both kingdoms. Bernard was to refer to this indispens-

[18] *Sensible and Irritable Parts of Animals,* pp. 2, 58. *Elementa,* 4.11, 2.12.

[19] For a brief historical resumé on irritability, see Max Verworn, *Irritability* (New Haven, 1913), pp. 1–17. For Glisson and his sources on this subject, see O. Temkin, "The Classical Roots of Glisson's Doctrine of Irritation," *Bulletin of the History of Medicine,* 38 (1964):297–328.

able reactivity now by the term irritability, now by the term sensibility. The central point in his scheme was that the fundamental response mechanisms of organisms have a common physicochemical basis. He cited in evidence his own experiments suggesting that anesthetics (ether, chloroform) reversibly inhibit not only consciousness but vital processes in general, even including those of plants (their germination, gas exchanges, fermentive activity, and the reactions of sensitive plants).[20]

Finally, as a fairly extreme and quite recent statement of the irritability idea, we may cite Verworn's observation (1913) that "if we could analyze the irritability of living substance to the essence, the nature of life itself would be fathomed."[21] Such a statement, coming when it does in the history of scientific ideas, raises a fundamental issue. Is irritability a collective abstraction for a whole ensemble of different vital responses or is it a particular molecular process common to all such responses?

The latter alternative had an active vogue from the latter nineteenth to early twentieth centuries. G. Hermann illustrates this position in his excellent article, "Irritabilité," in *La Grande Encyclopédie*. "We see, then," he says, "that we have not to do with a simple artefact of classification grouping under the one denomination *irritability* the diverse properties termed vital; [rather] we attach to this term the idea of an always identical molecular movement representing what is fundamental and constant in the mechanisms of life."

This statement wonderfully illustrates the estimable—but potentially misleading—ambition especially of many nineteenth-century theorists to find simple ways of organizing the vast new arrays of data that were becoming available then. There was much to attract a thinker of that era in "the idea of an always identical molecular movement," especially one that represents all that is "fundamental and constant in the mechanisms of life." But is there, in fact, such a movement? Certainly the biology of our time—at least by mid-century—has stopped looking for one. Our whole way of proceeding today is, in fact, less oriented than formerly toward the discovery of such vast but over-simple conceptions.

[20] Claude Bernard, "La sensibilité dans le regne animal et dans le règne végétal," in *La science expérimentale* (Paris, 1878), pp. 218–36. For a survey of the irritability doctrine to 1837, see John Fletcher, *Rudiments of Physiology* (Edinburgh, 1835–37), pt. 2, the long footnote on pp. 50–54.

[21] See Max Verworn, *Irritability* (New Haven, 1913), pp. 1, 6.

Very recent developments suggest the need for a reappraisal of Haller. A number of modern critics, imbued with the idea of irritability as a universal life principle, have found fault with Haller's effort to restrict it to the phenomena of contractility. Verworn, for example, considers that *"Haller's* theory represents a great regression in comparison to the correct fundamental thoughts of *Glisson."* From today's perspective, Verworn's judgment is debatable. In retrospect, Haller's restricted concept seems attractive precisely because of the absence from it of unwarranted generalization.

The notion that an identical transformation occurring in a unique molecular system accounts for all vital responses must be seriously questioned, if not outrightly rejected, in the mid-twentieth century. Present knowledge points, rather, to *varied* molecular systems responding *variously* to *varied* stimuli. Irritability as a collective term for this varied activity is not erroneous, exactly—but it is dangerous since it leads easily to the assumption of a single common process. As far as anyone knows, there is no such process, nor is the existence of one even probable. From this point of view Haller's original ideas of irritability and sensibility, with their basis in experimentation and their avoidance of generalization and theoretical pretense, have a certain appeal over the grander but less demonstrable ideas that both preceded and succeeded them.

Psychobiology in Haller's System

After Descartes. Post-Renaissance psychobiology concerned itself with classic questions and subquestions, mostly Greek in initial formulation but reformulated in the sixteenth century, especially by Descartes. Among the most vexed of these was that subsequently termed the "mind-body" problem. What Descartes had transmitted on this subject was an application of his own somatopsychic dualism to questions he encountered in his reading of the classics and the Scholastic critics thereof. For example, St. Thomas, an author to whom Descartes owed much, had inquired lengthily into such subjects as to whether soul is body or form or both; whether it is of the same nature as an angel; whether there are other souls than the intellective; whether a third entity is necessary to unite the soul to the body; whether the whole soul is everywhere in the body; and whether animals have souls.

Descartes' own goal, of eliminating the soul from animals, had opened a new phase in the debate about what came to be called the automatism of beasts. But Descartes, while depriving both men and animals of physiological life-souls, in the Greek sense, allowed man to retain a mind-soul that he saw as conscious and immortal. For those seventeenth- and eighteenth-century thinkers who agreed tacitly or openly with Descartes, the problem thus remained as to the nature, place, and extent of the mind-soul's interaction with the body (in the one species presumed to possess a mind).

Other classic psychobiological problems restated by Descartes had to do with: the nature and pathways of nervous conduction; the character of the excitative process (through which the nerves elicit action); the role of the brain in relation to its visible and subvisible organization; and others. As post-Cartesian physiological psychology developed, it made very slow progress in investigating these neuro-physiological problems, whilst the subtler and more subjective issues of mind-body relations remained mostly matters of speculation. In sum, Descartes' reformulation of the neurophysiological aspects of psychobiology had provided a program, little more, for interim studies of the nervous system. This fact may help us to place Haller's efforts in perspective, since these efforts amount in part to a thorough working through of the Cartesian program in the light of interim inquiry.

Haller's ideas. Haller saw the latent irritability of muscle as activated by an influx of juice from the nerves. He concluded this by ruling out other possibilities such as vibration (nerves are not elastic) or electrical conduction. From such considerations "the only probable supposition that remains is, that there is a liquor that comes from the brain, descends into the nerves, and flows out to the extreme parts of the body; . . . the same liquid being put in motion in an organ of sense, by a sensible body, transmits its motion upwards to the brain. . . ."[22]

He viewed the nerve fluid as motile, thin, invisible, tasteless, odorless, and reparable from the aliment.[23] He envisioned a continuous slow flow of it from the bloodstream into the nerves (each nerve starts in the brain at the surface of an artery) and an episodic, rapid flow associated with nervous conduction.[24]

[22] *First Lines,* 377.
[23] Ibid., 381.
[24] Ibid., 383.

Nerve-muscle action did not suggest to Haller the presence of soul in the nerve itself or at the nerve endings. He noted that nerves can do things more adequately simply by following the laws that govern this action than by submitting to conscious deliberations. Thus, the "motive cause which occasions the influx of the spirits into the muscle, so as to excite it into action, seems not to be the soul but a law established by the Creator. For, animals, newly born, without any attempt, or exercise, know how to perform compound motions, very difficult to be defined by calculation. But the soul learns these things which it performs, slowly imperfectly, and experimentally."[25] Haller's way of equating *Divine Law* with *motive cause* is representative of a model of thought that evolved in eighteenth-century physiology under the impact of Newton's theory of gravitation.

"Concerning the seat of the soul [another classic question], we must inquire experimentally," Haller said, and he added that experimental evidence suggests that soul is not in the nerves; nerves lose their power when severed. It seemed to Haller to follow from this —and other evidence to the same effect—that soul is limited to the brain only. One could even venture where, within the brain, soul resides, since "it appears from the experiment of convulsions arising when the innermost parts of the brain are irritated, that it lies not in the cortex but in the medulla; and not improbably in the crura of the medulla, the corpora striata, thalami, pons, medulla oblongata, and cerebellum. Finally, by another not absurd conjecture, it lies at the origin of every nerve, so that the concurrence of the first origins of all the nerves, makes up the sensorium commune."[26] This attempt shows more knowledge of brain structure than Descartes' attempt and is more experimentally oriented than the parallel efforts of such interim investigators of the nervous system as Willis, but it still leaves much to be desired.

Haller saw soul, in effect as (*a*) initiating an efflux of nervous juices (in motor action) and (*b*) responding to their influx (in sensation). He was a dualist in that mind and brain were for him distinct and had their separate properties. What, he asks, is the *color* (a physical property) of *pride* (a psychic property)? or the *magnitude* (physical) of *curiosity* (psychological)? Moreover, mind, un-

25 Ibid., 408.
26 Ibid., 372.

like body, can alter its "movements without the application of external forces," that is, unlike matter, it lacks inertia.[27]

Though mind and brain are distinct, they are intimately linked in Haller's system. Thought is an activity of soul, but to each idea there corresponds a differentiation of the brain—an "impression"[28]—as well as a differentiation of the sense organ. The perceptive act involves five components: (1) the object, (2) the response of the sense organ, (3) the brain's response, (4) the mind's response, and (5) the mind's awareness of its response.[29] Between visible object, retinal response, brain response, and idea there is a "lawful correspondence," but no identity. The same object presented to the mind through a different optical system would give rise to a different idea. The laws acting in these matters are unknown, however, and we should admit our ignorance of them.[30]

The mind has its own differentiations or faculties "distinct from any corporeal faculty." Among these are retention (*omnis scientiae mater*), judgment (*duarum idearum comparatio*), ingenuity (*ideis unicundis*), reason (*series judiciorum*), abstraction and generalization (these utilize symbols or *signa*), and finally wisdom (a "slow examination of ideas").[31]

Even these distinctly mental faculties, however, depend on a healthily constituted brain. Physical compression, irritation, blood shortage, or textural alteration can occasion profound psychic disturbance. So can external causes—air, food, habits, vocation.[32] Reciprocally, the emotions anger, grief, terror, love, hope, joy and shame produce physiological responses of the circulation, the secretions, the skin, and so forth. Perhaps these effects are nerve-mediated; perhaps the nerves control arterial "sphincters" (this is Haller's guess at what we now term vasoconstriction). The passions also effect the physiognomy.[33] A few psychic faculties—hope, curiosity, the quest

---

[27] Ibid., 569.
[28] Ibid., 558, 570.
[29] *First Lines,* 557; *Elementa,* 5.17.1.4.
[30] *First Lines,* 571.
[31] *First Lines,* 562; *Elementa,* 6.17.1.10–16. For Vives and other earlier exponents of a factorial treatment of mind, see Brett, *A History of Psychology* (London, 1921), 2:162.
[32] *First Lines,* 563.
[33] Ibid., 565.

for the good, for glory—are, finally, viewed as apparently entirely unconnected with matter.[34]

Generation

Haller and Bonnet (chapter 30) were the last important exponents of preformationism and of the idea that the preformed individual is the product of the mother. Having commenced as an epigenesist, Haller became convinced through his own study of the chick embryo and through corroborative theories of Bonnet that the individual exists preformed in the ovum. He acknowledged that the embryo *appears* to undergo a progressive differentiation: one sees, first, merely an apparently structureless worm and, later in succession, a heart, eyes, intestinal viscera, and lungs. But invisibility does not imply nonexistence. He supposed the problem to be one of transparency. The individual exists preformed as a constellation of transparent rudiments of future parts. The original nature of the rudiments is mysterious, but Haller insists that all are formed simultaneously, and become visible in a certain order. He seems to admit, however, that, once formed, these rudiments are subject to a certain amount of elaboration.[35]

Haller gives a readable, complete, and well-organized survey of existing knowledge concerning asexual and sexual reproduction.[36] He reviews in a detailed and critical way the three possibilities that the new individual is the product of (*a*) the father alone, (*b*) the mother alone, or (*c*) both of these through some sort of seminal combination.[37] He decides that the primary role of the semen is to activate rudiments already present in the ovum and to make them grow. He opposes the proposition that the seminal animalcules are "as it were, the first rudiments of the future animal"; their nature seems rather to Haller "to be the same with that of eels in vinegar or other infusorial animals." He also says of infusorial animals that they are "tenacious of their own genus [and] never grow up into an entirely different sort of animal possessing limbs."[38]

He "regards as conjectures everything said about them by their

[34] Ibid., 564.

[35] *Elementa*, 8.29.27.

[36] A. von Haller, *Sur la formation du coeur dans de poulet* (Lausanne, 1758), republished in Latin translation in Haller's *Opera Minora* (Lausanne, 1767), 2:54.

[37] *Elementa*, 8.2.2.1–8.

[38] *First Lines*, 825, 882.

author Loewenhoek, that there are spermatical animals of both sexes, that one detects a certain difference in the tail region, that they couple, that the females become full, . . . that they lose their tails, that they change their skins . . . that some have been seen with two heads . . . [that] they have the actual figure of a man." He notes with disbelief that "M. Gautier has given a marvelous picture of a fetus discovered in horse semen which was as big as a bean and had the distinct figure of a horse; he said, also that in the semen of an ass, he saw a fetus with big ears. . . ." Haller asks "how it is possible that so many other hard working men could have been blind enough not to have seen so large an animal. . . ."[39]

With his usual thoroughness Haller has read everything obtainable about *ova* and made a comparative study of them himself. In the ovary are vesicles discovered long before and termed *ova* by de Graaf. They are filled with a "clear humor, sometimes roseate or yellow, which is coagulated by alcohol and heat, and which forms strong white filaments like whites of egg." These vesicles contain, though it is invisible, that which, on fecundation, can produce a new individual. Fecundation takes place within the ovary itself. Ovulation follows, and the fecundated product is transferred through the Fallopian tubes to the uterus. After ovulation, the vesicle that previously held the rudiment of the new animal is converted into a *corpus luteum*.[40]

In this connection, Haller cleared up a point whose status had been clouded by Buffon's misinterpretation of what he saw in the ovary. "Although Buffon says that the corpus luteum is found in virgin animals and that other eminent men report having found it in virgin women . . . I have opened forty ewes and thirty bitches not to mention she-goats, cows, sows, dormice, doe rabbits, and finally the cadavers of six women who died during pregnancy or after abortion, or in confinement. Graaf similarly opened a hundred doe rabbits and forty ewes." In no case did Haller find a corpus luteum except where conception had occurred.[41]

Haller reviewed, in this connection, the whole debate as to whether there really is a typical egg in the ovary. He himself had used the term *ovum* for the entity within the ovary that is destined to form the future

[39] *Elementa*, 7.27.2.10.
[40] Ibid., 7.28.2.34, 8.29.1.24. Also, *First Lines*, 878.
[41] *Elementa*, 8.29.15.

individual, but he now acknowledged that there is much debate about the status of this body. He cited scores of authors on this subject, and gave specific page references to the literature of the subject. He himself sought for the egg without success in forty pregnant ewes without finding anything answering that description until the seventeenth day after conception, and he was skeptical of the reports of authors who claimed to see something sooner.

Whether there is a mammalian egg, he concluded, is partly a definitional problem. "If you so interpret *ovum* as to make it a hollow membranous receptacle within which there be a humor and within that a fetus, the ancient opinion can be accepted that obtains every animal out of an egg, except for a few very simple animals referred to before." Modern readers must be careful not to impose a modern interpretation on Haller's approach to this question. Twentieth-century biology sees the *ovum* as the organism itself in the single-cell stage before or after fertilization. Haller's approach was quite different. Extrapolating from the familiar bird's egg, he viewed the ovum as a surrounding shelter (*hospitium*) for the developing fetus. That an egg in this sense, as an envelope, seemed not to appear until the twentieth day did not in any way counter his fundamental assumption that the primordial individual is contained, preformed, in a vesicle within the ovary.[42] Acknowledging the conjectural state of the subject, Haller supposed that "in the male semen there is some power, which can influence the form of the soft substance of the very minute embryo."[43] He saw this power as permitting the individual, though preformed in the mother, to be in effect the product of both its parents (for an alternative of this idea, see Bonnet, chapter 30).

Before fecundation, he said, humors oscillate gently between heart and arteries. But semen excites the already irritable heart to greater contractions driving fluids to the body parts in such a way as to cause their development (*evolutio*).[44] Development does permit successive changes. The heart is in the form of a parabola; lacunae appear; its parts draw together; it shortens; it thus arrives at its state of perfection. These changes are occasioned by mechanical forces of expansion, attraction, pressure, repulsion, derivation, resorption of hu-

---

[42] Ibid., 8.29.25–29. For a short contemporary summary of Haller's experiments, see "Observations sur la génération et sur la non-existence de l'oeuf tout formé dans l'ovaire," *Histoire de l'académie royale des sciences 1753* (Paris, 1757), pp. 134–35.

[43] *First Lines,* 881.

[44] Ibid., 886; *Elementa,* 8.29.2.27.

mors, evaporation.[45] It would not be incorrect to think of Haller's theory as combining preformation with perfectibility.

Haller justified his preformationism logically by three assertions based on his study of the developing chick embryo. (1) The yolk-sac is continuous with the walls of the fetal intestine (*both* are parts of the embryo). (2) The yolk-sac could only have been formed by fetal blood vessels. (3) Since the yolk is present, so must fetal blood vessels be present (even before fecundation).[46] The first point is an understandably confused interpretation of what Haller saw (there *is,* later, a connection between the intestine and the yolk). The second point is essentially gratuitous. Hence, the third point—the conclusion of the syllogism—is unwarranted.

Haller likewise reviewed and rejected an ensemble of theories which we might term "animistic" or "vitalistic," Wolff's *vis essentialis,* Stahl's biomedical soul, and others.[47] He sought rather "the efficient causes of this beautiful animal machine." He acknowledged that there is "much difficulty concerning the means by which the rude and shapeless mass of the first embryo is fashioned into the beautiful shape of the human body. We readily reject such causes as the fortuitous concourse of atoms, the blind attractions of nutritive particles, and the action of ferments inconscious of their effects. The soul is certainly unequal to the task of producing such a beautiful fabric; and internal molds [see Buffon] of which I never could obtain a conceivably clear idea,[48] are to be referred to those hypotheses which the desire of explaining things, of which we are unwillingly ignorant, has produced."[49]

He protests that the "most beautiful frame of animals is so various, and so exquisitely fitted for its proper and distinct functions of every kind . . . that it is calculated by rules more perfect than any human geometry . . . ," and he concludes that "no cause can be assigned to it below the infinite wisdom of the Creator himself." But insofar as we can interpret these matters scientifically "the more frequently, or the

[45] *Elementa,* 8.29.2.28.

[46] *First Lines,* 883; *Elementa,* 8.29.2.7. See also F. J. Cole, *Early Theories of Sexual Generation* (Oxford, 1930), p. 86.

[47] Ibid., 8.29.2.15.

[48] See A. von Haller, *Reflexions sur le système de la génération de M. de Buffon* (Paris and Geneva, 1751), and *Elementa,* 8.29.2.4–6. See also C. W. Bodemer, "Regeneration and the Decline of Preformationism in Eighteenth Century Embryology," *Bulletin of the History of Medicine,* 38 (1964):20–31.

[49] *First Lines,* 884.

more minutely we observe the long series of increase through which the shapeless embryo is brought to the perfection necessary for animal life, so much the more certainly does it appear, that these things which are observed in the more perfect fetus, have been present in the tender embryo . . ." from the first.[50]

Preformation raises, Haller acknowledges, the thorny problem of encapsulation, a phenomenon which he for one is quite willing to embrace. He discusses the problem mathematically, assuming the age of the earth to be 6,000 years, or time enough to have permitted 200 generations. He calculates a total population of two hundred billion. The problems posed seem not insuperable to him. They would be insuperable if one were to insist that the volumetric ratio of mother-to-fetus was the same for all future mothers inside the mother that we see. But Haller seems to suppose the problem soluble by assuming that the future mothers inside the visible mother are merely involucres or shells.[51]

SUMMARY

Haller's microsystem was an elaboration of that developed by his teacher, Boerhaave: it postulated granules glued together to form linear or laminar "fibers" and these deployed in an alveolar, or "cellular," arrangement which may remain open or may become compacted. Haller's physiology was in part an exhaustive reformulation of classical physiological problems in terms of the microsystem he constructed. These problems included nutrition, motion, psychic action, and generation. Between his micro- and macrophysiology he by no means always achieved a precise consonance or perfect agreement, but that was a task not for one biologist but for generations of them. In terms of method, organization, completeness, and documentation, his achievement was remarkable. And there is no better guide to the biology—at least the animal biology—of his era than the exhaustive footnoted longer edition of the *Elements*. The repercussions of Haller's experimental approach to irritability will appear as our story unfolds (see below, especially, Bordeu, Brown, Broussais, Bichat, Lawrence, Bernard, Virchow, and Verworn).

[50] Ibid., 885 (translation from the edition of 1779).

[51] *Elementa*, 8.29.2.29. For important further information on epigenesis and preformation see H. B. Adelmann, *Marcello Malpighi and the Evolution of Embryology* (Ithaca, 1966), 2:871–86, and, in general, his treatments of Harvey, Haller, Bonnet, and C. F. Wolff.

Index to Volume I

Brackets are used around certain subitems to suggest that they deviate from or contradict, rather than exemplify, the item under which they are found

Assimilation: in Aristotle, 110–11; in
Boerhaave, 378; in Descartes, 260; in
Galen, 147; in van Helmont, 209;
in Plato, 97; in Willis, 317
Astrology, 168, 176–77
Atomism: in Democritus, 53–54; in Epicurus, 122–26; in Leucippus, 53; in
Lucretius, 122–26
Automata, 222–24, 256, 344
Avicenna, 179, 199
Avogadro, A., 269

Bacon, Francis, 56, 163, 214, 251, 272;
on air, role of, 236; on epistemology,
250–51; on fire and life, 235–36; on
generation, 237–38; on hylozoism,
232–33; on living vs. nonliving matter,
235; on matter, 230–34; on nutrition,
235; on particles, 231; on respiration,
236; on spirits, 232, 234; on vital spirit,
238
Bacon, Roger, 168
Baglivi, Georgio, sketch of theories,
373–74
Bailey, C., cited, 121 n, 129–30
Balsam, ideas concerning: in Anaximenes, 31; in Paracelsus, 177
Barometer, 221
Bartholin, C., 214, 254
Basso, Sebastian, 251, 315
Bastholm, E., cited, 333 n, 373 n, 397 n
Bauhin, C., 214, 254
Bayon, H. P., cited, 241 n
Beeckman, I., 251, 280
Being and not being, in Parmenides, 41
Berg, Alexander, cited, 373 n
Bernard, Claude, 36, 325, 398
Bichat, X., 5, 106, 142
Bile(s): in Galen, 148; in *Nature of
Man,* 73; in Paracelsus, 178. *See also*
Humors
Bios, 13–14
Birch, Thomas, quoted, 307
Blas, 210
Blend of the body. *See* Composition;
Krasis; Temperament
Blood.
—composition of: in Boerhaave, 374;
in Galen, 149; in Lower, 341
—formation of: in Galen, 147–48; in
van Helmont, 209
—heat of: in Harvey, 244
—as humor: in Galen, 149; in *Nature
of Man,* 73
—as locus of life and/or awareness: in
Empedocles, 48; in Harvey, 243; in

Lower, 340; in Willis, 321
—motion of: in Empedocles, 50; in
Descartes, 258; in Galen, 149, 155; in
Stahl, 364
Boerhaave, Hermann, 60, 171, 274, 280,
393; on alchemy, 388–89; on assimilation, 378; on blood and body fluids,
374; on composition of the body, 371,
385–87; on development, 383–84; on
fibers, 372; on flux of the body, 376;
on generation, 381–84; on matter,
384–85, 387; on mechanics of the
body, 369–84; on mind-body problem, 370; on muscles, 379; on nerve
action, 380; on nutrition, 375–78; on
plants, 385; on respiration, 374–75;
on secretion, 380–81; on somatopsychic dualism, 370
Bonnet, C., 383, 404, 406
Bordeu, T., 393
Borelli, Giovanni, 227–29, 333, 338, 352,
356, 367; biographical notes, 342; on
air, 347; on *anima,* 343; on fermentation, 344; on heart, 344; on mechanics
of the body, 343–48; on muscle action,
344–45; on regulation of body function, 346; on respiration, 344–47; on
soul, 343
Boundless, the, 26–28, 31, 41, 47, 54,
267. See also *Apeiron*
Boyle, Robert, 167, 198, 225, 270, 273,
274, 297, 330, 331, 356, 367; on
cosmology, 287; on disease, 289; on
life, 285–90; on matter, 280–83;
on nutrition, 288–89; on respiration,
275–76
Brahe, Tycho, 168, 214
Brain: in Aristotle, 138; in Galen, 138,
158–59, 162
Breathing. *See* Respiration
Brett, G. S., cited, 403 n
Bridges, C. B., cited, 77 n
Browne, John (seventeenth century),
342
Bruno, Giordano, 171, 214, 280
Buffon, G., 19, 467
Buridan, John, 224
Burnet, John, xiii, 52
Butterfield, Herbert, 244

Calcination, Stahl on, 355
Castration, Galen on, 154
Catamenia, 114; reproductive role of, in
Aristotle, 113–14; reproductive role
of, in Fernel, 201; reproductive role
of, in Galen, 151